Insulating Films on Semiconductors 1991

Conference Chairman and Proceedings Editor

Professor W Eccleston
Department of Electrical Engineering and Electronics
The University of Liverpool

Programme Committee:

I A Aizenberg (Moscow)
P Balk (Delft)
J Kassabov (Sofia)
W Eccleston (Liverpool)
O Engstrom (Gothenburg)
H Flietner (Berlin)
S Hall (Liverpool)
M Heyns (IMEC, Leuven)
F Koch (Munich)

P Migliorato (Cambridge)
W Milne (Cambridge)
P Rosser (STC, Harlow)
M Schulz (Erlangen)
J Simonne (Toulouse)
M J Uren (RSRE, Malvern)
— Co-Editor
W Weber (Siemens, Munich)
D R Wolters (Philips, Eindhoven)

Local Organisation Committee:

W Eccleston (Chairman)
S Hall
B Lussey
J Marsland

K I Nuttall
S Taylor (Secretary)
V Wilson

International Advisors:

D J DiMaria (IBM)

Li Zhi-Jian (Tsing-Hua, Beijing)

Conference Sponsors:

Siemens AG
GEC-Plessey Semiconductors Ltd
Pilkington plc
Department of Trade and Industry

STC Technology Ltd
University of Liverpool
B.T. Research Laboratories

Books are to be returned on or before
the last date below.

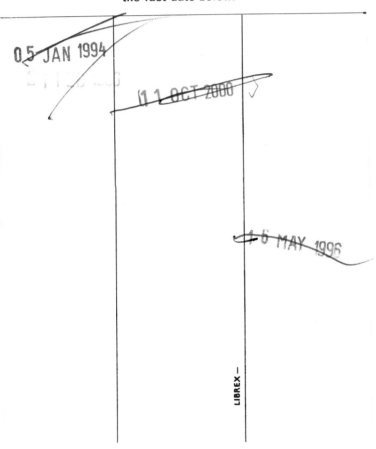

Insulating Films on Semiconductors 1991

Proceedings from the 7th Biennial European Conference, including Satellite Workshops on *Silicon on Insulator: Materials and Device Technology* and *The Physics of Hot Electron Degradation in Si MOSFETs* held at the University of Liverpool, 2nd to 6th April 1991

Edited by W Eccleston and M Uren

Adam Hilger
Bristol, Philadelphia and New York

© 1991 by IOP Publishing Ltd and individual contributors

All rights reserved. No part of this publication may be reproduced, stored in a retrieval system or transmitted in any form or by any means, electronic, mechanical, photocopying, recording or otherwise, without the prior permission of the publisher. Multiple copying is only permitted under the terms of the agreement between the Committee of Vice-Chancellors and Principals and the Copyright Licensing Agency.

British Library Cataloguing in Publication Data

Biennial European Insulating Films On
Semiconductors Conference (7th 1991 Liverpool)
  Insulating Films on Semiconductors 1991.
  I. Title   II. Eccleston, W.   III. Uren, M.
  621.3815

ISBN 0-7503-0168-6

Library of Congress Cataloging-in-Publication Data are available

Published under the Adam Hilger imprint by IOP Publishing Ltd
Techno House, Redcliffe Way, Bristol BS1 6NX, England
335 East 45th Street, New York, NY 10017-3483, USA
US Editorial Office: 1411 Walnut Street, Philadelphia, PA 19102

Printed in Great Britain by J W Arrowsmith Ltd, Bristol

Contents

xi **Preface**

Invited Papers

1–18 Discrete conductance fluctuations and related phenomena in metal–oxide–silicon device structures
K R Farmer

19–32 Oxidation of silicon
A M Stoneham

33–42 UV and plasma effects in the Si/SiO$_2$ system
J Kassabov

43–51 Hot-electron transport studies in SiO$_2$ using soft-x ray induced internal photoemission
E Cartier and F R McFeely

53–63 Interface and oxide engineering for high quality SIMOX devices
S Cristoloveanu

Workshop Papers

65–72 The relationship of trapping and trap creation in silicon dioxide films to hot carrier degradation of Si MOSFETs
D J DiMaria

73–82 Charge trapping and degradation of thin dielectric layers
M M Heyns and A V Schwerin

83–92 Hot carrier-induced degradation modes in thin-gate insulator dual-gate MISFETs
H Iwai

93–106 The impact of hot carrier degradation on scaling of sub-μm CMOS processes
H M Mühlhoff, M Steimle and J Dietl

107–116 SOI CMOS devices
M Haond and O Le Néel

117–125 Poly-Si thin film transistors
S D Brotherton

Contributed Papers

Section 1: Fundamental Electronic Processes and Measurements

127–130   Modelling of individual interface states in MOSFETs
M Schulz

131–134   Tunneling resonance by barrier symmetrisation tuning at the Si–SiO$_2$–interface
T Poppe, M Bollu and F Koch

135–138   Hole trap analysis in SiO$_2$/Si structures by electron tunneling
M Schmidt and H Köster jr

139–142   Interface trap measurements using 3-level charge pumping
N S Saks, M G Ancona and Wenliang Chen

143–146   Individual attractive defect centers in the Si–SiO$_2$ interface
A Karmann and M Schulz

147–150   Electron trapping-induced conductance- and noise-dynamics in ultra-thin metal–oxide–silicon tunnel diodes
M O Andersson, K R Farmer and O Engström

151–154   Interface state distributions and the center of reconstruction model
H Flietner

Section 2: Growth and Properties of Grown Oxides of Silicon

155–158   Dynamics of silicon oxide growth: molecular simulation and test
V V Pham, R Razafindratsita, G Sarrabayrouse and J J Simonne

159–162   Effect of preliminary atomic hydrogen treatment on Si oxidation
I A Aizenberg, A V Andrianov, S V Nosenko and V A Khvostov

163–166   An effective oxidation technique for the formation of thin SiO$_2$ at $\leqslant 500°C$
V Nayar and I W Boyd

167–170   Effects of cooling rate on MOS interface properties
K Heyers, A Esser, H Kurz and P Balk

171–174   Origin of reliability differences in oxides grown by RTO in HCl/O$_2$ and pure O$_2$ ambients
K Barlow and V Nayar

175–178   Low pressure thermal oxide applicable as bottom oxide in oxide–nitride–oxide triple layers
E P Burte and A Bauer

179–182   Electrical characteristics of Cl-implanted thin gate oxides
S Verhaverbeke, M M Heyns and R F De Keersmaecker

Contents

183–186 Noise and other electrical characteristics of CMOS FETs fabricated with furnace, Anodic and rapid thermal oxides
D C Murray, J C Carter, N Afshar-Hanaii, A G R Evans, S Taylor, J Zhang and W Eccleston

Section 3: Properties of Deposited Dielectrics

187–190 Thin TiO_2 as a film with a high dielectric constant
A Spitzer, H Reisinger, J Willer, W Hönlein, H Cerva and G Zorn

191–194 Diffusion of hydrogen in and into LPCVD silicon oxynitride films
C H M Marée, W M A Bik and F H P M Habraken

195–198 Characterization of SiO_2 films deposited by reactive excimer laser ablation of SiO target
A Slaoui, E Fogarassy, C Fuchs and P Siffert

199–202 Low pressure MOCVD of tantalum oxide
E P Burte and N Rausch

203–206 Epitaxial fluoride films on semiconductors: MBE growth and photoluminescent characterization
N S Sokolov, S V Novikov and N L Yakovlev

207–210 Improved surface treatments and passivation procedures of GaAs crystals controlled by photoluminescence measurements
S K Krawczyk, M Gendry, J Tardy, F Krafft, P Viktorovitch, P Abraham, A Bekkaoui, Y Monteil, R Schütz, R Riemenschneider, R Richter and H L Hartnagel

211–214 Low temperature chemical vapor deposition of Si_3N_4 thin films assisted by electrical discharge
B Balland, R Botton, J C Bureau, Z Sassi, J Prudon and M Lemiti

Section 4: Defects and Impurities in SiO_2

215–218 Depassivation of P_b sites by heat and electric field
E H Poindexter, G J Gerardi, F C Rong, W R Buchwald and D J Keeble

219–222 Information on the spatial distribution of P_b defects at the (111)Si/SiO_2 interface revealed by ESR observation of dipolar interactions
A Stesmans and G Van Gorp

223–226 Radiation damage in SiO_2/Si structures induced by high energy heavy ions
M C Busch, A Slaoui, E Dooryhee, M Toulemonde and P Siffert

227–230 The effect of zinc in silicon dioxide gate dielectrics
T Brozek, V Y Kiblik, O I Logush and G F Romanova

viii   Contents

Section 5: Poly and Amorphous Structures

231–234 Drain bias instability in polycrystalline silicon thin film transistors
N D Young, S D Brotherton and A Gill

235–238 Low temperature plasma oxidation of polycrystalline silicon
P K Hurley, J F Zhang, W Eccleston and P Coxon

239–242 The effect of phosphorus implantation dose on the gap-state density in polycrystalline silicon
P Vassilev, R Paneva, V K Gueorguiev and L I Popova

Section 6: Hot Carrier Phenomena

243–246 Purely electronic dielectric breakdown of thin SiO_2 films
J Suñé, E Farrés, M Nafría and X Aymerich

247–250 Spatially-resolved measurements of hot-carrier generated defects at the Si–SiO_2 interface
A Asenov, J Berger, P Speckbacher, F Koch and W Weber

251–254 Modeling and characterization of submicron P-channel MOSFETs locally degraded by hot carrier injection
A Hassein-Bey and S Cristoloveanu

255–258 Gate-controlled electroluminescence from reverse-biased Si-MOSFET drain contacts
A Kux, M Schels, F Koch and W Weber

259–262 Electron spin resonance study of trapping centers in SIMOX buried oxides
J F Conley, P M Lenahan and P Roitman

263–266 Oxide field dependence of bulk and interface trap generation in SiO_2 due to electron injection
A V Schwerin and M M Heyns

Section 7: Degradation and Trapping in SiO_2

267–270 The role of the two-step process in hot-carrier degradation
M Brox and W Weber

271–274 AC hot carrier degradation behaviour in n- and p-channel MOSFETs and in CMOS invertors
R Bellens, P Heremans, G Groeseneken and H E Maes

275–278 Correlation of generated electron traps in degraded silicon dioxide
J F Zhang, S Taylor and W Eccleston

Contents

279–282   Defects of oxidized <100> silicon surfaces induced by high electric field stress from low (100 K) to high (450 K) temperatures and relation with the trivalent silicon defect
D Vuillaume, R Bouchakour, M Jourdain, G Salace and A El-Hdiy

283–286   Homogeneous hole injection into gate oxide layers of MOSFETs: injection efficiency, hole trapping and Si/SiO_2 interface state generation
A V Schwerin and M M Heyns

287–290   Fast and slow interface state changes in Fowler–Nordheim stressed capacitors
M J Uren

291–294   Detrapping of trapped electrons in SiO_2 under Fowler–Nordheim stress
J F Zhang, S Taylor and W Eccleston

295–298   Characterization of hot-carrier effects in short channel NMOS devices using low frequency noise measurements
M J Deen and C Quon

299–302   A search for protons in irradiated MOS oxides
J T Krick, J W Gabrys, D I Semon and P M Lenahan

303–306   Prediction of hot electron degradation in MOSFETs: a comparative study of theoretical energy distributions
C C C Leung and P A Childs

Section 8: Silicon on Insulator

307–310   Anneal characteristics of E'_1 centres in buried oxide layers and oxide precipitates in silicon
R C Barklie, T J Ennis, K J Reeson and P L F Hemment

311–314   An application of a new chemical etching process for oxidation induced stacking faults in SIMOX structures; comparison with silicon
C Tsamis, D Tsoukalas, N Guillemot, J Stoemenos and J Margail

315–318   Infrared microscopic spectroscopy analysis of silicon on insulator bevelled samples
J Samitier, A Pérez-Rodríguez, B Garrido, J R Morante and P L F Hemment

319–322   Silicon-on-insulator waveguides
N Mohd Kassim, T M Benson, D E Davies and A McManus

323–326   Evaluation of SOI/SIMOX structures by Raman scattering measurements obtained at different excitation powers
A Pérez-Rodríguez, F Coromina, J R Morante, J Jiménez, P L F Hemment and K P Homewood

327–330   Investigation of hysteresis and floating-body effects in SOI-MOSFETs
T Ouisse, G Ghibaudo, J Brini, S Cristoloveanu and G Borel

331–334 A comparison of the relative merits of n+ or p+ polysilicon gates for ultra thin SOI MOSFETs
G A Armstrong and W D French

335–338 A study of heavy doping effects on the performance of thin film SOI MOSFETs
G A Armstrong and W D French

339–342 Determination of generation lifetime in thin film silicon-on-insulator (SOI) material using capacitance time, charge time and gated diode measurements
L J McDaid, S Hall, W Eccleston and J C Alderman

343–344 **Author Index**

Preface

The 1991 INFOS Conference was held in Liverpool, England from April 2nd to 5th. It was followed by two workshops on the *Hot Electron Degradation of MOS Devices* and *Silicon on Insulators: Materials Devices and Technology.* Key papers from the workshops have been included in this volume.

Approximately 70 papers were accepted for either full or poster presentations, an acceptance rate of 60%. The manuscripts were very fully refereed to the highest standards and we believe the resulting Proceedings represent work of a high calibre carried out in USA, Germany, UK, France, Japan and countries of Eastern Europe. Despite the travel problems that resulted from the conflict in the Gulf, a total of fifteen countries were represented at the meeting and papers covered all aspects of MOS science and technology.

We would wish to express our thanks to the Programme Committee and reviewers who did a difficult job in a very short time span. Particular thanks are owing to Werner Weber and Steve Hall who organised the workshops. A large number of staff at Liverpool were involved in the paper selection procedure, in particular Dr J S Marsland who prepared the spread sheets and Mrs B Lussey and Miss V Wilson who, alongside many other conference tasks, ensured minimum delay in distributing the relevant papers. Thanks are due to the staff at the publishers Adam Hilger (Institute of Physics), particularly Kathryn Cantley, who plans to have the Proceedings available and distributed in record time.

We look forward to seeing you at Delft in 1993 where Professor Balk has the pleasurable if onerous task of hosting the meeting.

W Eccleston
M Uren

Paper presented at INFOS '91, Liverpool, April 1991
Invited Papers

Discrete conductance fluctuations and related phenomena in metal—oxide—silicon device structures

K R Farmer

Dept. of Solid State Electronics, Chalmers Univ. of Technology, S-412 96 Göteborg, Sweden

ABSTRACT: This paper reviews the efforts both to understand and make use of discrete conductance fluctuation phenomena in metal-oxide-silicon device structures. The observations of simple two-level fluctuations and of more complex multi-level fluctuations by numerous groups are compared. The various models that have been proposed to explain the wide range of fluctuation behavior are presented and discussed. The applications that have grown out of an increased understanding of these phenomena are surveyed.

1 INTRODUCTION

The discovery of low-frequency discrete fluctuations in the channel conductance of small metal-oxide-silicon field effect transistors (MOSFETs) (Ralls *et al* 1984) and in the current through small area MOS tunnel diodes (Farmer *et al* 1987) has opened a new field of exploration into the nature of individual defects near the oxide-silicon interface. From studies in this field it has been possible to gain insight into the microscopic origins of ensemble processes found in larger devices such as $1/f$ noise, deep level transient phenomena, and oxide and channel degradation effects. In their simplest incarnation, the fluctuations appear as randomly occurring switching between two distinct conductance levels. The lifetimes of each level are usually found to be exponentially distributed, so the fluctuation time-dependence can be characterized entirely by two time constants. These time constants, which can be observed in the range from fractions of a millisecond to hours (limited mainly by the speed of the measuring system and the observer's patience), together with the fluctuation magnitude completely describe the signal for a given set of device bias conditions. An increasing body of data being collected in the literature reveals that these easily described fluctuations can display a surprisingly complicated range of behavior in response to changes in temperature and voltage, and among different fluctuators under the same bias conditions. In addition, anomalously large fluctuation amplitudes and complex multi-level fluctuations are often observed. Numerous models have been developed which generally attribute the simplest two-level fluctuations (TLFs) to discrete changes in charge or configuration of individual trap states or oxide defects. In the case of diodes, trap-assisted tunneling has also been suggested to explain the fluctuations. Experiments implicate all three processes: charging, defect motion, and tunneling.

In this paper I review the efforts both to understand these discrete fluctuations in MOS devices, and to make use of the information which they may be able to provide. In an attempt to discern common features, careful attention is paid to the voltage and temperature dependence noted by different research groups. The complexity of this behavior coupled with the low number of observables means that the fluctuations are naturally difficult to model. The various models which appear in the literature are presented in some detail, along with a discussion of their ability to describe the broad range of fluctuation phenomena. I also mention several models which have been developed outside the context of the MOS conductance fluctuations, and I note similar fluctuation behavior that has been observed in other device systems. While considerable progress has been made toward understanding TLFs in MOS devices, the question of their exact microscopic origin is still very much an open one.

© 1991 IOP Publishing Ltd

In the final section of this paper I survey what has been learned about the relationship between TLFs and such phenomena as $1/f$ noise, deep level transients, oxide and channel degradation, photoemission, low-frequency conductance, and "slow" states. In addition I describe some of the applications arising out of this knowledge. Realistic MOSFET noise models now utilize the connection between TLFs and $1/f$ noise. Individual fluctuators can be employed as internal probes in working sub-micron FETs. Oxide and channel degradation effects can be monitored as changes in the properties of extant TLFs, or by the sudden appearance of new ones. Finally, the conductance state of certain TLFs can be controlled electrically, suggesting the possibility that bistable devices which work on the atomic scale in MOS structures at 300 K may not be just a dream.

2 DISCRETE CONDUCTANCE FLUCTUATIONS: OBSERVATIONS AND MODELS

This section is divided into three parts. First I review the observations and models associated with fluctuations in the channel current of small MOSFETs. This is followed by a similar review for fluctuations in the tunnel current through small, thin oxide MOS diodes. It is clear that great energy has been put into measuring the properties of TLFs in these systems, and an equal or larger effort has been directed toward understanding the underlying microscopic processes. The resulting models vary widely in scope and believability, and explain the fluctuation phenomena with differing amounts of success. I will briefly discuss the TLF data and models in the final part of this section.

2.1 MOSFETs

The pioneering studies of discrete fluctuation phenomena in small MOSFETs were performed by a group at Bell Laboratories. Their initial paper (Ralls et al 1984) describes two-level current fluctuations in a 0.15 μm × 1 μm n-channel device biased mainly in strong inversion. They present additional measurements in a subsequent paper (Howard et al 1985). Because of the discrete nature of the TLFs, and because of their size and gate voltage dependence, the authors attributed the TLFs to random filling and emptying of individual traps in the oxide, adjacent to the channel. The time constants associated with a given state of the TLFs were seen either to increase or to decrease with increasing gate voltage, indicating that the traps could be either negative or positive scattering centers. Hence they were identified as the time constants for single electron capture and emission, τ_c and τ_e, respectively. The fluctuation magnitudes varied greatly, corresponding to a change in the number of carriers in the channel which could be either larger or smaller than one. Thus they concluded that the channel mobility was also affected by the fluctuations. By assuming that each trap was (1) located at a well-defined energy level E_T, (2) at a depth d_o inside the oxide, and (3) in equilibrium with the channel electrons at the Fermi level E_F, they used the principle of detailed balance (equal number of captures and emissions) to estimate d_o for the various traps, with the resulting values ranging from 0.2 to 2.0 nm. One of their most interesting observations was that at temperatures ranging from 4.2 K to 111 K, for all of the TLFs they studied, both τ_c and τ_e were thermally activated with activation energies in the range $E_{act} \sim 9\text{-}280$ meV. This implied that the simplest model of activated field emission over a potential barrier between the trap and the silicon conduction band could not be used to explain their data, because at distances ≤ 2 nm and for such low activation energies charge capture should occur mainly by direct tunneling. They suggested that the activated dynamics might be explained either through a detailed study of inelastic tunneling processes, or by a potential barrier which could be present if the capture of an electron at a given trap site were accompanied by the atomic motion of the surrounding lattice or network, analogous to the configurational coordinate models used to describe the behavior of charged defects in semiconductors (for example, Henry and Lang 1979). In a later paper the group investigated the possibility that lattice motion was involved in the fluctuation process (Jackel et al 1985). At low temperatures the electrons and phonons are only weakly coupled, so an applied source-drain voltage can raise the electron temperature, T_e, above that of the lattice, T_l. By independently controlling these two temperatures they showed for one trap that while τ_c and τ_e were strongly dependent on T_l, τ_c did not change as quickly at the corresponding values of T_e, and τ_e did not change at all with T_e. Since electron capture and emission coupled with motion of the lattice would be expected to show a stronger dependence on T_l than on T_e, the authors suggested that atomic relaxation did occur during capture of an electron from the high energy tail of the inversion layer, and that emission occurred when the lattice activated the trap system to an energy where the electron could escape back to the inversion layer.

While the Bell group does not work out a detailed theory for the fluctuations, they do suggest equations that would be expected to apply for a process involving charge capture and emission. In such a model, detailed balance leads to an expression for the ratio of the time constants

$$\tau_c/\tau_e = (1-f_T)/f_T = g \exp\left[(E_T-E_F)/k_BT\right], \quad \text{Eq. 1}$$

where f_T is the trap occupancy function if only single electrons are involved, and g is the trap electronic degeneracy. From this relation and the usual expression for the electron capture cross-section, σ_n, the two measured time constants would be related to σ_n by

$$\sigma_n = \frac{1}{\tau_c v_{th} n} = \frac{\exp\left(\Delta E_{CT}/k_BT\right)}{\tau_e v_{th} N_c}, \quad \text{Eq. 2}$$

where n is the density of electrons in the channel, v_{th} is the mean carrier velocity, $\Delta E_{CT} = E_C - E_T$ is the energy level of the trap relative to the conduction band edge E_C, and N_c is the effective density of states at E_C. These equations serve as the starting point for a number of models described below.

Perhaps the most extensive investigation and modelling of the discrete fluctuations in MOSFETs has been performed by Kirton, Uren, and coworkers who have studied devices mainly near 300 K. These researchers have reported the temperature dependent kinetics of TLF time constants in both small n-channel (Kirton and Uren 1986) and p-channel (Cobden *et al* 1990) devices. Like the Bell group, they find that the time constants for both levels of a TLF are thermally activated. To explain this behavior they develop a model similar to the one described above in which lattice relaxation accompanies the trapping of a single charge. For n-channel devices, the total energy, electronic + vibronic, of this trap-plus-electron system is illustrated in Figure 1(a) as a function of a normal configurational coordinate, Q. In this model, thermally activated capture occurs when the empty trap is excited to a crossover point, A, far away from equilibrium, where it can be filled by an electron from the conduction band. This leads to a thermally activated capture cross-section,

$$\sigma = \sigma_0 \exp(-\Delta E_B/k_BT), \quad \text{Eq. 3}$$

where ΔE_B is the barrier for capture and σ_0 is a cross-section pre-factor which depends on the probability for inelastic tunneling from the inversion layer to the trap. At the crossover point there is a strong overlap between the two possible states of the system, so the trap can make a transition from a thermally excited empty state to a highly excited filled state. The filled trap then relaxes to a new, lower energy state, dissipating the excess energy by multiphonon emission. This new state has a different barrier, $\Delta E_{CT}+\Delta E_B$ to overcome before the trap can make the transition back to its original (empty) vibronic state. Expressions for the capture and emission time constants can be obtained by combining Eqs 2 and 3. Kirton and Uren (1989b) relate n to $I_d(T)$, the channel current, and extract the temperature dependences associated with v_{th} and n to reduce these expressions to

$$I_d(T)T\tau_c = \frac{\exp(\Delta E_B/k_BT)}{\sigma_0\chi}, \text{ and} \quad \text{Eq. 4}$$

$$T^2\tau_e = \frac{\exp\left[(\Delta E_B + \Delta E_{CT})/k_BT\right]}{\sigma_0 g\eta}, \quad \text{Eq. 5}$$

where χ and η are numerical constants for the system. From these equations they could calculate two independent estimates of σ_0, which typically were found to differ by two or three orders of magnitude. They also calculated values of ΔE_{CT} which placed the filled trap level well below E_F, too low to allow the possibility of observing emission. To resolve these difficulties they assumed that the trap binding energy, ΔE_{CT}, was temperature dependent, to first order varying linearly with T. Thus they replaced ΔE_{CT} in Eq 5 with $\Delta E_{CT} = \Delta H_{CT}-T\Delta S$, interpreting ΔH_{CT} and ΔS as the enthalpy and entropy of ionization, respectively. This interpretation results from the work of Engström and Alm (1983), who noted that the activation energy measured in thermal experiments is an enthalpy which can be split into its entropy and

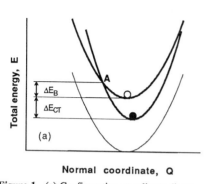
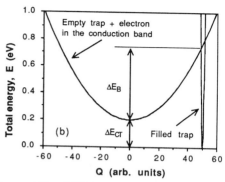

Figure 1. (a) Configuration coordinate diagram after Kirton and Uren (1989b), referenced to E_F, showing an empty trap with an electron at E_F (lower curve), the empty trap state with an electron in the conduction band (upper curve, open circle), and the filled trap state (solid circle). The barriers for capture and emission are ΔE_B and $\Delta E_{CT}+\Delta E_B$, respectively. (b) Approximate configuration coordinate diagram derived for $\Delta S = 7k_B$. This diagram is referenced to E_T.

Gibbs free energy components. (Thus Eq 1 is not strictly correct.) From their final expressions for the temperature dependence of τ_c and τ_e in n-channel devices, they calculated values of ΔE_B in the range 0.19-0.65 eV, ΔE_{CT} in the range 0.08-0.22 eV, and σ_0 in the range 10^{-20}-10^{-15} cm^2. They obtained similar values of these parameters for fluctuations in p-channel devices. They attributed the wide spread in ΔE_B to a continuous distribution of trap environments, and the large range of cross-section prefactors to the possibility that the traps were distributed over a range of distances into the oxide. One of their most interesting findings related to the values of ΔS which they extracted from their measurements. ΔS ranged from $4.4k_B$ to $11.8k_B$ for emission of an electron to the conduction band in n-channel devices, and from $1.5k_B$ to $12.9k_B$ for emission of a hole to the valence band in p-channel devices. Typical values were $5k_B$-$7k_B$, and ΔS was positive for *both* emission processes. They explained the large values of ΔS by the softening of the lattice near the trap upon emission, the change of the trapped electron in n-channel devices from a bonding state inside the the trap to an anti-bonding state in the conduction band, and perhaps a change in the trap degeneracy g, though for a process involving only a single charge this last contribution to ΔS would be expected to be relatively small. They suggested that such large values of ΔS may not be unreasonable, noting that the temperature dependence of the bandgap in silicon near 300 K leads to a surprisingly large entropy change of $\sim 3k_B$ for the creation of an electron-hole pair (Heine and Van Vechten 1976). Positive values of $\Delta S \sim 5k_B$ for emission from traps in both n-channel and p-channel devices implied that different types of transitions were occurring in the two systems, since if the same trap state were involved, the sum of the entropies would be equal to the silicon bandgap entropy, $\sim 3k_B$ rather than $\sim 10k_B$.

In addition to studying the temperature dependence of τ_c and τ_e, Kirton, Uren, and coworkers have also reported the gate voltage dependence of the time constants (Kirton *et al* 1989a) and the magnitudes (Uren and Kirton 1989a, Kirton and Uren 1989b), the observation of anomalous signals (Uren *et al* 1988, Uren and Kirton 1989a), the relationship between the simple TLFs and $1/f$ noise in larger devices (Uren *et al* 1985, Kirton *et al* 1987), and the observation of these slow states in low-frequency conductance measurements (Uren *et al* 1989b, Uren *et al* 1989c). Much of this work is discussed in two review articles (Kirton and Uren 1989b, Kirton *et al* 1989c). Analogous to the Bell group, they used the dependence of τ_c/τ_e on the applied gate voltage, V_g, for a number of TLFs to estimate the trap depths d_0 into the oxide. They extended this technique to argue that most of the fluctuations they observed could not correspond to a multi-electron capture and emission process at a defect site in the oxide. In the single electron model, for devices biased in strong inversion where the trapped charge is fully screened by the inversion layer charge, they obtained values of $d_0 \sim 1$-2 nm. In this regime ΔS was noted to be nearly independent of the gate voltage. In contrast, TLFs near threshold were found to lead to unreasonably large estimates of d_0, as high as 20 nm. Here, as the inversion layer charge was being built up, ΔS for one TLF was shown to increase strongly with increasing gate voltage, while ΔE_B, the barrier for electron capture hardly changed at all. These anomalous findings were considered to reflect the sensitivity of the bonding and dynamical properties of the trap system to the rapidly changing local environment during the growth of screening by the inversion layer charge, but the anomalous ΔE_B findings were noted to be not fully consistent with this explanation.

The voltage dependence of the electron capture rate in strong inversion was generally found to be much stronger than expected from theoretical calculations, which assumed that the fluctuations could modify both the number and mobility of the charge carriers in the inversion channel. The strong dependence of τ_c on V_g was found to be caused by a voltage dependent capture cross-section. In turn, the dependence of σ on V_g was accounted for by a voltage dependent σ_o, with ΔE_B essentially independent of V_g. In all cases, τ_c decreased and σ_o increased with increasing V_g. These results were attributed to changes in the equilibrium inversion layer charge distribution with changing V_g that significantly modified the capture cross-section for a trap removed slightly into the oxide. The electron emission rate was generally found to exhibit a wide variation in its sensitivity to V_g, from no sensitivity to a response equal and opposite that of the capture rate.

They performed a systematic study of the step-height magnitudes as a function of V_g, from weak to strong inversion. Using a simple model which assumed that TLFs could affect only the number of carriers in the channel, they found, after testing 356 distinct fluctuators, that the fractional TLF amplitude $\Delta I_d/I_d$ tended toward a value corresponding to one trapped charge in the oxide. The effect of this charge on the change in the number of inversion layer electrons, Δn ranged from $\Delta n \rightarrow 0$ in weak inversion to $\Delta n \rightarrow 1$ in strong inversion. Analogous to the Bell group, they found that $\Delta I_d/I_d$ varied greatly, in their model suggesting changes in trapped charge which could be ten times less or ten times greater than unity.

While studying the simple TLFs they observed that a fraction of the fluctuators displayed anomalous switching behavior. Three types of such behavior were described: (I) the modulation of a clear TLF by an envelope of the same magnitude which was seen in ~4% of the TLFs, (II) the modulation of a TLF by an envelope of a different magnitude, and (III) fluctuations between three distinct levels, 3LFs. This classification scheme is my own for the sake of comparison with other work, not that of the authors who group type II and type III fluctuators together. From simulations they showed that type I behavior could not be explained by Coulombic interactions between *randomly* placed defects; pairs of defects would have to be grouped and spaced no more than ~2 nm apart. Measurements of the gate voltage dependence of $\Delta I_d/I_d$ for the three-level fluctuations exhibiting type II and type III behavior revealed that while the associated Δn approached 1 in strong inversion for two of the three steps, which is the usual behavior for simple TLFs, it approached 0 for the third step. This third step was the actuating step from the middle to the low conductance level in one type II fluctuator, where steps between the high and middle conductance states apparently never were resolved. The anomalous step was between the high and middle conductance levels in one type III fluctuator; in this case transitions could be seen occurring between all of the levels. The time constants associated with the high and low levels of the type II fluctuator, τ_{high} and τ_{low}, varied with V_g in a manner consistent with charge carrier capture and emission: τ_{high} decreased and τ_{low} increased with increasing V_g. The time constant τ_{middle} of the intermediate level increased with increasing V_g (and thus n, also), suggesting that τ_{middle} was not a capture time. However, the channel current always decreased upon transition from this level, inconsistent with τ_{middle} being an emission time, since at 300 K in simple TLFs they never observed emission fluctuations which decreased I_d. Two possible equivalent explanations for this behavior were that electron capture could occur either to one of two metastable states of the same defect or to one trap in an interacting defect pair, shifting the occupancy level of the second trap away from the Fermi level. They argued that while either of these explanations could be used to model the type I behavior, for this type II fluctuator the transitions between the high and low levels would have to occur so quickly that they would never be observed. An alternative model was suggested which involved the filling of a trap which then periodically flipped between two metastable configurations *without* changing its charge state. In this model the change in trap configuration without a change in charge modifies the channel conductance by approximately the same amount as the change in charge alone.

Kandiah et al (1989) report a detailed study of TLFs and related signals in small MOSFETs, bipolar junction transistors, and in the drain current of junction field-effect transistors. In MOSFETs the TLF magnitudes, $\Delta I_d/I_d$, varied greatly from fluctuator to fluctuator, but generally were found to be independent of temperature, gate voltage and source-drain voltage in weak inversion. The characteristic times for a given TLF, τ_i, usually changed in opposite directions when V_g was varied at constant V_{sd}. The same behavior was observed when V_{sd} was varied at constant V_g. Both time constants were observed always to decrease with increasing temperature, though the authors did not state explicitly that this change was thermally activated. In addition to "normal" TLFs, they mention the observation of all of the types of anomalous fluctuations described above. For all of their device structures they attribute the simple fluctuations to charge state transitions of traps, where the current modulation is due not to the *presence or absence* of the charge, but to the *motion* of the charge upon trapping and de-trapping. Citing Ramo's theorem (Ramo S 1939) and its extensions (for example, Pellegrini B 1986), they suggest that the *process* of filling a given trap induces a *dc step* in the current to the biasing electrode. The current is then restored to its original value when the charge is released. The low current state would correspond to the more negative charge state of the trap in n-channel devices, the more positive state in p-channel devices. In MOSFETs the traps were considered to be either recombination or generation centers located in the oxide, within ~2 nm of the oxide-silicon interface. It was argued though, that if the filling and emptying of a trap occurred by emission (generation), then the characteristic time constants at 300 K for such a trap located at midgap, where the τ_i would be about equal, would be expected to be only on the order of milliseconds, not seconds or hundreds of seconds as was often observed. Hence it was concluded that only capture processes (recombination) were important for explaining TLFs. In this model, they were able to explain all of the observed behavior of their fluctuators. For example, the invariant $\Delta I_d/I_d$ in weak inversion follows directly from Ramo's theorem. The fact that the time constants usually change in opposite directions as a function of gate and source-drain voltage reflects the dependence of the capture rates, $1/\tau_i = n_i \sigma_i v_{th,i}$, on the changing concentrations, n_i, of holes and electrons in the vicinity of the trap. For small values of V_{sd}, assuming temperature-independent cross-sections, the temperature dependence of the product of the time constants is predicted to be essentially that of the bandgap, $\sim \exp(E_g/k_B T)$ where $E_g \sim 1.1$ eV. They show data consistent with this prediction, but do not comment on the possible thermal activation of the individual time constants.

Ohata et al (1990) report a study of anomalous fluctuations in 0.1-0.2 µm wide n-channel MOSFETs. While they usually observe simple TLFs with values of $\Delta I_d/I_d \sim 0.01$, occasionally they see fluctuators with relative amplitudes as large as 30%, even at 300 K. These TLFs can be observed at any bias level, from sub-threshold to strong inversion, and are found to have the customary exponentially distributed dwell times in each state. Both lifetimes are strongly thermally activated, with, for example, $E_{act} \sim 0.7$ eV for each state of one TLF. The high and nearly equal values of E_{act} suggested that, in the activated charge trapping model, the trap level is closer to the conduction band edge than to midgap. This was taken to imply that strong lattice displacement accompanies the charging process. The time spent in the low current state increases with increasing V_g, while the high-state time constant decreases, similar to the behavior of most of the smaller TLFs described above. In addition, the values of $\Delta I_d/I_d$ are found to decrease with increasing gate bias, again the usual behavior. $\Delta I_d/I_d$ decreases with increasing temperature. The authors show that a 30% current modulation is inconsistent with both mobility degradation due to a single scattering center and a number fluctuation of one in the channel carrier concentration. They estimate that the channel area affected by a 30% TLF would be ~80 times the expected screening area due to one charge. Besides seeing the large TLFs, they also observe the same multi-level fluctuations described by other researchers, namely 3LFs and TLFs modulated by an envelope of the same or different magnitude. They suggest that lattice reconstruction at the oxide-silicon interface may explain their anomalously large TLFs, since areas appear to be influenced which are too large to be explained by simple Coulombic interactions, and lattice displacement seems to be involved. From this they infer that the lattice reconstruction with a long-range correlation via interacting double-well potentials may be responsible for their complex signals, though the interaction mechanism is not described. The double-well potentials act such that the capture of an electron at one defect site modifies the capture and emission probabilities at a distant site.

Nakamura et al (1989) report a study of TLF step-heights and 3LF time constants in 0.1-0.2 μm wide n-channel MOSFETs. They observe the usual phenomenon of decreasing fractional step-height with increasing gate voltage, and develop a model for the behavior based on screened Coulomb potentials, similar to the approach taken by a number of other groups. They describe the V_g-dependence of the time constants for the states of one typical 3LF: τ_{high} decreases exponentially, τ_{middle} increases exponentially, and τ_{low} does not change with increasing V_g. They noted that τ_{low} in simple TLFs is usually strongly dependent on V_g, thus the unusual behavior of τ_{low} in this 3LF implied that the signal arose from a doubly charged oxide trap. The nearly equal step sizes connecting the three levels was interpreted as additional evidence that the trap was doubly charged.

Schulz and Karmann have studied TLFs in both n and p-channel MOSFETs (Karmann and Schulz 1989, Schulz and Karmann 1990, 1991a, 1991b). The fluctuations were generally found to exhibit the usual behavior: a wide spread in fractional step-heights; a rapid increase in the fractional step-height with increasing V_g as the device is swept from the sub-threshold bias regime to weak inversion; a decreasing fractional step-height as the silicon is swept from threshold to strong inversion; τ_c (the time constant for the high current state) decreasing with increasing V_g; τ_e (for the low current state) showing a range of behavior with increasing V_g, from not changing to decreasing as strongly as τ_c; and thermally activated time constants with a wide spread in activation energies in the range ~0.1-0.7 eV. Also, they make special note of an unusual type of TLF found in p-channel devices below room temperature whose amplitude seems to be always on the lower side of the wide distribution, whose time constants are only weakly thermally activated, again at the lower end of the spread, and whose dependence on V_g is generally opposite to the "normal" behavior. That is, τ_{high} increases with increasing V_g, while τ_{low} decreases. TLF behavior similar to this in MOSFETs has been reported only in the original paper by Ralls et al (1984). Other researchers (Kirton and Uren 1989b, Restle and Gnudi 1990) make a point to note that they have never seen behavior such as this, but they worked mainly at 300 K, while the anomalous behavior is reported predominantly for devices at lower temperatures. For this fluctuator, the rate of change of the time constants with V_g does not seem to be purely exponential, but instead appears to show symmetric fine-structure.

They attribute the two types of TLFs, normal and anomalous, to repulsive and attractive defect centers in the oxide, respectively. Repulsive centers are said to fluctuate between a neutral state when the center is empty, and a charged state when the center is filled, thus modulating the channel conductivity. This is completely analogous to the process described in many of the models above. In contrast, attractive centers are said to have charge of the opposite sign of the channel carriers when empty, becoming neutral when filled. The researchers develop two entirely different models for the two sets of behavior.

They use a thermionic emission model to describe normal TLFs. In this model, a trap is located in the oxide ~<2 nm from the oxide-silicon interface. It is described by a potential well which, by superposition with the usual interface potential barrier, forms a barrier between the trap and carriers in the inversion layer which can be as high as 3.2 eV. Charges are thermally excited in both directions over this barrier, thus inducing the fluctuations by modulating the number and mobility of carriers in the channel. Expressions describing the capture and emission rates are derived and shown to describe the data. One key hypothesis undergirds this model: "Direct tunneling into the level seems to be an unlikely process, possibly because the probability for inelastic tunneling of free carriers in the channel into a localized level below the band edge is too low" (Schulz and Karmann 1990). For the purpose of calculating the TLF time constants, they model the charge density in the region of the channel near the trap as being shared by the channel and the trap, so that the filled trap removes one carrier from the channel. This "depleted" charge is described by an effective density of states, N_{eff}, to and from which tunneling takes place. As a result of the depletion, the occupancy function f_T for the trap is referenced, not to the silicon Fermi level E_F, but to a local Fermi level $E_{F,eff}$ in the vicinity of the trap. They note that this concept is in conflict with the notion that the charge in the channel would be expected to establish equilibrium quickly after each fluctuation step.

For the case of anomalous TLFs the density of holes in the p-channel is said to be increased at the interface near the defect center, increasing the probability of tunneling to the trap. One of these holes is bound in a potential well in the channel, spatially separated from the trap, never free to contribute to the conductivity of the channel. The fluctuations occur when the hole tunnels between its bound position in the channel and its bound position in the trap, modulating the mobility of the rest of the holes in the channel by changing the scattering strength of the hole-trap system. The system is said to be a weaker scatterer when the hole is in the trap where it can more effectively screen the trap from the channel.

2.2 MOS Tunnel Diodes

The first experiments with TLFs and related phenomena in MOS tunnel diodes were reported by Farmer et al (1987) and Rogers et al (1987) who studied discrete fluctuations in the current through ~1 µm^2 sized, aluminum gate devices. The oxides were ~20 Å thick and grown on p-type substrates which were insulated from the gate by a thick field oxide/nitride layer. Three later works extended this investigation (Farmer et al 1988, Farmer and Buhrman 1989, Farmer 1990). TLFs were observed at negative gate voltages, when the silicon surface was either depleted or accumulated. The relative magnitudes of the TLFs, $\Delta I/I$, varied greatly from fluctuation to fluctuation, in the range 0.1-10%, with 0.1% being the limit of resolution of the measuring system. It was noted that the values of $\Delta I/I$ were as much as 2000 times too large to be explained as a modulation of the tunnel current through a homogeneous oxide by the trapping of a single charge; large-scale barrier inhomogeneities, while not ruled out as a possible explanation for the large amplitudes, were not noted during current-voltage (I-V) characterization of the oxides. $\Delta I/I$ was strongly voltage dependent, increasing sharply as V_g was swept through the depletion plateau, peaking near the flatband bias condition at V_{fb}~-1.1 V, and then decreasing more slowly as the silicon surface became accumulated. One TLF was noted which showed an unusually strong, superlinear decrease in magnitude (on a Log($\Delta I/I$) versus voltage scale) as $|V_g|$ was increased. While $\Delta I/I$ for a given TLF decreased as $|V_g|$ increased above V_{fb}, at any bias level in this regime TLFs could be found whose amplitude was ~1%. The absolute magnitudes, ΔI, varied approximately linearly with increasing temperature and showed all possible types of behavior, either increasing, decreasing or not changing at all. The lifetimes, τ_H (for the high current state) and τ_L, were found to be exponentially distributed in time. These time constants usually varied roughly exponentially with V_g and displayed nearly every possible combination of behavior; the only behavior not observed was the simultaneous increase of τ_H and τ_L as $|V_g|$ was increased. In general, the lifetimes were found to be thermally activated, but they showed a wide variation in activation energies, with $0.007 \leq E_{act} \leq 0.442$ eV in the temperature range $50 \leq T \leq 270$ K. The higher values of E_{act} were found at the higher temperatures and for lower values of $|V_g|$. Occasionally τ_i would not show pure thermally activated behavior, but rather would display curvature on an Arrhenius plot, decreasing in slope with decreasing T.

Since single charge trapping seemed to be ruled out as a strong candidate to explain these phenomena, a model was developed based on the possibility that a defect in the oxide may experience thermally activated transitions between two metastable configurations, a well-known concept in amorphous systems (Galperin et al 1989). As a consequence of this structural change, trap states near the silicon quasi-Fermi levels may either be created or, more likely, shifted in energy, significantly changing their occupation probabilities. These changes in the distribution of trap levels are reversible as the defect fluctuates between its two metastable states. In this model, the change in oxide charge associated with the defect is a *consequence* of the change in the defect configuration, not its cause as in, for example, the multiphonon emission model of Kirton and coworkers. Furthermore the net change in charge during one fluctuation is not restricted to unity; the reconfiguration of a strong, perhaps extended bistable defect could affect the distribution of a number of trap levels, particularly if the traps occur in clusters as has been suggested from preliminary scanning tunneling microscope measurements of oxide on silicon (Welland and Koch 1986, Koch and Hamers 1987).

Invited Papers

The large fluctuation magnitudes and their wide distribution in size were attributed to a change in the number of trapped charges $\Delta Q/q > 1$, coupled with the possibility of local barrier inhomogeneities. The voltage dependence of the relative magnitudes was explained to be consistent with changes in the amount of positive charge trapped in the oxide, which to a first approximation would be equivalent to a change in the voltage drop ΔV_{ox} across the oxide. Such a change would have its largest relative effect in depletion. The decreasing values of $\Delta I/I$ with increasing $|V_g|$ above V_{fb}, particularly for the TLF which showed the anomalously strong effect, were attributed to a decreasing value of ΔQ, since $\Delta I/I$ was seen to fall off more sharply than the fastest decrease that would be observed in the limiting case that the fluctuation acted as an intermittent metallic short, $\sim 1/I$. The voltage dependence of ΔQ would result from either a bias-dependent change in the number of traps influenced by the fluctuating defect, or a bias dependent change in the number of interacting defects which could participate coherently in a fluctuation. The latter explanation was based on observations of multi-level fluctuation behavior to be discussed below. The temperature dependence of ΔI was also attributed to a variable value of ΔQ because ΔI could vary with T more strongly than could the total diode current. The voltage and temperature dependence of the lifetimes were accounted for in the phenomenological equation

$$\tau_i = \tau_{o,i} \exp\left[(E_{act,i} \pm \xi_i q V_g)/(k_B T + g_i(V,T))\right], \qquad \text{Eq. 6}$$

where E_{act} describes the activated transitions between the two metastable defect configurations, ξ relates the applied voltage to the resultant tilt of the two-well potential explaining much of the wide variation of τ_i with V_g, and $g_i(V,T)$ accounts for the curvature occasionally seen in thermal activation plots and the fact that both fluctuation rates often increase with increasing $|V_g|$. This last term describes the possibility that the applied bias can provide a random energy input to the fluctuating defect. The researchers argued that this energy could not be ohmic heating, but rather was due 1) to a localized excitation of the defect which was caused by inelastic scattering of the tunnelling electrons by the defect, and 2) to energy exchange resulting from the change in the total energy of traps associated with the fluctuating defect upon capture and emission of charge. The second process is the multiphonon emission effect described by Kirton *et al.* It was noted that defect excitation or "heating" by such mechanisms has already been established to be an important aspect of TLF behavior in metallic nanoconstrictions, where to first order, $g_i(V,T) \propto V^2$ (Ralls and Buhrman 1988, Ralls *et al* 1989).

In addition to describing the dynamics of individual TLFs, this group also comments on the sensitivity of the fluctuations to thermal and voltage cycling. They find that the fluctuations are stable provided that V_g and T are kept relatively constant. But wide excursions of T, for example from T<100 K to T>270 K and back, or V_g, for example from depletion to strong accumulation and back, often cause the TLFs either to change in character or to disappear altogether. This behavior is in contrast to that of TLFs in MOSFETs which have been reported to be stable for months at 300 K (Restle 1988). The group also reports the observation of a wide variety of multi-level fluctuations at relatively low biases, and extremely complex intermittent fluctuations for $|V_g| \sim > 2.9$ V. They note four different types of multi-level phenomena: A) 3LFs in which transitions between the highest and lowest conductance levels always occur via an intermediate level, B) 3LFs in which transitions to the intermediate level seem to occur only via the lowest level, C) four-level fluctuations in which the duty-cycle *and* amplitude of one TLF seem to be modulated by the transitions of another TLF, and D) a fluctuator which shows TLF behavior at low temperatures, but at higher T, the conductivity pauses at several intermediate levels between the two extremes. Type A and type B fluctuations are similar to the type II fluctuations seen in MOSFETs. In the metastable defect reconfiguration model, these multi-level fluctuations would be interpreted as interactions between two local or extended defects, the state of one defect and its associated trap or traps determining the kinetics of another defect-trap system. The mechanism governing this serial process was postulated to be some strain-related interaction, or the result of the band-like rather than molecular modes of vibration which can be found in amorphous SiO_2 (Sen and Thorpe 1977). The sensitivity of fluctuations to excursions in T and V_g was explained as changes in the detailed potential seen by a particular defect due to changes in the configuration of other defects which become active during the excursion. The intermittent fluctuations at $|V_g| \sim 2.9$ V are found to be composed of periods of stable TLF and multi-level fluctuation phenomena which evolve in time, giving the complex fluctuations the character of a random walk in a random fluctuating potential. At higher voltages, these fluctuations superimpose leading to a gradual and noisy, irreversible increase in the tunnel current. This charging or "wear-out" effect will be discussed in the next section.

Andersson et al (1990) report a detailed study of one TLF in the current through a 0.13 μm² aluminum gate MOS tunnel diode biased in accumulation. The oxide was ~25 Å thick, grown on a p-type silicon substrate. The relative magnitude $\Delta I/I$ of this TLF decreased with increasing $|V_g|$, analogous to the behavior of TLFs seen in this regime by Farmer and coworkers. Furthermore, $\Delta I/I$ was observed to be independent of T in the temperature range 100-150 K. In this same range, the lifetimes, τ_L and τ_H, of each state of this single TLF exhibited nearly the full spectrum of voltage and temperature dependence described above for the collective behavior of several dozens of TLFs. Below ~130 K, the lifetimes were either weakly activated (τ_L) or independent of temperature (τ_H). In this temperature regime, both time constants decreased with increasing $|V_g|$. Above ~130 K τ_H crossed over to strongly thermally activated behavior with E_{act} ~ 0.5 eV (by my calculations), while τ_L remained only weakly activated. At the crossover, the voltage dependence of τ_H became inverted, thus above ~130 K τ_H increased while τ_L decreased with increasing $|V_g|$. The values of τ_L appeared to be merging as T increased, perhaps to a crossover at ~170 K.

To explain these data, a model was developed based on trap-assisted tunneling through two different states of a single oxide defect. The ground state of this defect, D_0, is located in the oxide a certain distance from the aluminum gate, at an energy level positioned ΔE_e below the silicon conduction band edge. The defect has an excited state, D^*, which is located in the oxide at a different distance from the gate, at an energy level situated between the metal Fermi level and the silicon conduction band edge. Electrons are rapidly captured from the metal to D^* with a characteristic time constant τ_{tc}, and emitted from D^* to the silicon with a different time constant τ_{te}. This process provides the current path for the high conductance state of the TLF, and the magnitude of this current is given by $\Delta I = q/\tau_{te}$ if the excited state is located closer to the aluminum (which they find to be the case in order to explain this fluctuator). Occasionally, with a time constant $\tau_c = \tau_L$, an electron is captured from the metal to D_0. This event closes the tunneling path through D^*. Capture to D_0 is as much eight orders of magnitude slower than than capture to D^* mainly because, unlike capture to D^*, it is thermally activated at higher temperatures, indicating the presence of relaxation phenomena associated with the defect ground state. Emission from D_0, also much slower than its excited state counterpart, occurs either by thermal activation to the silicon conduction band or by tunneling to the top of the valence band where holes are accumulated. The time constant for emission is $\tau_e = \tau_H$. After emission, the trap relaxes and is ready to accept another electron, either to D_0 or to D^*.

Quantitative calculations are performed which show that this model can be used to explain the voltage dependence of the fluctuation step-height; in their devices TLFs even as large as 16% can be explained by trap-assisted tunneling through a single defect. The weakly activated behavior of τ_e is attributed to the energy ΔE_e required for emission from D_0. The strongly activated behavior of τ_c above ~130 K is attributed to the lattice relaxation accompanying capture to D_0. The temperature independent behavior of τ_c below ~130 K is explained either by a local temperature increase at the trap site due to extra power dissipation during the high-current intervals, or by the possibility that at low temperatures capture can occur without thermal activation via direct tunneling from states well below the aluminum Fermi level. In this model, the curvature displayed in the Arrhenius plot of τ_e cannot be explained by local heating, since for the defect in this state, no current is flowing through D^*. They conclude that for emission, two concurrent activation energies must be be involved, one to the silicon conduction band, the other possibly associated with recombination with the holes accumulated near the valence band.

Stroh (1990) has shown that discrete conductance fluctuations can be seen in MOS tunnel diodes other than those with aluminum gates. He has observed TLFs in the current through small area polycrystalline-silicon gate devices.

2.3 Discussion

Two of the above models are debatable. First, it seems to me that in using Ramo's theorem, Kandiah *et al* have applied an expression for the generation of a displacement current to describe a persistent effect. They do not explain why the current induced by the motion of a charge to a trap does not decay once the charge stops moving after it is captured. If the persistence is due to the charge's Coulombic effect on the channel, then there is no need to cite Ramo's theorem at all, and the basis of the model becomes identical to the charge trapping models of a number of the other researchers. In addition, in a model where only capture processes are important, if the temperature dependence of the product of the time constants, $\tau_c\tau_e$, is determined by the temperature dependence of the square of the intrinsic carrier concentration as suggested by the authors, then it is carried mainly by only one of the time constants. Thus it is difficult to explain how both time constants can have large, nearly equal activation energies, as is sometimes observed. In this model it is also difficult to explain how the sum of the activation energies can be much less than E_g, which has been seen. This said, it is interesting to note that while the activation energy for the product $\tau_c\tau_e$ for 16 TLFs reported by Kirton *et al* ranges from 0.29 to 1.59 eV, the values cluster around the mean which is $<E_{act}>$ = 1.09 ± 0.17 eV (~E_g). The error is one standard deviation.

The second model which may be brought into question is the thermionic emission model proposed by Karwath and Schulz. The underlying assumption is that tunneling will not be the dominant capture process. As pointed out in the seminal paper by Ralls *et al* (1984), direct tunneling *should* occur to traps situated ≤ 2 nm from the channel. In fact, it is well known from studies of MOS tunnel diodes that quantum tunneling can easily be observed up to distances of ~3 nm through a 3.2 eV oxide barrier. Thermionic emission may still be applicable for traps located deeper than this inside the oxide, or if some as of yet unspecified mechanism inhibits the tunneling.

The hole-trapping model used by Schulz and Karmann to explain the dynamics of one particular type of fluctuation behavior is an interesting proposal. Quantitative calculations are needed to determine if the trap-hole system can modulate the channel mobility to the extent suggested.

It is clear that any successful model must adequately account for the charging, defect motion, and tunneling which accompany the fluctuations. The photoemission observed by Restle and Gnudi (see the next section) is direct evidence that charge emission is involved in at least a fraction of TLFs. Also the reasonable success of numerous groups in predicting the qualitative behavior of TLF step-heights as a screened Coulombic effect provides additional evidence that charge almost certainly is involved. (We note that in this model, quantitative estimates for the step-heights in MOSFETs have been obtained by at least six groups: Jäntsch and Kircher (1989), Kirton and Uren (1989b), Nakamura *et al* (1989), Hung *et al* (1990a), Restle and Gnudi (1990), and Schulz and Karmann (1990)). The wide distribution of step-heights remains to be explained. The earliest, most convincing evidence that defect motion occurs during the fluctuation process was provided by the electron heating experiments by Jackel *et al*. Tunneling obviously is involved in the diode experiments, though it is not clear if the TLFs are due to the modulation of a trap-assisted tunneling path, a changing amount of trapped charge, or both.

In the trap-assisted tunneling model proposed by Andersson *et al*, at least two questions must be answered if it eventually is to be proven successful: (1) What constraints does this model place on the number, spatial, and energetic distribution of traps that would be required in order to explain the transition from single-defect Lorentzian noise behavior in small devices to pure $1/f$ noise in larger devices? It is probable that if the sort of trap they propose were present in numbers typical of intrinsic oxide defects, distributed uniformly in both energy and space, then the resulting $1/f$ noise would be orders of magnitude larger than what is observed. (2) Why would little or no relaxation be expected to be observed for trapping in the excited state of the defect as compared to very strong relaxation for the ground state? These questions aside, it is conceivable that TLFs in tunnel diodes ultimately will be associated with some type of trap-assisted tunneling process. For example, the voltage dependence $\Delta I/I$ noted by Farmer *et al*, and attributed to changes in the amount of trapped positive charge, can be described just as well by electrons tunneling to states at a well-defined energy level in the band gap, followed by recombination with holes in the silicon valence band.

The key issue in the multiphonon emission model by Kirton et al is the values they obtain for the change in entropy, $\Delta S \sim 5\text{-}7k_B$, for the emission of one electron. No detailed calculation has yet been presented which can justify such a large change in entropy for a single ionization. Figure 1b shows the configuration coordinate diagram that one would obtain for a value of $\Delta S = 7k_B$, assuming that the curvature of each well is determined by the ground state energy of a one-dimensional harmonic oscillator. In this approximation

$$E_0/E_0^* \sim (M \omega_0^2 Q^2)/(M^* \omega_0^{*2} Q^{*2}) \text{ and} \qquad \text{Eq. 7}$$

$$\Delta S \sim k_B \ln (\omega_0/\omega_0^*), \qquad \text{Eq. 8}$$

where M and M* are the effective masses of the defect system in its filled and empty state respectively, and ω_0 and ω_0^* are the associated ground state vibrational frequencies. Assuming that the masses and ground state energies of both oscillators are nearly equal gives

$$Q^* = Q \exp (\Delta S/2k_B), \qquad \text{Eq. 9}$$

the expression used to generate Figure 1b. Typical values of $\Delta E_{CT} \sim 0.2$ eV and $\Delta E_B \sim 0.5$ eV determine the relative positions of the oscillator minima. Despite all of its approximations, this diagram is perhaps more realistic than Figure 1a. In this simple picture it is clear just how significant the role of the trapped electron would have to be in binding the defect to the lattice. Of course the configurational coordinate does not represent excursions in real space, but if oscillations in the filled trap potential well corresponded to the motion of a single defect atom on the Ångström distance-scale, then the atom would move over tens of Ångströms when not filled. Obviously the trap would have to extend over several atoms, all of which would be strongly bound to the lattice by the capture of one electron.

Another challenge facing the multiphonon emission model is the consideration of multi-level fluctuations. As can be seen above in the summaries of various groups' efforts, a different model could be developed for each type of complicated behavior. The complexity of multi-level fluctuations eventually led Kirton et al to suggest that trap charge state changes could be coupled with metastable behavior, where a change in configuration *without* a change in charge could alter the channel conductance. Fluctuations would be made between two or more of these metastable configurations as well as between the two different charge states. It is not clear why a change in the internal state of a defect would modify the channel conductance to approximately the same extent as a change in charge alone, but a metastable model can certainly be developed to explain any multi-level fluctuation.

As an alternative to the multiphonon emission model, the metastable defect reconfiguration model was proposed. In this model, the change in oxide charge associated with the defect is a *consequence* of the change in the defect configuration, not its cause. Here complex fluctuations are the result of interactions between defects. At this point, a weakness of this model is its inability to specify the exact cause of the interaction. In addition some question might be raised as to the amount of charge ΔQ that might be required to explain typical TLF step-heights. Evidence has recently been presented for fluctuations in *p-n* junctions that suggests that even "simple" TLFs can be a succession of randomly delayed small steps (Knott 1990).

It is important to note that a model different from the previous two has recently been developed to explain perhaps a related charging effect in MOS capacitors. Hysteretic charging was observed in band to trap tunneling experiments by Zavnut et al (1988, 1989). This phenomenon was interpreted in terms of a process analogous to optical Franck-Condon transitions (Fowler et al 1990a). Here transitions between two potential wells are most likely to occur at a particular vibrational level where the vibrational wave-function overlap between the initial and final states is a maximum. Thus in this model, a change in the charge state of a trap is induced by the excitation of the trap to an appropriate vibrational state. Upon transition, relaxation to the ground state may take place via multiphonon emission. Hence this effect has characteristics of both the metastable defect reconfiguration model and the multiphonon emission model: the defect mediates the charging process, and relaxation can occur through the emission of phonons. It remains to be seen if the Franck-Condon concept can be applied to discrete fluctuations.

Defect fluctuations such as those proposed in the metastable defect reconfiguration model are now a familiar concept in amorphous materials where tunneling transitions occur between two-level systems at low temperatures and activated transitions occur at higher temperatures (Galperin *et al* 1989). Both Farmer *et al* and Andersson *et al* noted the weak temperature dependence of the time constants at low T, the curvature displayed by τ_i on an Arrhenius plot, and strong activation at higher T. Andersson *et al* suggested that two different models might be needed to explain the low T behavior, one for each time constant. Farmer *et al* attributed the curvature to scattering-induced excitation of the defect analogous to the "heating" described by Ralls *et al* (1989) for defects in metal nanoconstrictions. Tunneling transitions between two-level systems provide a natural alternative explanation for the weak temperature dependence. It is possible to describe all of the data acquired to date by suggesting that TLFs in MOS tunnel diodes experience a crossover from strongly activated to weakly or non-activated behavior at ~150 K, but it may be impossible to prove that the fluctuators exhibit such a general characteristic because most TLFs can be observed only over a limited range in temperature. Comparable crossover behavior has been observed for TLFs in a number of other systems including Nb_2O_5 (Rogers and Buhrman 1985), a-Si (Rogers *et al* 1986), AlGaAs (Judd *et al* 1986), Al_2O_3 (Jiang and Garland 1991), and a quantum-well diode in an InP/InGaAs structure (Cavicchi and Panish 1990). The anomalous fluctuators at lower temperatures noted by Schulz and Karmann (1991a) and by Ralls *et al* (1984) may indicate that a similar crossover exists for TLFs in MOSFETs. Additional evidence for this possibility may be provided by the studies of Wong and Cheng (1989) who observed reproducible structure in the noise amplitude in *n*-channel transistors versus temperature in the range 77-375 K, particularly near 200 K. Clearly, additional studies of the temperature dependence of TLFs in MOSFETs are needed.

Comparisons such as this suggest that it may be possible to describe the fluctuation behavior in MOSFETs and MOS tunnel diodes simultaneously, using a single model. However, there are two obvious reasons why this may be too much to expect. First, the devices are sensitive to traps in different regions of the bandgap. In the MOS tunnel diodes biased in strong accumulation, the electron quasi-Fermi level is very near the valence band edge, thus in principle, tunneling electrons can be trapped at energy levels anywhere in the gap. Also, the tunnel current is sensitive to traps above the conduction band edge, between E_C and the gate Fermi level. In contrast, MOSFETs are biased in inversion, so the channel carriers can be captured most easily by traps near the band edges. Second, the gate oxide in a MOSFET experiences fields only up to 1 MV/cm, while the tunnel oxide fields can be as high as 10 MV/cm. This difference may account for the observed contrast in TLF stability. Despite these differences, the two systems do display remarkably similar fluctuation behavior which does encourage speculation that they are related. The study of TLFs associated with identifiable defects, the P_b center for example, may be the next step in discerning the relationship. Recent theoretical calculations of defect processes for such known defects in silicon dioxide and near the oxide-silicon interface have led to predictions of bistability, metastability and strong relaxation effects (Chu and Fowler 1990, Fowler 1990b). Much has been learned about the discrete fluctuation phenomena in MOSFETs and MOS tunnel diodes, but the problem is far from being solved.

3 RELATED PHENOMENA AND APPLICATIONS

3.1 $1/f$ Noise

The connection between the discrete fluctuations in many small device systems, including tunnel junctions and MOSFETs, and bulk $1/f$ noise in larger devices is now well established. Rogers and Buhrman (1984) showed that individual TLFs active at low temperatures in a metal-insulator-metal tunnel junction gave rise to distinct Lorentzian noise power spectra. These spectra, distributed randomly over a broad frequency range, superimposed to create a spectrum with $1/f$ character. As the temperature was raised, an increasing number of fluctuators became active until no single TLF could be distinguished, and the noise power spectrum showed nearly perfect $1/f$ behavior. Similar experiments confirming the relationship between discrete fluctuations and $1/f$ noise in MOSFETs have been reported (Uren et al 1985, Bollu et al 1987a, Bollu et al 1987b, Kirton et al 1987, Karmann and Schulz 1989, Kirton and Uren 1989b, Wong and Cheng 1989). As a result of the increased understanding of the microscopic processes underlying $1/f$ noise, improved MOSFET noise models are being developed (Surya and Hsiang 1986, Hung et al 1990b, Hung et al 1990c). For additional information on $1/f$ noise, the interested reader is referred to three recent reviews (Kirton and Uren 1989b, Weissman 1988, Van Vliet 1991).

3.2 Deep Level Transient Spectroscopy

Karwath and Schulz (1988, 1989) have pioneered the study of deep level transient spectroscopy (DLTS) on small area MOSFETs. After a device is switched from strong inversion to a bias level near threshold, the changing channel voltage V_{sd} (at constant I_d) is seen to be comprised of discrete steps of varying magnitude. These steps, which normally appear as decreases in $|V_{sd}|$ in n-channel devices and increases in $|V_{sd}|$ in p-channel devices, were explained to be due to re-emission of electrons (n-channel devices) or holes (p-channel devices) from traps near E_F which had been filled while the device was biased in strong inversion. In repeated measurements of DLTS transients in which only a few steps can be observed, a given jump can often be attributed to a particular defect because the individual step-heights do not change from trace to trace. The decay times associated with each step appear to be exponentially distributed in time, thus the mean decay time is interpreted as an emission time constant, τ_e. For one identifiable step, τ_e was found to have an activation energy of 0.06 eV in the temperature range 160-270 K. The relative magnitude of another identifiable step was shown to be the largest for low values of V_g (weak inversion), and to decrease with increasing V_g. Occasionally, comparatively large steps were observed which were described as being made up of several smaller steps that were correlated; one clear example was shown where the correlation occurred on a 0.1 second time-scale. These "giant" steps were attributed to the simultaneous emission of charge from a cluster of traps. The study of the statistical properties of τ_e associated with a single fluctuator, compared with the properties of ensembles of fluctuators having a distribution of time constants, represents an interesting opportunity to test the ergodic hypothesis.

3.3 Conductance Measurements

Uren et al (1989c) report that in relatively large area (>>1 μm²) MOS capacitors, the parallel conductance G_p shows a distinct plateau at very low frequencies (≤ 100 Hz). They attribute this plateau to the $1/f$ noise spectrum arising from the "slow" states at the interface. $1/f$ noise requires an exponential distribution of time constants which translates into a frequency-independent ac loss, i.e., a frequency-independent G_p.

3.4 Photoemission

Restle and Gnudi (1990) report the observation of photoemission from a single trap. In three of ~100 MOSFETs studied for sensitivity to 830 nm laser light, they found that the emission time of an active TLF was dependent on the intensity of the light. In contrast, the capture time was independent of the light intensity, indicating that the sample was not being heated by the laser. By measuring τ_e as a function of temperature at three different light intensities, they showed that the emission rate could be expressed as the sum of the known thermally activated rate without light and a temperature-independent photoemission rate. It was not understood why most fluctuators did not show measurable photoemission effects. Light sensitive TLFs have also been reported in a GaAs-AlGaAs single-barrier tunneling device (Snow et al 1990).

3.5 Quantum Effects

The group at Bell Labs has manufactured novel, nanometer scale MOSFET devices with conductance channels as narrow as 30 nm, having a number of closely spaced voltage probes between the source and the drain (Mankiewich et al 1985). In these structures they observe that at low temperatures (<10 K), a single TLF defect can give rise to correlated switching in sev-

Invited Papers

eral adjacent regions along the channel, indicating that its spatial range of influence can extend hundreds of microns along the inversion layer. They interpreted the long-range correlations as being due to quasi-classical or quantum mechanical effects of the scatterer on the source-drain conductance (Howard *et al* 1986, Skocpol 1986).

3.6 Probes Inside Active Submicron MOSFETs

A procedure has been developed for using individual TLFs as probes inside active submicron MOSFETs (Restle 1988, Restle and Gnudi 1990). It requires few assumptions about the fluctuation mechanism, only (1) that TLFs are due to a defect located in the oxide in communication with the inversion layer, (2) that the defect is situated a fractional distance $X \cong L_{ts}/L$ from the source region (where L is the channel length and L_{ts} is the distance between the source and the defect), and (3) that τ_c and τ_e are determined by the potential difference between the gate and the quasi-Fermi level below the defect, V_{gt}. With these assumptions, the voltage dependence of the time constants can be written

$$V_{gt} = V_{gs} - V_{ts} = F_{\tau_c}(\tau_c), \text{ and} \qquad \text{Eq. 10}$$

$$V_{gt} = V_{gs} - V_{ts} = F_{\tau_e}(\tau_e), \qquad \text{Eq. 11}$$

where V_{gs} is the gate-source voltage and V_{ts} is the quasi-Fermi potential difference along the channel between the defect and the source. The functions $F_{\tau_i}(\tau_i)$ are determined from cubic spline fits to plots of τ_c and τ_e versus V_{gs}, measured in the linear regime of operation, where $V_{sd} << V_{gs} - V_{th}$ and the entire channel is in strong inversion. In this regime, since V_{sd} is small, V_{ts} can be neglected in obtaining the functions $F_{\tau_i}(\tau_i)$. Once these functions are known, the trap is said to be "calibrated" because now, for any measured value of τ_i, under any bias condition where Eqs 10 and 11 are good approximations, V_{ts} can be determined. Eq 10 would not be expected to be valid for large values of V_{sd}, assuming that the fluctuations were due to charge capture and emission at a trap site, because the charge density near the trap and τ_c are determined only by local potentials. The location of the defect along the channel can be determined from the relation $V_{ts}/V_{sd} \cong L_{ts}/L$. Using symmetric FETs Restle and Gnudi show that this technique is consistent in its prediction of L_{ts}, particularly for higher gate voltages. At this point the defect is ready to be used as a voltage probe inside the device; changes in the oxide field or changes in the inversion layer charge density can be measured as changes in the time constants with bias. One of the most interesting applications of this technique is to the study of device degradation. Changes in the properties of a pre-existing TLF can be monitored at different bias conditions as a function hot carrier stressing, in order to determine the effects of stress on the local potential and carrier density. Alternatively, defects can be created by hot carrier stressing (Bollu *et al* 1987a, Ohata *et al* 1990), and the properties of the resulting TLFs studied. Restle and Gnudi noted that stressing sometimes led to the creation of stable TLFs, but other times led to TLFs which lasted only a few seconds. Occasionally, the number of observable active TLFs decreased after stressing.

3.7 Oxide and Channel Degradation

Bollu *et al* (1987a) have compared the noise power observed in the channel conductance of undamaged MOSFETs to that which results from intentional hot carrier stressing. In undamaged samples at low temperatures they observed the usual Lorentzian peak structure corresponding to active TLFs, which made a smooth transition to $\sim 1/f$ noise at higher temperatures, as more TLFs became active. The room temperature noise power spectrum for a degraded sample did not exhibit pure $1/f$ behavior. Instead it showed a peak in the 1-100 kHz region of the trace which they interpreted as an interaction effect between traps.

Ohata *et al* (1990) indicate that TLF defects can be created by hot carrier stressing at values of the source-drain voltage as low as 2.9 V, i.e., for electron energies less than the Si-SiO$_2$ interface potential barrier height.

The relationship between TLFs and oxide degradation in MOS tunnel diodes has been reported by a number of groups. Using oxides in the range ~ 50-100 Å thick, Nguyen *et al* (1987), Neri *et al* (1987) and Olivo *et al* (1988) study Fowler-Nordheim (FN) tunneling-induced degrada-

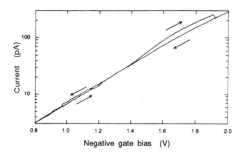
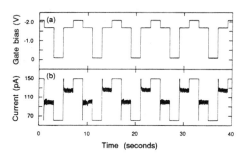

Figure 2. A current-voltage curve for a 0.8 μm² MOS tunnel diode at 300 K showing discrete hysteretic behavior.

Figure 3. (a) Three-level signal to control the state of the device in Figure 2: $V_g=-1.7$, read; $V_g=-0.1$, switch to high current state; $V_g=-2.1$, switch to low current state. (b) The tunnel current in response to the applied signal.

tion. They observe low-field leakage induced by the FN stressing, and discrete complex fluctuations in the tunnel current as precursors to catastrophic breakdown. As noted above, Farmer *et al* (1987, 1988,1989) report the observation of intermittent complex fluctuations in direct tunnel oxides biased at $|V_g| \sim 2.9$ V, and show that at higher voltages, these fluctuations superimpose leading to a gradual and noisy, irreversible increase in the tunnel current. The voltage threshold for this effect corresponds to an ~2 eV electron energy at the oxide-silicon interface, similar to a threshold reported for the generation of electron traps by hot electrons in the oxide conduction band, accelerated towards the interface (DiMaria and Stasiak 1989). Studies of the dynamics of the increasing tunnel current as functions of oxide quality, thickness, bias conditions and temperature are leading to new insights into the origins of oxide degradation (Farmer *et al* 1991).

3.8 Scanning Tunneling Microscopy

The scanning tunneling microscope (STM) may prove to be a valuable tool in finally understanding the microscopic dynamics of discrete fluctuation phenomena. The first attempts to apply scanning tunneling microscopy to the study of defects at the oxide-silicon interface were reported by Welland and Koch (1986) and Koch and Hamers (1987). They observed the familiar TLF structure in the constant voltage tunnel current between a tungsten scanning tip and an oxidized silicon sample. The oxide was ~1.5-2 nm thick. As mentioned above, their findings indicated that traps sometimes appear in clusters. As STM techniques have improved, it has become possible to map out the morphology of the oxide-silicon interface over relatively large areas (Niwa and Iwasaki 1989). With this advancement, it may soon be possible to intentionally incorporate or create interface defects of a known origin, and study their evolution over time. This is particularly interesting when one notes the recent progress made in the field of ballistic electron emission microscopy (BEEM) applied to Schottky barriers, where silicon surface potential variations can be mapped out by changing the voltage on an electrically "transparent" surface electrode (Kaiser and Bell 1988, Bell and Kaiser 1988). It has recently been reported that subsurface modifications in the interfacial electronic properties can be induced by the BEEM tunnel current (Fernandez *et al* 1990).

3.9 Bistable Devices

Occasionally TLFs are found which exhibit hysteretic behavior. Such an effect was noted recently for the tunnel current through a double-barrier normal-metal tunnel junction (Jiang and Garland 1991). As the bias voltage V_b was increased from 0 mV, the time constants τ_i changed monotonically. Suddenly, at $V_b \sim 80$ mV the τ_i changed abruptly by a factor of 2-3, one time constant increasing, the other decreasing. As the bias was decreased, the τ_i remained in their altered states until $V_b \sim 60$ mV. The TLF was attributed to single electron transitions, but the hysteretic modulation of the τ_i was attributed to metastable defect transitions.

Similar, but more dramatic behavior can be observed in MOS tunnel diodes in which positive charge has been created by high field stressing at 100 K. Details of the charging process are described by Farmer et al (1990). The hysteretic behavior for such a device is illustrated in Figure 2 which plots the current-voltage characteristic for a 0.8 µm^2 tunnel diode at 300 K. Notice that as the negative gate voltage is increased, the current switches to the "low" state near $|V_g| = 1.9$ V. As $|V_g|$ is decreased the current switches back to the original state at $|V_g| \sim 1$ V. Successive I-V traces show that this behavior is reproducible, thus the device is bistable in the region $\sim 1 < |V_g| < \sim 1.9$ V. With the applied signal shown in Figure 3(a), it is possible to toggle the device current between the two states, Figure 3(b). The extrema in Figure 3(b) are due to amplifier saturation. The maximum power dissipation of this device is 0.3 nW. Bistable behavior similar to this has been seen in a number of positively charged MOS diodes. It is interesting to speculate that an atomic scale defect is the source of this static memory effect.

4 CONCLUSION AND ACKNOWLEDGMENTS

This paper has been an attempt to collect the key observations associated with fluctuation behavior in MOS devices, to describe and evaluate a number of models which have been proposed to explain the wide range of behavior that can be exhibited, and to survey the applications that have grown out of an increased understanding of these phenomena. A broad theoretical and experimental foundation now exists on which to build studies covering a wide range of topics, from fundamental physics to new applications. It is clear that as device sizes shrink, the effects of the fluctuators will become more important, and the need both to understand and to utilize them will increase. Seven years after the first report of fluctuations in MOSFETs, the work associated with TLFs in MOS systems is just beginning.

I wish to thank R. A. Buhrman at Cornell University for getting me started in this field, and M. O. Andersson and O. Engström, my colleagues at Chalmers. I have benefitted greatly from their support and long conversations. This work was sponsored by a grant from the Swedish National Board for Technical Development.

REFERENCES

Andersson M O, Xiao Z, Norrman S and Engström O (1990) *Phys. Rev. B* **41** 9836
Bell L D and Kaiser W J (1988) *Phys. Rev. Lett.* **61** 2368
Bollu M, Koch F, Madenach A and Scholz J (1987a) *Appl. Surf. Sci.* **30** 142
Bollu M, Madenach A J, Koch F, Scholz J and Stoll H (1987b) *Noise in Physical Systems* ed C M Van Vliet (Singapore: World Scientific) pp 217-20
Cavicchi R E and Panish M B (1990) *J. Appl. Phys.* **67** 873
Chu A X and Fowler W B (1990) *Phys. Rev. B* **41** 5061
Cobden D H, Uren M J and Kirton M J (1990) *Appl. Phys. Lett.* **56** 1245
DiMaria D J and Stasiak J W (1989) *J. Appl. Phys.* **65** 2342
Engström O and Alm A (1983) *J. Appl. Phys.* **54** 5240
Farmer K R (1990) Ph.D. Thesis Cornell University
Farmer K R and Buhrman R A (1989) *Semicond. Sci. Technol.* **4** 1084
Farmer K R, Andersson M A and Engström O (1990) *Proc. 20th Int. Conf. Phys. Semicond.* eds E M Anastassakis and J D Joannopoulos (Singapore: World Scientific) pp 391-4
Farmer K R, Andersson M A and Engström O (1991) *Appl. Phys. Lett.* in press
Farmer K R, Rogers C T and Buhrman R A (1987) *Phys. Rev. Lett.* **58** 2255
Farmer K R, Saletti R and Buhrman R A (1988) *Appl. Phys. Lett.* **52** 1749
Fernandez A, Hallen H D, Huang T, Buhrman R A and Silcox J (1990) *Appl. Phys. Lett.* **57** 2826
Fowler W B (1990) *Rev. Sol. St. Sci.* **4** 551
Fowler W B, Rudra J K, Zavnut M E and Feigl F J (1990) *Phys. Rev. B* **41** 8313
Galperin Yu M, Karpov V G and Kozub V I (1989) *Adv. Phys.* **38** 669
Heine V and Van Vechten J A (1976) *Phys. Rev. B* **13** 1622
Henry C H and Lang D V (1977) *Phys. Rev. B* **15** 989
Howard R E, Jackel L D, Mankiewich P M and Skocpol W J (1986) *Science* **231** 346
Howard R E, Skocpol W J, Jackel L D, Mankiewich P M, Fetter L A, Tennant D M, Epworth R W and Ralls K S (1985) *IEEE Trans Electron Devices* **12** 1669
Hung K K, Ko P K, Hu C and Cheng Y C (1990a) *IEEE Electron Dev. Lett.* **2** 90
Hung K K, Ko P K, Hu C and Cheng Y C (1990b) *IEEE Trans. Electron Devices* **37** 654
Hung K K, Ko P K, Hu C and Cheng Y C (1990c) *IEEE Trans. Electron Devices* **37** 1323
Jackel L D, Skocpol W J, Howard R E, Fetter L A, Epworth R W and Tennant D M (1985) *Proc. 17th Intern. Conf. on the Physics of Semiconductors* ed J D Chadi (New York: Springer) pp 221-4

Jäntsch O and Kircher R (1989) *The Physics and Chemistry of SiO$_2$ and the Si:SiO$_2$ Interface* eds C R Helms and B E Deal (New York: Plenum) pp 349-56
Jäntsch O (1989) *Appl. Surf. Sci.* **39** 486
Jiang X and Garland J C (1991) *Phys. Rev. Lett.* **66** 496
Judd T, Couch N R, Beton P H, Kelly M J, Kerr T M and Pepper M (1986) *Appl. Phys. Lett.* **49** 1652
Kaiser W J and Bell L D (1988) *Phys. Rev. Lett.* **60** 1406
Kandiah K, Deighton M O and Whiting F B (1989) *J. Appl. Phys.* **66** 937
Karmann A and Schulz M (1989) *Appl. Surf. Sci.* **39** 500
Karwath A and Schulz M (1988) *Appl. Phys. Lett.* **52** 634
Karwath A and Schulz M (1989) *The Physics and Chemistry of SiO$_2$ and the Si:SiO$_2$ Interface* eds C R Helms and B E Deal (New York: Plenum) pp 327-33
Kirton M J and Uren M J (1986) *Appl. Phys. Lett.* **48** 1270
Kirton M J and Uren M J (1989b) *Adv. Phys.* **38** 367
Kirton M J, Uren M J and Collins S (1987) *Appl. Surf. Sci.* **30** 148
Kirton M J, Uren M J and Collins S (1989a) *The Physics and Chemistry of SiO$_2$ and the Si:SiO$_2$ Interface* eds C R Helms and B E Deal (New York: Plenum) pp 341-7
Kirton M J, Uren M J, Collins S, Schulz M, Karmann A and Scheffer K (1989c) *Semicond. Sci. Technol.* **4** 1116
Knott K F (1990) *Solid-St. Electron.* **33** 1347
Koch R H and Hamers R J (1987) *Surf. Sci.* **181** 333
Nakamura H, Yasuda N, Taniguchi K, Hamaguchi C and Toriumi A (1989) *Jap. J. Appl. Phys.* **28** L2057
Neri B, Olivo P and Riccò B (1987) *Appl. Phys. Lett.* **51** 2167
Nguyen T N, Olivo P and Riccò B (1987) *Proc. IEEE Int. Rel. Phys. Symp.* (New York: IEEE) p 66
Niwa M and Hiroshi I (1989) *Jap. J. Appl. Phys.* **28** L2320
Ohata A, Toriumi A, Iwase M and Natori K (1990) *J. Appl. Phys.* **68** 200
Olivo P, Nguyen T N and Riccò B (1988) *IEEE Trans. Electron Devices* **35** 2259
Pellegrini B (1986) *Phys. Rev. B* **34** 5921
Ralls K S and Buhrman R A (1988) *Phys. Rev. Lett.* **60** 2434
Ralls K S, Ralph D C and Buhrman R A (1989) *Phys. Rev. B* **40** 11561
Ralls K S, Skocpol W J, Jackel L D, Howard R E, Fetter L A, Epworth R W and Tennant D M (1984) *Phys. Rev. Lett.* **52** 228
Ramo S (1939) *Proc. IRE* **27** 584
Restle P (1988) *Appl. Phys. Lett.* **53** 1862
Restle P and Gnudi A (1990) *IBM J. Res. Develop.* **34** 227
Rogers C T and Buhrman R A (1984) *Phys. Rev. Lett.* **53** 1272
Rogers C T and Buhrman R A (1985) *Phys. Rev. Lett.* **55** 859
Rogers C T, Buhrman R A, Kroger H and Smith L N (1986) *Appl. Phys. Lett.* **49** 1107
Rogers C T, Farmer K R and Buhrman R A (1987) *Noise in Physical Systems* ed C M Van Vliet (Singapore: World Scientific) pp 293-302
Schulz M and Karmann A (1990) *Proc. 5th Cong. of the Brazilian Society of Microelectronics* ed V Baranauskas (SPIE #1405) p 2
Schulz M and Karmann A (1991a) *Appl. Phys. A* **52** 104
Schulz M and Karmann A (1991b) *Physica Scripta* in press
Sen P N and Thorpe M F (1977) *Phys. Rev. B* **15** 4030
Skocpol W J (1986) *The Physics and Fabrication of Microstructures and Microdevices* eds M J Kelly and C Weisbuch (Berlin: Springer-Verlag) pp 255-65
Snow E S, Campbell P M, Glembocki O J, Moore W J and Kirchoefer S W (1990) *Appl. Phys. Lett.* **56** 117
Stroh R J (1990) Ph.D. Thesis, University of Cambridge
Surya C and Hsiang T Y (1986) *Phys. Rev. B* **33** 4898
Uren M J and Kirton M J (1989a) *Appl. Surf. Sci.* **39** 479
Uren M J, Collins S and Kirton M J (1989c) *Appl. Phys. Lett.* **54** 1448
Uren M J, Day M J and Kirton M J (1985) *Appl. Phys. Lett.* **47** 1195
Uren M J, Kirton M J and Collins S (1988) *Phys. Rev. B* **37** 8346
Uren M J, Kirton M J and Collins S (1989b) *The Physics and Chemistry of SiO$_2$ and the Si:SiO$_2$ Interface* eds C R Helms and B E Deal (New York: Plenum) pp 335-40
Van Vliet C M (1991) *Solid-St. Electron.* **34** 1
Weissman M B (1988) *Rev. Mod. Phys.* **60** 537
Welland M E and Koch R H (1986) *Appl. Phys. Lett.* **48** 724
Wong H and Cheng Y C (1989) *Appl. Surf. Sci.* **39** 493
Zavnut M E, Feigl F J and Zook J D (1988) *J. Appl. Phys.* **64** 2221
Zavnut M E, Feigl F J, Fowler W B, Rudra J K, Caplan P J, Poindexter E H and Zook J D (1989) *Appl. Phys. Lett.* **54** 2118

Paper presented at INFOS '91, Liverpool, April, 1991
Invited Papers

Oxidation of silicon

A.M.Stoneham AEA Industrial Technology Harwell Laboratory
Didcot Oxon OX11 0RA, UK

ABSTRACT The continued dominance of silicon in semiconductor technology stems in part from its easily-formed, stable, insulating oxide. As technical demands grow with the need for more compact devices and higher performance, so it becomes more important to understand oxide growth and what might be possible by optimised processing. Moreover, new types of experiment (isotope data, pulsed laser atom probe, noise data, scanning tunnelling and atomic force microscopy) provide demanding tests on models. The current unified model for silicon oxidation in wet and dry conditions goes beyond the traditional descriptions of kinetic and ellipsometric data by explicitly addressing the issues raised in isotope experiments. The framework is still the important Deal-Grove model, in which diffusion and interfacial reactions, operating in series, limit growth. Support comes from theory as well as experiment, and I shall note some of the self-consistent calculations of transport mechanisms for oxidation. However, there are substantial deviations from Deal-Grove behaviour, and some for which that model is clearly inapplicable. Several features emerge as important. First is the role of stress and stress relaxation. Second is the nature of the oxide closest to the Si, where there is evidence of differences from the amorphous stoichiometric oxide further out. These differences may be in composition, in network topology, or otherwise, but they appear to result in a "reactive layer": whereas molecular oxygen diffuses interstitially in the outer dioxide, in the reactive layer there is isotope exchange between diffusing and network oxygens. This interaction can be stimulated electronically or by the use of atomic oxygen. In dry oxidation, we believe that it is this reactivity plus altered diffusion in the altered region close to the Si which lead to deviations from Deal-Grove. For wet oxidation, the reactive layer is less significant, and there remain many problems of interpretation. Thirdly, we must consider the charge states of both fixed and mobile species. In thin films with very different dielectric constants, image terms can be important; these terms affect interpretation of spectroscopies, the injection of oxidant species, and more mundane phenomena like whether water wets a growing oxide film. A more exotic possibility is that self-trapping of injected carriers may occur, with the image term supplying the extra energy gain on localisation.

1 Introduction

Silicon has dominated the semiconductor industry for so long that the two have become almost synonymous. Partly this position is the result of excellent basic properties of silicon; partly it is the wealth of experience in materials control, so that obstacles can be overcome with confidence; partly too, it is ancilliary properties like an easily-formed, stable, insulating, oxide which allows ingenuity in processing and high performance. It is the oxide which gives an edge (among other factors) over, say, diamond, or II-Vs, or some of the more complex systems, like Si/Ge strain layers, despite the specialist advantages they have.

© UKAEA

For the major areas of semiconductor technology, a range of factors come in: breakdown of the oxide, noise associated with traps in the oxide, charges in the oxide affecting transport near to the interface, etc. The volume of research literature on oxidation and the Si/oxide interface continues to surprise, though some is concerned with new methods: buried layers, ion-beam methods, and so on. Nevertheless, the range of unresolved problems continues, and it is these that I shall address in this survey. I shall give only a brief discussion of the oxidation phenomenon itself, since a major review has been given recently. I shall also limit myself to thermal oxidation, omitting anodic oxidation and the several promising (but still expensive) ion-beam methods, like SIMOX. Yet there are still new features to highlight, and new phenomena to discuss.

2 Basic Ideas of Oxidation.

The key ideas of oxidation were laid down over half a century ago. They involve a small number of key ideas (Mott & Gurney 1940; for a more recent review in the context of defect processes, see Hayes & Stoneham 1985). The first is transport: either Si ions must move out, or oxygen ions must move in, or perhaps both may occur. If ions are transported, rather than neutral atoms, then electronic charges must move too. The transport phenomenon has many possible complications, but clearly the simplest assumption is that the chemical potential difference which drives transport is independent of film thickness, so the flux (proportional to the gradient of that chemical potential) is inversely proportional to the oxide thickness, x. This, when diffusion is rate-determining, leads directly to the famous "parabolic" law of oxidation, with the square of the thickness linear in time ($x^2 \approx At$). The second key idea is an interfacial rate process, whether the adsorption or reaction of gas at the outer oxide surface, or some interface process between silicon and its oxide. It is tempting to say this rate is independent of oxide thickness (so giving a linear rate, $x \approx Bt$), but again there are possible complications. The Mott-Cabrera mechanism is just one: here electrons tunnel from substrate (e.g. Si) to adsorbed species at the outer oxide surface (e.g. adsorbed oxygen molecules give oxygen ions) until a particular potential difference is established; this would provide a thickness-dependent field which affects injection of ions at the Si/oxide interface. All these models give particular kinetics x(t), and there is considerable - usually unjustified - optimism in attempting to deduce subtleties of mechanism from measurements of oxidation rate (not always x(t) directly).

2.1 A framework for Silicon Oxidation

Here the classic description is that of Deal and Grove(1965), whose model forms the framework within which almost all generalisations are analysed. The idea is that oxidation is controlled by the two main processes operating "in series". This yields characteristic kinetics, with the thickness x at time t governed by two parameters. The "parabolic" constant describes the diffusion-limited regime, when the square of the thickness is linear in time; the "linear" constant describes the regime (at short times, if indeed it is seen) when interface reactions are controlling. If the Deal-Grove description were complete, with the two parameters independent of oxide thickness, oxidation could be described by an equation in which the inverse oxidation rate (dt/dx) is linear in thickness, x. The slope is controlled by the reciprocal diffusion rate, and the intercept at small thicknesses x by the reciprocal reaction rate.

Linearity of (dt/dx) with x is observed for thick oxides, but there is a substantial deviation for thin oxides (for dry oxidation, this means thicknesses less than 10-20nm). It is this deviation and its implications which account for much of the interest and the plethora of models. Broadly speaking, explanations rely on either a reduction in diffusion constant close to the Si/oxide interface or an enhanced reactivity at early times. Both descriptions give broad qualitative accord with the kinetics of dry oxidation. The kinetics (i.e. x(t)) for a real growing oxide put limits on the processes involved, but do not indicate unique microscopic processes. To go further, detailed calculations and experiments are necessary, these going beyond kinetics alone. In particular, isotope experiments (using sequences of oxidation in which different oxygen isotopes are used in turn, followed by determinations of where within the film each isotope lies, for example) have been especially revealing, and experiments which measure composition close to the interface indicate several features of great interest. It is the deviations from Deal-Grove and the results of isotope experiments which point clearly to potential problem areas for any application of thin oxides.

2.2 General features of Silicon Oxidation.

At the risk of repeating much that is well-known, the results which are of direct interest here are summarised below. A fuller discussion, with extensive references is given by Mott et al (1989); that review is based, in turn, largely on the papers collected in volume B55 of the Philosophical Magazine.

First, the oxide formed is largely silicon dioxide. There are serious qualifications for the region close to the Si/oxide interface, notably from measurements with the pulsed laser atom probe which show a limited region of stoichiometric SiO. With such limited exceptions, the oxide appears to form a continuous random network, with each Si linked to four further silicons via an oxygen "bridge". The oxide is usually amorphous, again with the exception of local regions at the interface where steps favour epitaxy, or impurity features like SiC, or else on very flat Si, where some crystalline form of silica appears to be stable. The local regions of crystallinity appear to have no effect on oxidation kinetics.

Secondly, there is a major difference between dry and wet oxidation; even minor levels of water enhance the rate. Just as in quartz, where there are at least six forms of hydrogeneous species, there may be many distinct hydrogenic species. It is known from theory and from muon and other resonance studies that atomic hydrogen is stable in quartz (contrary to widely-expressed views) and H atoms are presumably stable in thermal oxide away from defects. Clearly, however, a mobile reactive species like atomic hydrogen will rapidly seek defects. Water can have a range of effects on oxide properties, especially on mechanical properties and plasticity. Some of these differences are a natural consequence of the continuous random network, in that less energetic processes can split a water molecule and insert a proton and hydroxyl than are needed to split an oxygen molecule and insert it into the network (possibly at a defect site). A fuller analysis of atomic mechanisms is given by Heggie and Jones (1986, 1987).

Thirdly, in the diffusion-limited regime (thicker oxides) the transport mechanism is mainly interstitial molecular oxygen (dry case) or interstitial molecular water (wet case). Other species can contribute, especially in special regimes, or in the presence of electronic excitation. Silicon motion is negligible (Murrell et al 1990) contrary to the oxidation of silicides and alloys of Si. The dominant transport processes identified lead to oxidation rates which show good agreement between theory and other experiments (Hagon et al 1985). In particular, the parabolic rate constant agrees well with independent measurement of diffusion rates in macroscopic silica (Atkinson 1987), and the activation energies observed agree well with thise predicted theoretically. Theory also makes it clear that the water molecule diffusing in quartz would have a lowest-energy path with rotation, and this is presumable more structure-sensitive than the oxygen molecule motion for which barriers will be relatively similar from one oxide structure to another. It is widely believed (and probably true) that most transport is by neutral species, but there are special cases (e.g. in the presence of electron beams) when charged species may be important; some workers strongly favour charged species. In most cases, diffusion of oxygen into the silicon substrate is not rate-determining, though the consequences (like enhanced impurity diffusion from Si interstitial injection) can be important.

As oxidation proceeds, the interface between oxide and silicon moves into the silicon. Fig 1 shows how this can be envisaged if there is an intermediate layer of oxide. The figure is general, in that it need not presume any special properties of the intermediate layer. However, if it were the reactive layer proposed by Stoneham Grovenor & Cerezo (1987) one would expect the initial oxide (to a thickness similar to that of the reactive layer) to grow by reaction at the outer surface. When the oxide is much thicker, oxygen molecules will diffuse to the outside of the "reactive layer", and there react, exchanging with network oxygens. This simple picture is sufficient to explain many of the isotope experiments (which are not to be understood by most models based on kinetics alone) and to suggest further isotope oxidation sequences.

2.3 Aspects of the thermal oxide.

As noted above in discussing composition, there is evidence for an "intermediate layer" between the silicon and the amorphous silicon dioxide. The intermediate layer appears to be from 0.3 to 3nm thick, depending on preparation and on the type of experiment used to detect it. Thus the layer can differ in composition, for the pulsed laser atom probe shows several layers of stoichiometric SiO; it may differ in local coordination, for spectroscopic methods appear to indicate charge states of Si other than those expected from Si/silicon dioxide alone; it may appear to differ simply because the different dielectric constants of Si and oxide lead to image terms affecting energies of transitions with change in localised charge states (Browning et al 1989). The intermediate layer may differ too in refractive index, suggesting altered local density, or perhaps local topology. Differences in structure are deduced by Brügemann et al (1990). In special circumstances, the oxide can be crystalline (Ourmazd et al 1987; Rochet et al 1986, Fuoss et al 1988, Bevk et al 1990) but this seems not to be important for the present discussion. Oxidation of gas-evaporated Si particles down to 10nm diameter shows thermal oxide growing to about 2nm thickness after anneal in air at 400°C, with an O:Si ratio of about 1.2 (Hayashi et al 1990); this is in line with data from the pulsed laser atom probe, and also suggests an influence of small-scale roughness.

What is important in understanding isotope experiments in dry oxidation conditions (reviewed by Mott et al 1989), is the reactivity of the layer. Whilst oxygen molecules can penetrate most of the silicon dioxide network without significant reaction (subject to some qualifications) they do react within the oxide close to the interface; only at this stage (and to a lesser extent at the outside surface of the oxide) is there exchange between network and diffusing oxygen isotopes.

Some properties of the thermal oxide result from stress, and the consequences are at least part of the explanation of deviations from Deal-Grove and the differences between dry and wet oxidation. Note that the plasticity of the silicon (e.g. from interstitial generation) is important, not just that of the oxide. One view is that the oxide has an altered local topology in response to the stress, e.g. smaller (Si-O) rings. There is a wealth of papers on these aspects (surveyed by Mott et al 1989), though the term stress is used in several distinct ways, and the consequences claimed are sometimes naively large. The existence of stress is no surprise, since the number of silicons per unit area of a surface of silicon is nearly twice that for its oxide. (this is true for real surfaces or even for geometric planes intersecting the infinite solid; clearly, the precise ratio depends on the plane and on the form of the oxide). Relaxation of the stress with time has been inferred from a variety of experiments. Landsberger and Tiller (1987) find two characteristic times are needed: one appears to correspond to the rapid local relaxation of compacted silica, and the other to the viscous relaxation of silica, similar to that seen in bulk experiments.

The amorphous structure and the possible intermediate layer between silicon and fully-oxidised silica may be another consequence of stress. Suppose there is a layer of thickness L and higher energy e per unit volume, and that this is a consequence of a mismatch δ between the silicon and its standard dioxide. The average strain in the layer will be δ/L, leading to a strain energy $\frac{1}{2}c(\delta/L)^2$ per unit volume, or $\frac{1}{2}cL(\delta/L)^2$ per unit area, with c an elastic constant. The sum of this energy (proportional to $1/L$) and the energy eL has a minimum which defines the layer thickness. Of course, if the interfacial energy obtained omitting the layer is low enough, there will be no layer at all.

Related to these interfacial energies, however, are substantial Coulomb energies which are often conveniently ignored. If the Si has a planar surface, and if all the oxide were to start with oxygens attached to it (rather than Si) then one would have a plane of charged oxygen ions, with a divergent energy for an infinite interface; such a polar interface would be a good example of the type which does not occur without extra conditions (e.g. facetting into anionic and cationic regions, or impurity compensation); here steps at the interface and a nearby region deficient in oxygen are possibilities. The excess Coulomb energy puts constraints on the way the interface develops and on its structure; indeed, it may be this term which drives the creation of an intermediate layer more than stress itself. The same Coulomb term may well influence the roughness of the interface and step formation; crystalline interfaces are only noted for very small zones. For discussions of observed roughness, see Goodnick et al (1985), Honda et al (1988) and Roos et al (1989).

The properties of the silicon dioxide produced in different ways are similar for many of the several forms. Thus we can use for guidance on thermal oxides data on the various amorphous silicas and even the crystalline silicas like quartz. This is especially important for stress response, plasticity and viscosity, etc. In all the basic network topology (oxygens twofold coordinated to silicons, silicons fourfold coordinated to oxygens) holds, and other properties vary systematically. Navrotsky has summarised data on enthalpy and volume per (Si + 2O) unit for many such silicon dioxides (fig 2 is based on this work). An examination of her data summary shows two main features: (i) for crystalline forms, the lowest energy corresponds to quartz, and more compact forms have an energy higher by an amount comparable with that expected from the elastic properties of quartz, and (ii) for amorphous forms, the lowest energy corresponds to a larger unit volume (by about 25%) and a higher energy (by about 0.25eV/unit) than quartz, consistent, inter alia, with the large stored energies reported for radiation-amorphised quartz (Tinivella 1980).

2.4 Further comments on Kinetics.

Experimentally, it is natural to plot thickness x against time t. Yet when thinking of mechanisms, it is natural to plot rate (dx/dt) against thickness. It is the thickness which determines the chemical potential gradient driving diffusion, or which fixes the electric fields which may affect injection at interfaces. But what does the oxide thickness x mean? Most measurements infer a thickness. The direct measurement might be the number of oxygens per unit area; if so, the thickness comes only after an assumption about the density. Or it might be an optical phase shift, as in ellipsometry; if so, there is an assumption about refractive index in the near-surface region. But we know the density can vary widely for silicas, and that the refractive index normally scales roughly with the density, so there are uncertainties. These uncertainties matter when we need to look at stress response and perhaps more when one turns to electrostatics and those processes influenced by adsorbed ions (e.g. if the Cabrerra-Mott mechanism applies).

One illuminating way to analyse the kinetics (Stoneham & Tasker 1987) is to look at the data $x(t)$ close to some reference time τ, and to evaluate the logarithmic derivative of the growth rate from data close to some reference thickness $X \equiv x(\tau)$:

$g(X) = - d[\log(dx/dt)]/d[\log(x)]$

where the minus sign is present because oxidation is slower for thicker films. The advantage is twofold. First, one can identify those regimes for which Deal-Grove theory fails, in that when $g > 1$ there is no combination of simple diffusion and thickness-independent interfacial reaction which is acceptable. Secondly, if an extended model is used, e.g. Mott-Cabrera (see below) one can see from the data near $t = \tau$ how plausible the model is and over what range of times that model holds.

As an example, data have been compiled from various sources (mainly Massoud et al 1985 and Adams et al 1980) by Wolters and Zegers-van Duynhoven for a wide range of oxidation conditions: temperature, oxygen pressure, etc. Wolters' discussion has a quite different thrust, but values of g can be obtained readily from these lists ($g = \beta = (1-\alpha)/\alpha$ in their notation). What emerges (fig 3) are systematic failures of Deal-Grove: lowering temperature and, above all, lowering gas pressure, takes one outside the Deal-Grove regime. Thus the model of diffusion and thickness-independent interface reactions in series fails for the thinner (slower-growing) films. It is exactly these regimes which become important if a controlled thin oxide is to be grown. Kinetics alone do not tell one the mode of failure, but remain an important test of models. The failure of simple kinetics also suggests one should look carefully at oxide defects and their consequences.

A similar analysis (Stoneham & Tasker 1987) for oxidation data taken in the presence of an electron beam (Collot et al 1985) indicates (fig 4) strikingly non-Deal-Grove behaviour. The data have several novel features. For very small thicknesses, oxidation is slower at higher temperatures; as noted elsewhere, this seems to be because the sticking probability is rate-determining (Stoneham Grovenorand Cerezo 1987, Miatello & Toigo 1987). In this regime the rate-limiting process is adsorption or reaction of gas molecules with the surface, and re-evaporation leads to a rate which falls as temperature rises. The characteristic activation energy is about 0.25 eV. A second feature is the dependence on electron energy, where dissociation of interstitial molecular oxygen is inferred. Recent work on interstitial oxygen in silica glasses (Awazu & Kawazoe 1990) shows reactions of this type can be controlled by ArF laser excitation; these authors identify several probable reactions giving ozone and atomic oxygen, as well as the molecular form. Dissociation requires more than about 5.1eV.

3 What could be done better?

At this meeting, many may be asking what can be done better as a result of this understanding? I cannot provide all the answers, but several of the issues can be identified.

First, the processing parameters might be optimised. The temperatures, the timing, sensitivity to gas dryness, and the selectivity of which region is oxidised, are all targets for optimisation. Under control are temperature, the nature of gas (e.g. the use of ozone instead of - or as well as - oxygen) and extra factors like electrons, optical stimulation, or the use of ion beams.

Secondly, the oxide reproducibility standards need improving in line with miniaturisation (Fair & Ruggles 1990). The thin, highly-reliable dielectric layer needed for 0.25µm technology has to be controlled in the range 70±3.5Å. The control necessary is not just of time, but also of temperature uniformity, water levels, any doping, and the existence of native oxide, of contamination, or of left-overs from the cleaning process. The control of water is a severe requirement: at 980°C, 125ppm of water will result in oxide outside the specification for sub 0.5µ technology.

Thirdly, the quality of the oxide matters: what is the defect density, where are the fixed charges? How can these be monitored? The defects will influence breakdown and other phenomena like stability against change of stoichiometry and the way reduction occurs. What defect reactions occur under irradiation? Can one devise treatments to make the system harder against radiation?

The possible effects of contamination in cleaning are shown in the way growing oxide is wetted by water. Immediately after removal of thermal oxide, water does not wet the surface: the wetting angle is close to $\pi/2$. As the oxide grows, wetting develops, being essentially complete by the time the thickness has reached 4 nm. The simplest explanation is that there are adsorbed charged ions (perhaps fluoride; see e.g. Ermollett et al 1991) from the etch, and that these remain close to the outer oxide surface; the change in wetting angle then follows from simple electrostatics and the different dielectric constants involved for reasonable charge densities and depths. The key factor is the image potential, since both water and Si are far more polarisable than oxide (Stoneham & Tasker 1983). The image interaction shows up in other ways too, and is evident in Si XPS and Auger spectra measured for thin oxides (Browning et al 1989). It cannot be emphasised too strongly that electrostatic energies are a fact of life and of a magnitude which becomes increasingly significant for thinner oxides.

Whilst most studies of the oxidation process conclude that neutral species dominate (typically the interstitial oxygen and water molecules) any model which includes charged species must allow for image effects. These are probably most significant in the injection processes of oxidation, i.e. when ions move from Si to oxide or from gas to oxide. Stoneham and Tasker (1987) have analysed effects on kinetics.

4 Effects of Oxide Defects.

Since an aim of oxide growth understanding must be how one can control charged defect concentrations in the oxide near to the Si, I now turn to some aspects of these defects. It cannot be stressed too strongly how large are the gaps in our knowledge; even in quartz, where optical and spin resonance studies have clarified the natures of many defects, there are some major uncertainties. The many forms of hydrogen are one instance; the excited states involved in optically-enhanced oxidation, or in the consequences of injected charges, are another. Such difficulties are compounded by the way in which the oxide changes close to the growing interface. Thus, when one turns to a problem like noise, or breakdown, it is an achievement to know what are or what are not controlling factors!

4.1 Fixed Charge.

Fixed charges may be intrinsic or extrinsic. The extrinsic charges are often associated with Na or with H, though one should not forget that the cleaning process can leave surface charged species. The random potential has a range of effects, one relating to mobility close to Si/oxide interface (see Jain et al 1990 for a review).

4.2 Pb centres and similar defects.

Spin resonance and other methods show the existence of singly-occupied silicon dangling bonds. These are presumed to be strictly at the interface; their number is small (less than one per hundred silicons; for a recent analysis see Stesmans & Van Gorp 1990) and there is no reason to believe they play a major role in oxidation. The number depends on temperature, but varies little with oxygen pressure. These defects account for all the fast interface states (for recent views on these centres and their properties, see the special issue of Semiconductor Science and Technology, 1990).

Hydrogen passivation appears to work for Pb centres. There are certainly other traps (some surely intrinsic) which can be passivated to a useful degree. One striking result is the significant difference in capture cross-section (a factor 10) between H associated and D-associated defects (Gal et al 1989). The larger value for H suggests a multiphonon process in which fewer, larger, local mode phonons are involved.

4.3 Noise and associated Traps.

The 1/f noise observed in capacitor structures is associated with charge trapping. Systematic studies (Kirton & Uren 1989; Cobden et al 1990) show there are both "fast" and "slow" trap states. What is less clear is the nature of the traps involved. For example, is it the case that the electron and hole transitions (which can be written formally as:

(1) Hole transition: valence electron + [P] becomes [Pe]
(2) Electron transition: [Ne] becomes [N] + conduction electron with [N], [P] representing defects of unspecified charge state which can capture electrons (so [P] could be a neutral acceptor and [Ne] a neutral donor). However, the trap is presumably in the oxide, and the carriers may well be in the silicon, so there is charge transfer across the interface. In consequence, the recent measurements giving both enthalpies and entropies associated with the traps raise interesting features.

Both the forward reactions (1,2 above) above have substantial positive entropies, of order 5k. It is unlikely that degeneracies could give such large values; it is certainly not plausible that soft (low frequency) vibrations could be effective either (this issue also arises in diffusion problems; note that experiment finds a constant-pressure entropy, not a constant-volume entropy - the two can even have opposite signs - and thermal expansion will ruin any precise compensation of force constants needed to give a high entropy). However, there is a simpler explanation. Suppose that the energy difference between the middle of the oxide band gap and the middle of the Si gap is almost independent of temperature (this is not necessary, but simplifies explanation). Then the known strong temperature dependence of the Si bandwidth (which falls by 0.6eV when kT rises by about 0.1eV) means that there will be a significant positive entropy (perhaps 3k or more) from a deep level (assumed to stay at mid-gap) for both the electron and hole channels.

What might the slow traps be? They appear to be intrinsic; they show large lattice relaxation after capture by tunnelling. The traps appear to be discrete, that is as well-defined as is expected for an oxide which is non-crystalline. Could they be associated with self-trapping?

Invited Papers

4.4 Self-Trapping in Silicas

In self-trapping, a free carrier - unassociated with any defect - can lower its ground-state energy by causing a lattice distortion; the localised carrier is then self-trapped. The phenomenon was first predicted by Landau, who also realised that there could be a thermal barrier to the self-trapped state. The first unambiguous observations were on KCl, where Kanzig observed the formation of Cl molecular ions, essentially a hole localised on two chloride ions. Later observations (some inferred from transport measurements) suggest self-trapping is widespread, and the barrier to self-trapping was verified too in some cases (for a review see Stoneham 1989).

In the several forms of silica, there is evidence for self-trapping. In quartz, the well-known blue luminescence comes from the self-trapped exciton: in effect, the hole component of the exciton is localised on an oxygen, which is substantially displaced, and the electron is localised partly at that oxygen and partly at a neighbouring silicon (Fisher et al 1990). Since vitreous silica shows similar luminescence, self-trapped excitons are presumably involved there too.

Whilst the evidence for self-trapped excitons is clear, the position for single carriers (electrons or holes) is less certain. Indeed, in quartz neither seem to be trapped, though a modest perturbation, like substitutional Ge, leads to trapping of both electrons and holes. In amorphous oxide, where there is a site-to-site variation of properties, the situation is different, and Griscom(1989) has spin resonance evidence for self-trapping in silica glasses. Mott & Stoneham (1977) suggested delayed self-trapping as a partial explanation of transport in thin oxide films on silicon. What is the situation in thin oxide films? Clearly, self-trapping cannot be far away: not only is it present or close in other silicas, but the non-crystalline nature of the oxide will help too. The situation is further complicated by the fixed charges known to be present, since these will also encourage charge localisation (e.g. an electron would trap preferentially close to an Na+ interstitial).

Yet there is a further term favouring localisation too: self-trapping could occur through the extra stabilisation because of the electrostatic image terms, for Si has a relatively large dielectric constant (Stoneham & Tasker 1987). If self-trapping does occur, it would be favoured very close to the silicon, so that self-trapped electrons or holes would be prime candidates for the slow states giving $1/f$ noise.

This possibility, still conjecture, has implications for radiation effects. The first is for radiation-enhanced adhesion, for charged defects lead to an extra term in the interfacial energy when adjacent to a medium with higher dielectric constant. The second aspect is radiation stability. Ionisation damage is well-known to occur in silicas, when purely electronic excitation suffices to create vacancies and interstitials. To create damage, energy localisation is needed. In cases where mechanisms are understood, it is normal for one intermediate stage to be a self-trapped species. Thus four key questions arise. First, do carriers self-trap at the Si/ thermal oxide interface? Secondly, are these self-trapped forms related to the slow traps giving $1/f$ noise? Thirdly, are these same species formed during exposure to ionising radiation? Fourthly, what defect reactions are possible involving these self-trapped states? This fourth question brings us back to the oxidation process, and to the substantial series of experiments on optical and electronic excitation. The many models and rationalisations still leave an incomplete picture. This is, of course, not the only aspect of charge transfer between silicon and its oxide which remains unsettled. Thermionic emission ideas (Irene & Lewis 1987) suggest electron flux can be a rate-limiting step. Other open questions relate to the energy needed for an electron to dissociate an interstitial oxygen molecule.

4.5 Remarks on Electrical Breakdown

One of the aims of oxide growth must be to limit breakdown. It is natural to assume that there is a critical field for breakdown, so that thicker films withstand larger voltages. Yet it is well known (e.g. Dearnaley et al 1970) that more complex "forming" processes can occur at sub-breakdown fields. Such forming appears to involve defect generation processes and can lead to longer-term degradation, though up to several hundred switches from high- to low-current states can be managed. Moreover, recent results (Sofield 1990) show that oxide films of 42Å thickness seem to withstand higher fields than the thicker films (up to 104Å) studied. How thin can a useful film be made?

Several models of breakdown have been noted (Rowland & Hill; Hill & Dissado). At present it seems likely that there is an initial degradation till a threshold, then a new conduction mechanism emerges which leads to breakdown; it seems to be a local phenomenon, and it may be associated with defect generation (Suñé et al 1990). The low field breakdown may be the result of oxide decomposition at defects (Si + oxide going to volatile SiO) and can be suppressed by adequate oxygen. This process relates to reduction, and is the opposite to the oxidation behaviour of relevance here. Defect dynamics are thus a natural component of the complex range of phenomena which have come to be known as "wear-out". Whether or not there is the link one might envisage with the defect sites which nucleate oxide decomposition reactions (Rubloff et al 1987) is not clear to me.

5 Other Systems: What happens for Si/Ge?

The presence of Ge completely alters behaviour. In Si-Ge random alloys (where compositions of 14% Ge and 50% Ge have been studied), the Ge is effectively stuck at the semiconductor/oxide interface. As oxidation proceeds, the outer zone of the semiconductor (next to the oxide) becomes Ge-rich (Ge piles up), preventing access of Si to oxygen. Thus Fathy et al (1987) produced epitaxial layers of Ge on Si(100) by steam oxidation of Ge-implanted Si. In related experiments, the effect of a marker layer of Ge on a silicon substrate has been studied. It appears that "Si interstitials cannot get through"; presumably it is also true that oxygen cannot get through either, since there is no significant Si motion in the normal oxidation of pure silicon. It is also true that there is no longer oxidation-enhanced diffusion of B associated with Si interstitial generation in the Si/Ge (LeGoues et al 1989). A distinction between "dry" and "wet" oxidation remains. Dry oxidation is relatively unaffected by Ge; wet oxidation is enhanced by Ge: under wet conditions, Ge/Si oxidises faster than Si, and the rate increases with Ge content. For the wet oxidation of a commensurately-grown Ge/Si strain layer on a silicon substrate, oxidation is both faster than for pure Si and its rate increases with Ge content. The density of fixed charge at the oxide/(Ge/Si) interface increases with the Ge content too (Nayak et al 1990).

Acknowledgements. I have had invaluable discussions with many colleagues, and would mention in particular Professor Sir Nevill Mott, Professor S C Jain, Mr M Murrell, Professor S Rigo, Dr F Rochet, Dr C Sofield and Dr P W Tasker. Some of this work was supported by the Underlying Research programme of the UKAEA, and later by its Corporate Research programme.

A C Adams, T E Smith and C C Chang, 1980 J Electrochem Soc **127** 1787
A Atkinson 1985 Rev Mod Phys **57** 437
A Atkinson 1987 Phil Mag **B55** 641

Invited Papers

K Awazu and H Kawazoe 1990 J Appl Phys **68** 3584
J Bevk L C Feldman T P Pearsall G P Schwarz and A Ourmazd 1990 Mat Sci & Eng **B6** 159
R Browning M A Sobolewski and C R Helms 1989 p243 of Helms & Deal 1989
L Brügemann R Bloch W Press P Gerlach 1990 J Phys Cond Mat **2** 8869
D H Cobden M J Uren M J Kirton 1990 Appl Phys Lett **56** 1245
B E Deal & A S Grove 1965 J Appl Phys **36** 3770
G Dearnaley A M Stoneham and D Vernon Morgan 1970 Rep Prog Phys **33** 1129
A Ermollett F Martin A Maouroux S Marthon & J F M Westendorp 1991 Semic Sci & Tech **6** 98
R B Fair & G A Ruggles 1990 Sol State Technol (May issue) 107
D Fathy, O W Holland C W White 1987 Appl Phys Lett **51** 1337
A J Fisher W Hayes and A M Stoneham 1990 Phys Rev Lett **64** 2667; J Phys Cond Matt **2** 6707.
P H Fuoss W Norton S Brenner A Fischer-Colbrie 1988 Phys Rev Lett **60** 600
R Gale H Chew F J Feigl C W Magee 1989 in Helms & Deal (1989).
S M Goodnick D K Ferry C W Wilmsen Z Lilental D Fathy O L Krivanek 1985 Phys Rev **B32** 8171
D Griscom 1989 Phys Rev **B40** 4224
J P Hagon, A M Stoneham and M Jaros 1987 Phil Mag **B55** 241, 257.
S Hayashi S Tanimoto and K Yamamoto 1990 J Appl Phys **68** 5360
W Hayes and A M Stoneham 1985 Defects and Defect Processes in Non-Metallic Solids (New York: John Wiley).
M Heggie & R Jones 1986 Phil Mag **A53** L65 and 1987 Phil Mag Letts **55** 47
C R Helms & B E Deal (editors) 1989 Physics and Chemistry of Silicon Dioxide (New York: Plenum Press).
R M Hill L Dissado 1983 J Phys **C16** 2145, 4447
K Hofmann G W Rubloff and D R Young 1987 J Appl Phys **62** 925
K Honda R Takizawa A Ohsawa 1988 J Appl Phys **63**
E A Irene and E A Lewis 1987 Appl Phys Lett **51** 767
S C Jain K H Winters and R van Overstraten 1990 Adv Electronics Electron Physics **78** 103
M J Kirton, M J Uren 1989 Adv Phys **38** 367
F K LeGoues R Rosenberg and B S Meyerson 1989 Appl Phys Lett **54** 644, 751; J Appl Phys **65** 1724
H Z Massoud J D Plummer E A Irene 1985 J Electrochem Soc **132** 2693, 2685
N F Mott and R W Gurney 1940 Electronic Processes in Ionic Crystals (Oxford: Oxford University Press).
N F Mott, S Rigo, F Rochet & A M Stoneham 1989 Phil Mag **B60** 189
N F Mott & A M Stoneham 1977 J Phys **C10** 3391
M Murrell, C Sofield 1990 to be published
A Navrotsky 1987 Diffusion and Diffusion Data **53/54** 61-66
D K Nayak K Kamjoo J S Park J C S Woo K L Wang 1990 Appl Phys Lett **57** 369
A Ourmazd D W Taylor J A Rentschler and J Bevk 1987 Phys Rev Lett **59** 213.
F Rochet S Rigo M Froment C d'Anterroches C Maillot H Roulet & G Dufour 1986 Adv Phys **35** 237
A Roos M Bergkvist C Ribbing 1989 Appl Optics **28** 1360
S M Rowland R M Hill L Dissado 1986 J Phys **C19** 6263
G W Rubloff K Hofmann M Liehr and D R Young 1987 Phys Rev Lett **58** 2379.
C J Sofield 1990 ESPRIT Basic Research Activity No 3109, Progress Report No 1.
A Stesmans and G Van Gorp 1990 Appl Phys Lett **57** 2663
A M Stoneham 1990 J Chem Soc Farad II **85** 505
A M Stoneham, C R M Grovenor & A Cerezo 1987 Phil Mag **B55** 201
A M Stoneham and P W Tasker 1986 Semicond Sci & Technol **1** 93
A M Stoneham & P W Tasker 1987 Phil Mag **B55** 237

J Suñé et al 1990 Thin Sol Films **185** 347
GTinivella 1980 Harwell Report AERE TP-877, citing e.g. A Antonini, A Manara and P Lonsi 1978, in "Physics of SiO2 and its interfaces" (edited S T Pantelides) Academic Press.
D R Wolters and A T A Zegers-van Duynhoven 1989 (Philips preprint)

Invited Papers

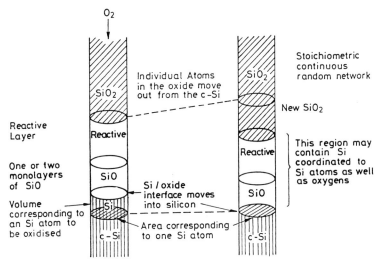

Fig 1. The oxidation process: the structure before and after oxidation by one oxygen molecule is shown for a column corresponding to one Si atom in the silicon substrate. The reactive layer and SiO layer (if present, since these will depend on conditions) move deeper into the Si, but the outer oxide surface moves further out.

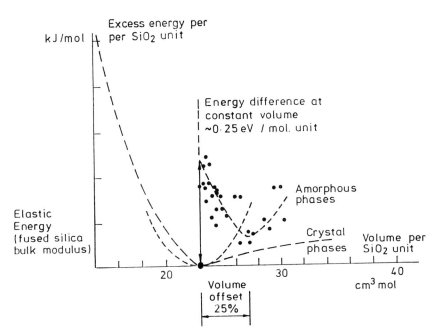

Fig 2. Data for various forms of silicon dioxide (after Navrotsky 1987) compared with elastic energies (with elastic constants for vitreous silica). Note amorphous phases have typically a 25% larger molecular volume and about 0.25eV more energy per molecular unit at constant volume.

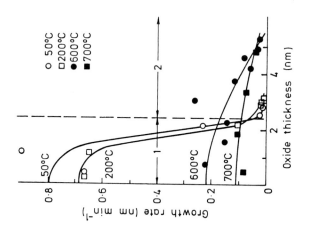

Fig 4. Data for electron-beam assisted oxidation from Collot et al replotted (Stoneham Grovenor & Cerezo 1987) showing clear deviations from Deal-Grove. The regime at low thicknesses where rates fall as temperature rises are controlled by the sticking probability. The data for thinner oxides are consistent with a reactive layer.

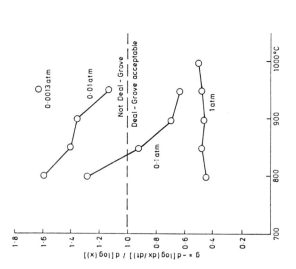

Fig 3. Data from Wolters et al 1989, Massoud et al 1985 and Adams et al 1980 analysed to identify conditions under which the Deal-Grove model fails (g > 1). Failure means that the rate is not determined by a combination of thickness-independent diffusion and interface reactions in series.

Paper presented at INFOS '91, Liverpool, April, 1991
Invited Papers

UV and plasma effects in the Si/SiO$_2$ system

Jordan Kassabov

Institute of Solid State Physics - Bulgarian Academy of Sciences 1784 Sofia, Bulgaria

1. INTRODUCTION

In the modern microelectronic technology many plasma processes are applied to implement integrated circuits with micron and submicron features. The prevalent amount of VLSI IC production is based on MOS structures. At the same time there is not enough knowledge concerning the influence of different plasmas on the fundamental properties of the silicon/silicon dioxide structures.

This paper is an attempt to clarify in a more general way the effects of different plasma treatments which are used or it is likely to be used in the microelectronic technology. For the purpose the effects of oxygen, argon, helium and hydrogen plasmas are studied. Taking into account that the UV radiation is an inevitable satellite of all plasmas special attention is paid to the separate effect of the UV radiation during different plasma treatments. Having also in mind that in all photolithographic processes UV light is used the effect of standard UV lamps is also studied.

Normally oxygen plasma is applied in photoresist stripping and recently also for silicon oxidation. Argon plasma is used in sputtering of many materials, ion milling, plasma etching and other processes. All these processes normally cause degradation of the silicon/silicon dioxide interface. It is interesting whether a suitable plasma treatment could restore the parameters of the Si/SiO$_2$ structures. If such plasma treatments are feasible a complete plasma technology in a closed vacuum line may seem very attractive. For the purpose some effects of helium and hydrogen plasma treatments are also studied.

2. OXYGEN PLASMA EFFECTS

Some investigations on oxygen plasma effects have been performed and published recently [1] on silicon/silicon dioxide structures with oxide thickness in the range of 20 to 100 nm.

The samples are p-type Si(100) 15-17 Ω cm wafers, MOS-grade cleaned and oxidized in dry oxygen at 1000°C, without high temperature annealing in N$_2$ (no POA). So prepared samples are exposed to a RF (13.56 MHz) O$_2$ plasma in a parallel plate reactor for different durations (t_e=1-40 min), at O$_2$ pressure 1 Torr and RF applied power density 0.1 W cm^{-2}. The samples are placed on a grounded electrode whose temperature is 300°C. The measurements were carried out on "gateless" inversion layer structures using the "Inversion layer I/V characteristics" method proposed recently, [2]. For the samples studied both the effective electron mobility in the inversion channel (μ_n) and the total interface charge (Q_T) were calculated.

In Fig.1 mobility dependence on oxide thickness is shown for as-grown structures and for structures after different oxygen plasma treatments. A strong degradation of electron mobility is seen for all samples. A low temperature annealing in hydrogen is not always sufficient to restore the values before plasma treatment.

© 1991 IOP Publishing Ltd

The plasma influence on Q_T is different depending on the initial values of Q_T. When the initial values of Q_T are in the range from 7×10^{10} to 1.5×10^{11} cm^{-2}, in O_2 plasma Q_T increases and reaches a saturation value Q_T^{sat}, O_2 plasma effect being much stronger for the thinner oxides.

When the initial value of Q_T is $(2-10)\times 10^{11}$ cm^2 O_2 plasma treatment reduces the value of Q_T with a saturation of the effect again. It is remarkable that in both cases, a LTA in H_2 (450°C) increases the value of Q_T i.e. some kind of negatively charged interface states are annealed. It is interesting also that saturation values of Q_T after H_2 LTA for both kinds of samples (low and high initial Q_T) tend to almost same value. For example the values of Q_T^{sat} for d=105 nm tend to about 1.5×10^{11} cm^{-2}. Shortly O_2 plasma brings about an increase of Q_T when it is initially low and a decrease in the opposite case.

In conclusion it is evident, that oxygen plasma effects are quite strong and are to be taken into account in microelectronic technology.

3. UV RADIATION EFFECTS

As mentioned above UV radiation is unavoidable in plasma and photolitography processing. In order to study the UV effect by itself [3], gateless silicon/silicon dioxide structures are investigated before (as-grown) and after irradiation by a mercury lamp with a spectrum in which all the lines have energies below 5 eV. Samples are irradiated directly from a distance of 15 cm without any focusing. During the irradiation no electric field across the oxide is applied.

The samples are prepared as above with an oxide thickness of 1000 Å and POA in N_2. The structures under investigation include also MOS capacitors. To form MOS capacitors and contacts to the diffused source and train a 0.5 micron Al+1% Si layer is sputtered. After deposition of metal layer a standard hydrogen PMA is carried out.

Main parameters obtained by the inversion layer I/V characteristics as mentioned above are inversion electron mobility total charge in the oxide and at the interface Si/SiO$_2$, surface potential barrier and Fermi level position at the silicon surface. Since equilibrium Fermi level is temperature dependent, by varying the temperature it is possible to measure also the density of interface states D_{it} near silicon conduction band making use of the expression

$$D_{it} = \frac{\Delta Q_T}{\Delta E_F} = \frac{\Delta(Q_f + Q_{it})}{\Delta E_F} = \frac{\Delta Q_{it}}{\Delta E_F}$$

A well pronounced effect of UV irradiation on the interface state density can be seen in Fig.2a. The plots shown are for an as-grown sample and for the same sample after UV irradiation for 20 min. No matter whether interface states are donor-like or acceptor-like, the Q_T vs.(E_C-E_F) curves have slopes of like signs; that is the reason why it is impossible to distinguish donors from acceptors by C/V measurements only. However, it can be seen in Fig.2a that the increased density of interface states (i.e. of the plot slop) due to the UV irradiation is accompanied by a decrease in the total charge Q_T for the whole (E_C-E_F) range under investigation. This fact together with the fact that positive fixed oxide charge Q_f, as measured by HF C/V curves, remains constant or even increases after UV irradiation suggest that UV radiation creates negatively charged traps at the interface. UV radiation effect on inversion layer electron mobility is shown in Fig.2b.; it can be seen that electron mobility is increased by UV radiation, despite of the almost twice increased interface trap density in the range 0.13-0.18 eV below E_C.

4. ARGON PLASMA EFFECTS

Experimental samples are prepared in the same way as for the oxygen plasma study. Exposure of the samples to RF (13.56 MHz) Ar plasma was carried out in the same planar plasma reactor. The purity of the argon used was 99.996%. The samples were placed also on a grounded electrode whose temperature was 300°C. The RF power density was the same as for the oxygen, 0.1 W/cm^2. The chamber was initially pumped out to a pressure of 5×10^{-4} Torr. Argon was introduced into the chamber under a pressure of 1 Torr. The plasma exposure times (t_e) were 5 and 20 min, [4].

After each treatment of the wafers μ_n, Q_T, Q_f and D_{it}, were determined. Q_f was evaluated by the standard HF C/V technique. Special MOS capacitors for these measurements were prepared by vacuum evaporation of aluminum dots 0.5 mm in diameter after wafer plasma treatment. Fig.3 shows a typical dependence of μ_n on the electron concentration in the inversion channel N_{inv} at room temperature with the oxide thickness as a parameter. N_{inv} is controlled by V_{BS} (V_{BS}-bulk-source bias). The highest values of μ_n for each oxide thickness correspond to $V_{BS}=0$. In Fig.3 for the first time a dependence of μ_n on d_{ox} at room temperature for very low inversion electron densities and for oxide thicknesses smaller than 100 nm is shown.

In general the effect of argon plasma is a reduction of the carrier mobility, proportional to t_e and relative weakly dependent on d_{ox}, Fig.4.

All results mentioned above concern the total effect of Ar plasma, without separating the effects of the different plasma components such as neutral atoms, ions, electrons and vacuum ultraviolet radiation. Further an attempt is made to study the net effect of a part of UV radiation of the argon plasma on the carrier transport and on the charges in the Si/SiO$_2$ structures, [5]. For the purpose during argon plasma treatment a number of structures on the silicon wafer are covered by a 2 mm sapphire filter (α-Al$_2$O$_3$) transparent for vacuum ultraviolet up to 8.7 eV. In this way the covered structures are subjected to the action of photons with energies less than bandgap of SiO$_2$. The sapphire was specially chosen as a filter because of its chemical stability and purity. In Fig.5 inversion electron mobility vs. N_{inv} is shown for as-grown and plasma-treated samples with and without α-Al$_2$O$_3$ filter respectively. It is seen, that the total argon plasma effect (i.e. without filter) is a drastic reduction of the carrier mobility. In opposite, when the plasma acts only by means of its UV radiation penetrating through the filter μ_n increases to values greater than those of the as-grown samples. This increase is similar to that described above after illumination with UV light with energy less than 5 eV.

As can be seen in Fig.6 as a result of the action of UV only, D_{it} increases with an amount of about 3×10^{11} cm^{-2}eV^{-1} in the scanned range of the silicon gap. At the same time the total charge for a fixed Fermi level position decreases by about 2.5×10^{10} cm^{-2}. A comparison of these data suggests that the argon plasma UV penetrating through the filter leads mainly to a generation of negatively charged interface traps. In Fig.6 it is seen also that total plasma effect is expressed in a strong reducing in D_{it} below values under the sensitivity of our technique. The results for different oxide thicknesses before and after total Ar plasma treatment, calculated from high frequency (1 MHz) C/V measurements show an 10-20% increase of Q_f.

It is found also that for the as-grown samples, Q_T (for $E_C-E_F=0.18$ eV) is about 1.3×10^{11} cm^{-2} and is weakly dependent on d_{ox}. A 5 min exposure in Ar plasma decreases Q_T to a practically constant value of about 1.1×10^{11} cm^{-2} in the investigated thickness range. A further increase of t_e does not change Q_T any more.

Taking into account that Q_T after plasma treatment is lower than that of as-grown samples and, also, the significant increase of Q_f after plasma treatment, it is clear that at the interface a quite large negative charge is trapped. In the samples investigated this charge $Q_{it}=Q_f-Q_T$ is in the range of $(1-4) \times 10^{11}$ cm^{-2} and is distributed somehow in the silicon bandgap from 0.25 eV up to 0.83 eV below silicon conductance band.

5. HELIUM PLASMA EFFECTS

Preparation of the samples and experimental techniques are same as above.

The main purpose to study helium plasma effects [6] is to clarify whether a low temperature He plasma will have annealing effect similar to high temperature annealing effect of Si/SiO$_2$ structures in helium ambient or will degrade the structures like argon plasma.

Two types of samples were investigated: group A - thermal Si/SiO$_2$ structures without high temperature annealing in N$_2$ and group B - same structures but subjected to a low temperature argon plasma treatment for a time enough to cause some degradation of the Si/SiO$_2$ structure. Helium plasma treatment conditions are the same as described above, for 5 and 20 min. The effective mobility in the inversion layer at room temperature versus the oxide thickness before and after helium plasma treatment is plotted in Fig.7 for a fixed value of the electron concentration in the inversion channel. It is seen in Fig.7 that a 5-minute helium plasma exposure of as-grown thermal oxide does not change the channel mobility for all oxide thicknesses. For 20 min, μ_n decreases and its degradation is a function of d_{ox}. In contrast to group A the initial wafers of group B are characterized by much smaller values of the mobility as a result of Ar plasma created defects. For comparison argon plasma brings about a considerably stronger decrease in μ_n (about 4 times) than helium plasma. A helium plasma treatment of B samples for 5 min restores μ_n to its initial values typical for as-grown structures.

For wafers from group A two ranges of thicknesses can be distinguished - from 16 nm to about 45 nm and from 45 nm to about 72 nm. Helium plasma reduces the density of interface states 1.5-5 times for thinner oxides and 100 times for thicker oxides. In general, a slight increase in Q_f (with about $(1-5) \times 10^{10}$ cm^{-2}) is observed for thicker oxides. At the same time, the helium plasma effect on samples from group A is a decrease of Q_T up to 5×10^{10} cm^{-2} in the range 0.15-0.25 eV below E_C i.e. He plasma creates also negatively charged interface states.

Initial wafers from group B (after argon plasma treatment) have a low density of interface states in the above mentioned energy range. For all oxide thicknesses five-minute helium plasma treatment gives rise in D_{it} of about $(2-3) \times 10^{11}$ cm^{-2}eV^{-1}. These values are near to those for as-grown oxides. Q_f decreases $(3.5-6) \times 10^{11}$ cm^{-2} without any clear dependence on d_{ox}. Q_T increases to values in the range of $(1.4-1.5) \times 10^{11}$ cm^{-2} which are near to those for as-grown oxides. These results show that helium plasma has different effects on A and B samples.

The net effect of the He plasma UV radiation was also studied [7] using a sapphire filter as described above for Ar plasma.

It was found that the effects of total and UV filtered He plasma are in general similar to those of Ar plasma concerning μ_n, D_{it}, Q_f and Q_T.

6. HYDROGEN PLASMA EFFECTS

Taking into account wide application of hydrogen in MOS technology for improving MOS parameters one could expect that a hydrogen plasma treatment would be beneficial as was mentioned in the introduction. That is the reason to carry out also some experiments with hydrogen plasma [8].

Experimental techniques and samples preparation are the same as above including the conditions of plasma treatments.

The dependence of the inversion channel mobility on the oxide thickness for irradiated and nonirradiated samples and also recovery of Ar plasma treated samples is shown in Fig.8. The total charge in the samples before and after H$_2$ treatment for d_{ox}=72 nm is shown in Fig.9.

H_2 plasma effects in the thin oxide samples are very strong. Fig.10 shows the change in μ_n, N_{inv}, Q_f and Q_T ($d_{ox} = 16$ nm) for as-grown and irradiated through sapphire UV filter samples. In Table 1 D_{it} (for E_C-E_F=0.15-0.25 eV), Q_T (for E_C-E_F=0.17 eV) and Q_f for different oxide thicknesses and plasma treatment condition are given.

Electrical breakdown of as-grown and hydrogen plasma treated Si/SiO$_2$ structures are also studied [9]. It can be seen in Fig.11 & Fig.12 that plasma introduces defects in Si/SiO$_2$ structures similar to the defects introduced by current-voltage stress.

Neither He nor H_2 plasma can restore in the necessary extent Si/SiO$_2$ parameters after a 5 or 20 min Ar plasma treatment concerning μ_n, Q_f, Q_T and D_{it}. Total H_2 plasma effects are more beneficial, [10].

The data about the charges in the Si/SiO$_2$ prepared by thermal oxidation in dry oxygen with HCl addition and the breakdown voltage given in Fig.13 show that the "chlorine" oxides are more sensitive to He and H_2 plasma treatments.

The breakdown voltage after different H_2 treatment, Fig.13, is measured by corona discharge technique. It is seen that the breakdown voltage of "chlorine" oxides drops drastically after plasma treatment depending considerably on plasma power and exposure time, [12].

7. DISCUSSION

Regarding experimental results presented in this paper, one have to remember that I/V inversion current measurements are slow. Time intervals between every two measured points are of the order of seconds. Hence it could be assumed that such kind of measurements are carried out in steady-state conditions. Therefore it is reasonable to accept that the interface traps under investigation reside not only at the Si/SiO$_2$ interface by itself but are distributed in a tunneling depth of 15-25 Å in the oxide near the silicon lattice. In this sense a "deeper" information about the interface structure and defects could be obtained. We believe that combining the information concerning the charges in the Si/SiO$_2$ system with the data about inversion electron transport (also for very low electron densities, 10-100 times lower than that achievable by MOS FET measurements) and using additionally other conventional techniques some new viewpoints on the Si/SiO$_2$ physics could be anticipated.

UV radiation from a lamp or from a plasma (O$_2$, Ar, He, H$_2$) in the energy range below 5 eV creates negative charge at the Si/SiO$_2$ interface and gives rise in Q_f. At the same time electron mobility is much higher than that in the as-grown samples also for very low inversion electron densities. This could be explained only assuming that UV radiation below 5 eV creates negatively charged interface traps (in the above mentioned 15-25 Å depth) which are in close vicinity to the Q_f centers and compensate their Coulomb scattering potential.

Total plasma treatments (without UV filter) cause strong reduction of inversion electron mobility, strong reduction of the shallow interface states (except O$_2$ plasma) and an increase in the interface traps in the midgap. It is reasonable to be assumed that interface charges caused by total plasma action are not "paired" with Q_f centers. In all plasmas these charges at the interface are negative. After all plasma treatments Q_f is normally higher.

H_2 plasma affects seriously the breakdown behavior of SiO$_2$ drastically decreasing breakdown voltage of dry oxides grown with addition of chlorine in the oxidation ambient.

Taking into account the complex character of the experimental results obtained (and the shortage of place) unequivocal explanation and defect models could hardly be given here. As it is mentioned by many authors [14-24] apparently trivalent silicon, H, H$_2$, H$^+$, OH, H$_2$O and other complexes are essentially engaged in the observed effects.

8. CONCLUSIONS

All plasma treatments introduce negative charges in the Si/SiO$_2$ system. Because of the steady-state conditions by I/V inversion layer characteristics measurements as "interface" one have to understand a 15-25 Å thickness of SiO$_2$ near the Si interface. Not all defects can be restored by H$_2$ LTA. He and H$_2$ plasmas don't recover completely the Ar plasma created defects. Ar, He and H$_2$ plasmas reduce the interfaces states density near silicon conductance band. Ar, He and H$_2$ UV radiations through a sapphire filter transparent for vacuum UV up to 8.7 eV and a direct irradiation by standard UV lamps give rise to electron mobility and interface charge density near silicon conductance band. H$_2$ plasma treatment and a current flowing through SiO$_2$ have similar effects on the breakdown voltage. "Chlorine" oxides are much more sensitive to plasma effects than pure dry SiO$_2$.

9. ACKNOWLEDGEMENTS

The author is highly indebted to Dr.E.Atanassova and Dr.D.Dimitrov for their help in the preparing of the paper and for the useful discussions during our common work.

10. REFERENCES

[1] J.Kassabov, E.Atanassova and E.Goranova, Plasma Report, April 1986, 43; "Advances in low-temperature plasma chemistry, technology, applications", Vol.2, ed.H.Boenig, pp.249-264, 1988
[2] J.Kassabov and D.Dimitrov, Solid State Electronics, 29 (1986) 477
[3] J.Kassabov, D.Dimitrov and A.Grueva, Solid State Electronics, Vol.31, No 31, pp.49-51, 1988
[4] J.Kassabov, E.Atanassova, D.Dimitrov and E.Goranova, Solid State Electronics, Vol.31, No 2, pp.147-154, 1988
[5] J.Kassabov, E.Atanassova, D.Dimitrov and E.Goranova, Microelectronics J., Vol.18, No 2, pp.21-27, 1987
[6] J.Kassabov, E.Atanassova, D.Dimitrov and E.Goranova, Microelectronics J., Vol.18, No 5, pp.5-12, 1987
[7] J.Kassabov, E.Atanassova, D.Dimitrov and J.Vasileva, Thin Solid Films, Vol.169, p.43, 1989
[8] J.Kassabov, E.Atanassova, D.Dimitrov and J.Vasileva, Semicond.Sci. Technology, 3, (1988) 686
[9] D.Dimitrov and J.Kassabov, ISPPEME'87, p.439, 1987
[10] J.Kassabov, E.Atanassova, E.Goranova, D.Dimitrov and J.Vasileva, Solid State Electronics, Vol.32, No 7, pp.535-540, 1989
[11] D.Dimitrov, J.Kassabov, T.Dimitrova and J.Halianov, ISPPEME'89, p.315, 1989
[12] J.Kassabov, St.Georgiev and D.Dimitrov, Thin Solid Films, Vol.199, 1991
[13] J.Kassabov, ISPPEME'89, p.1, 1989
[14] C.T.Sah, IEEE Trans.nucl.Sci., NS-23 (1976) 1563
[15] A.G.Revesz, IEEE Trans.nucl.Sci., NS-18 (1971) 113
[16] C.M.Svensson, The Physics of SiO$_2$ and its interfaces (Ed. S.T.Pantelides), p.328, Pergamon Press, New York (1976).
[17] M.Schulz, Surf.Sci., 132 (1983) 422
[18] W.Schmitz and D.R.Young, J.Appl.Phys., 54 (1983) 6443
[19] S.K.Lai, J.Appl.Phys., 54 (1983) 2540
[20] P.M.Lenahan and V.Dressendorfer, J.Appl.Phys., 54 (1983) 1457
[21] M.L.Reed, Semicond.Sci.Technol., 4 (1989) 980
[22] K.L.Brower, Semicond.Sci.Technol., 4 (1989) 970
[23] E.H.Poindexter, Semicond.Sci.Technol., 4 (1989) 961
[24] C.T.Sah, Solid State Electronics, 33 (1990) 147

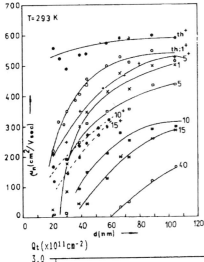

Fig.1. 293 K effective channel mobility vs. oxide thickness for different oxides: as-grown thermal oxide (th) and after LTA in H_2(th$^+$); numbers indicate plasma-exposure time; numbers with a sign (+) indicate plasma-treated oxides after LTA in H_2.

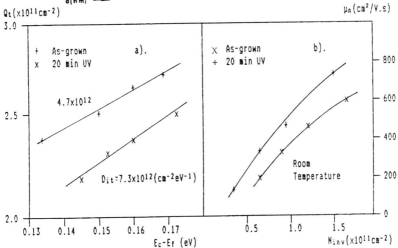

Fig.2. a) Total interface charge Q_T vs. (E_C-E_F) at the Si/SiO_2 interface before and after UV irradiation; b) Inversion electron mobility of as-grown samples before and after UV irradiation.

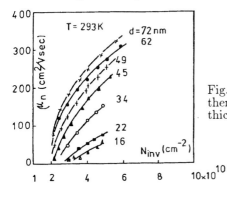

Fig.3. Dependence of μ_n on N_{inv} for thermal, as-grown oxides with different thicknesses.

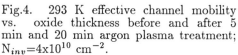

Fig.4. 293 K effective channel mobility vs. oxide thickness before and after 5 min and 20 min argon plasma treatment; $N_{inv}=4\times10^{10}$ cm^{-2}.

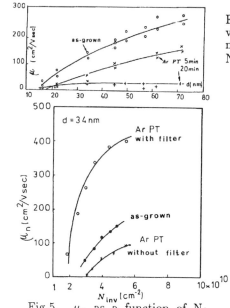

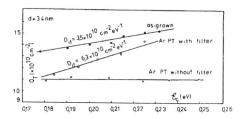

Fig.6. Q_T as a function of E_C-E_F before and after argon plasma treatment with filter and without filter; exposure time 5 min.

Fig.5. μ_n as a function of N_{inv} before and after argon plasma treatment with filter and without filter; exposure time 5 min.

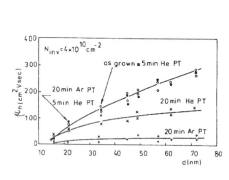

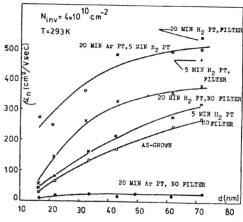

Fig.7. 293 K effective channel mobility vs. oxide thickness before and after 5 min He plasma treatment —o—; He PT 20 min —x—, Ar PT 20 min —+—; after both 20 min Ar PT and 5min He PT —●—.

Fig.8. 293 K effective channel mobility vs. oxide thickness before and after hydrogen plasma treatments.

Invited Papers

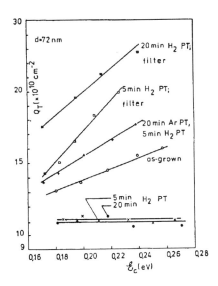

Fig.9. Q_T as a function of E_C-E_F for oxide thickness 72 nm before and after different plasma treatments.

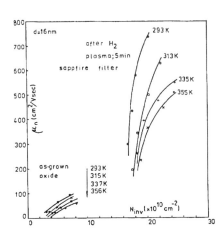

Fig.10. Dependence of μ_n on N_{inv} (d=16 nm) for different temperatures before and after H_2 plasma treatment with filter. For as-grown samples $Q_f = 3.5 \times 10^{11}$ cm^{-2} $Q_T^{0.17eV} = 1.4 \times 10^{11}$ cm^{-2}. After UV treatment $Q_f = 4 \times 10^{11}$ cm^{-2}, $Q_T^{0.17eV} = 1.65 \times 10^{11}$ cm^{-2}.

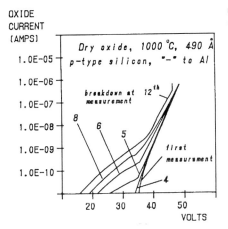

Fig.11. Successive measurements of oxide current, prior to breakdown for as-grown samples. Breakdown occurs at the twelfth measurment; (POA N_2 10 min, 1000°C).

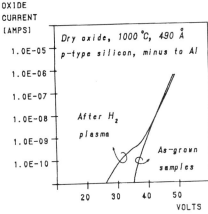

Fig.12. Prior to breakdown oxide current measurment for as-grown and H_2 plasma treated samples for 5 min (POA N_2 10 min, 1000°C)

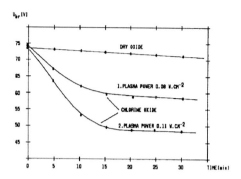

Fig.13. Breakdown voltages for dry oxide samples and "chlorine" samples (O_2+3%CHL). Oxide thickness 1000°C POA N_2 30 min, same temperature.

Table 1

d (nm)	As-grown oxide	H_2 Plasma Treatment			
		5 min		20 min	
		no filter	with filter (sapphire)	no filter	with filter (sapphire)
		D_{it} (×10¹⁰cm⁻²eV⁻¹)			
a). 16	20	5	230	3	200
22	48	5	120	<0.2	100
43	55	3	130	5	150
72	38	<0.2	100	<0.2	80
		Q_f (×10¹⁰cm⁻²) from HF C-V for E_c-E_f=0.83eV			
b). 16	35	52	40	48	40
22	42	61	43	62	43
43	48	65	45	60	50
72	36	57	40	55	42
		Q_t (×10¹⁰cm⁻²) for E_c-E_f=0.17eV			
c). 16	14.0	9.8	16.5	10.3	19.0
22	12.5	11.4	15.2	12.0	18.2
43	13.8	10.4	14.2	10.0	18.0
72	12.7	10.9	13.6	11.0	17.5

D_{it} (in the range 0.15 eV - 0.25 eV below Silicon conductance band). Q_f and Q_t for oxides with different thicknesses before and after H_2 plasma treatment (RF power density 0.1 W/cm²; pressure - 1 Torr; T_{sub}=300°C; 13.56 MHz planar plasma reactor).

Hot-electron transport studies in SiO_2 using soft-x ray induced internal photoemission

E. Cartier and F.R. McFeely

IBM Research Division Thomas J. Watson Research Center Yorktown Heights, NY 10598

The results of a series of zero field internal photoemission experiments designed to probe hot carrier dynamics in the range of 8-20 eV are reviewed. We illustrate how spectral reconstruction using classical Monte Carlo simulations may be employed to extract energy dependent scattering rates. We briefly discuss the relationship between the derived rates an current theory.

I. Introduction

Owing to its importance in electronic devices and high power laser optical coatings, hot-electron transport in wide gap insulators has been the subject of intense study.[1] In all these applications, the insulators are subject to high electric fields which may cause carriers to gain sufficient energy to induce bond breaking[2,3] and electron-hole pair generation, resulting in device degradation.[4] If impact ionization is sufficiently strong, electron multiplication is expected to lead to destructive dielectric breakdown. These problems are controlled by the hot electron dynamics. At kinetic energies $E_{kin} < E_g$, hot electrons can interact with the solid only by phonon processes (apart from scattering by defects, impurities, trapped charges, or disorder). Understanding hot electron dynamics therefore requires the knowledge of the hot-electron-phonon interaction over a wide energy range.

This understanding is currently incomplete. Difficulties arise from an inadequate knowledge of conduction band structures, and some electron-phonon coupling constants are poorly known in many solids.

Over the past few years, some novel experiments have been developed to study hot electron dynamics.[5-10] These have been based on hot electron transmission through thin insulating films on semiconductor or metal substrates, in which the electrons are injected into the film by UV, VUV, or soft x-ray induced photoemission from the substrate. The wide range of photon energies allows injection over the whole energy range, from the conduction band edge to energies much larger than the energy of the band gap. By studying the evolution of sharp features in the electron energy distribution as a function of overlayer thickness important transport parameters and their energy dependence can be derived.

In this paper we summarize recent progress made in the understanding of hot-electron transport in SiO_2 using soft X-ray induced internal photoemission.[10,11] This is the first technique capable of yielding detailed information on electron-phonon scattering rates and electronic scattering rates in the transition regime where both are important for the transport dynamics.

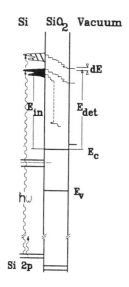

Fig. 1: Schematic energy band diagram illustrating the principle of the soft X-ray induced internal photo emission experiment used to study hot-electron transport in SiO_2

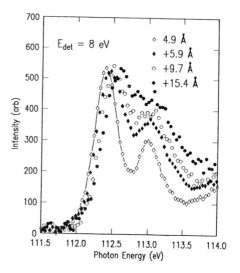

Fig. 2: Comparison of Si-2p substrate core level line shapes as measured after electron transport through SiO_2 overlayers of different thickness. The data is measured in the CFS-mode with a fixed electron analyzer energy of 8 eV with respect to the bottom of the SiO_2 conduction band.

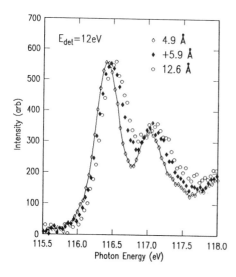

Fig. 3: Same as Fig. 2 but for a electron analyzer energy of 12 eV.

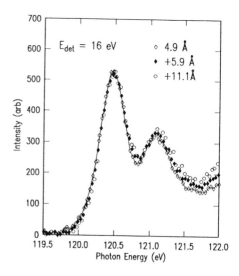

Fig. 4: Same as Fig. 3 but for a electron analyzer energy of 16 eV.

III. Experimental Techniques and Results

The principle underlying the experiments is illustrated in Fig. 1. Photoelectrons from the sharp substrate Si 2p core-level are excited by soft X-rays and injected into the oxide overlayer where they undergo scattering during transport to the outer surface. The line shape after transport through oxides of various thickness is obtained by measuring the energy distribution of the vacuum emitted electrons. Three typical electron trajectories are illustrated. One electron undergoes a deep inelastic event (dashed line) and gets removed from the spectral region of the Si-2p core level. Only electrons interacting exclusively with phonons can contribute to the Si-2p spectrum measured after transport. Photoelectron spectra are accumulated in the constant final state (CFS) mode. In this mode, the photon energy is scanned while the detection energy, E_{det}, with respect to the bottom of the oxide conduction band, is fixed.

Figures 2-4 show the results for overlayers of varying thickness at kinetic energies of 8 eV, 12 eV, and 16 eV. At 8 eV we observe drastic broadening of the bulk Si 2p spectrum along with a shift in the maximum of the distribution to higher photon energy, the direction corresponding to electron energy loss. This effect is still evident at 12 eV (fig. 3), but is substantially reduced, and by 16 eV (fig.4) it is barely evident.

To summarize the evolution of the photoelectron energy distributions with oxide thickness, we have defined scalar parameters, the linewidth and intensity of the substrate Si 2p3/2 component, to characterize the evolution of the distributions. In Fig. 5a, the broadening is shown as a function of oxide thickness for energies ranging from 8 to 18 eV. The broadening parameter, ΔB, is defined as the increase of the line width (FWHM) over the value obtained in the thinnest oxide measured ($t_0 = 7.4 \text{\AA}$). The broadening increases with oxide thickness at a rate strongly dependent on the detection energy. The decay of the intensities is shown in Fig. 5b. In sharp contrast to the line shape, the signal attenuation has little energy dependence. This arises from the interaction of a strongly increasing deep inelastic scattering rate and a decreasing phonon scattering rate.

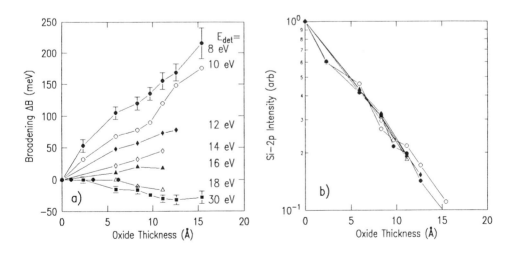

Fig. 5: Observed Si-2p 3/2 line broadening (a) and line intensity (b) as a function of oxide thickness and for various detection energies, E_{det}.

II. Transport Analysis

The unknown scattering rates and their energy dependence are extracted from the data via the theoretical reconstruction of the transport induced changes in the Si-2p core line, as a function of both energy and oxide thickness. Transport through the oxide and subsequent detection by the electron energy analyzer can be characterized by a detection probability function $P_{det}(E_{in}, E_{det}, t_{ox})$ giving the probability that an electron injected at energy E_{in} arrives at the detector at the detector pass energy, E_{det}, after traversing an oxide of thickness t_{ox} (Appropriate integrattions having been performed over the input angular distribution and angular acceptance of the electron analyzer.) P_{det} is calculated from the transport equation, which is solved for transport through oxides with thickness, $+t_{ox}$, on top of the thinnest oxide measured ($t_0 = 7.4$Å). This is achieved by a Monte Carlo solution of the classical Boltzmann transport equation.[12] The effect of transport on the spectrum is then given by the convolution of the detection probability function with the injection energy distribution, which is taken as the measured distribution for the thinnest oxide studied.
Longitudinal optical (LO) and acoustic (ac) phonon scattering are included in the Monte Carlo calculation.

The phonon scattering rates are calculated under the assumtion of unit effective mass, and using an isotropic parabolic band.

The LO-phonon scattering rate is calculated using the Fröhlich approximation, including the two dominant LO phonons at 0.063 and 0.153 eV and using the coupling constants given by Lynch.[13] At high energies, LO-phonon scattering is strongly forward directed and weak.

The acoustic phonon scattering rate, $1/\tau_{ac}(E)$, is calculated from the equation,[14]

$$1/\tau_{ac}(E) = \frac{3m_{eff}C_{ac}^2}{4\pi\rho\hbar k} \times \int_{q_c}^{q_{max}} \frac{q^3}{\hbar\omega(q)} \left(\frac{1}{2} \pm \frac{1}{2} + n_q\right) dq,$$

where $q_c = 2m_{eff}c_s$, $c_s = 4.6 \times 10^5$cm/s is an effective sound velocity obtained by averaging over longitudinal and transverse polarizations, $q_{max} = 2k - q_c$, C_{ac} is the nonpolar electron-acoustic phonon coupling constant, ρ the SiO_2 density, n_q the Bose function at wave vector q and temperature T. The phonon dispersion is approximated by $\hbar\omega(q) = (2/\pi)\hbar k_{BZ}c_s[1 - \cos(\pi q/2k_{BZ})]^{1/2}$ for ($q < k_{BZ}$) and by $\hbar\omega(q) = (2/\pi)\hbar k_{BZ}c_s$ for ($q \geq k_{BZ}$). $k_{BZ} = 1.208 \times 10^8$cm^{-1} is the Brillouin zone edge wavevector.

The integration over q is performed numerically and accounts for the temperature dependence of the emission and absorption rates.

To reconstruct the experimental spectra we allow C_{ac} to be energy dependent. We thereby maintain the phonon occupation for emission and absorption processes as dictated by the phonon dispersion and the Bose function. Acoustic scattering favors large angle scattering and is essentially isotropic.

Inelastic scattering can be treated in terms of an integral deep inelastic rate. $1/\tau_{inel}(E)$. Any deep inelastic process leads to large energy loss, $\delta E_{inel} \gtrsim 2.5 eV$, completely removing the scattered electron from the spectrum. We therefore need not consider the electron after scattering, nor do we require any knowledge of the nature of the deep inelastic processes.

The deep inelastic rate is also allowed to be an energy dependent free parameter, adjusted in conjunction with (C_{ac}) until the experimental line shapes and intensities are reproduced for every thickness studied.
The results are shown in Fig. 6 and 7 with the 8 eV data as an example. Fig. 6 shows the change of the detection probability function $P_{det}(E_{in}, E_{det}, t_{ox})$ with oxide thickness P_{det} for zero additional thickness ($t_0 = 7.4$Å) reproduces simply the rectangular detection window. With increasing thickness, P_{det} rapidly broadens towards

Invited Papers

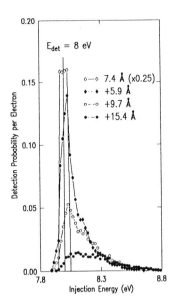

Fig. 6: Evolution of the detection probability with oxide thickness as obtained from the Monte Carlo simulation for a detection energy, E_{det}, of 8 eV.

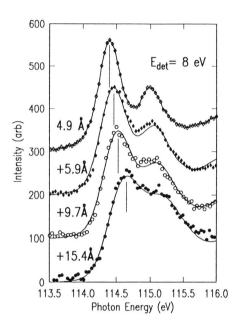

Fig. 7: Evolution of the Si-2p line shape (solid lines) as obtained by convoluting with the detection probabilities shown in Fig. 6 in comparison with the experimental data (symbols).

higher injection energies because electrons efficiently lose energy to the lattice by phonon emission and therefore may downscatter into the detector window after having been at energies, E_{in}, larger than E_{det}. The absolute values of P_{det} rapidly decrease with thickness. This is a result of the deep inelastic scattering. (A small contribution to the attenuation is the result of phonon backscattering into the substrate, where the carriers are assumed lost.)

Fig. 7 finally shows the line shape evolution imposed by the detection probabilities shown in figure 6. Theoretical (lines) and experimental (dots) distributions at each thickness are scaled to facilitate the comparison of the line shapes, independent of intensity variations.

Our simulations reproduce not only the line shapes but also accurately reproduces the experimental absolute intensities at every measured thickness, using only two ajustable parameters. Only a single combination of the two scattering rates simultaneously yields correct line shapes and attenuations.

We present In Figs. 8 and 9 a series of simulations, in which we have turned off certain scattering mechanisms and/or modified their scattering properties. This affords a deeper insight into the importance of these mechanisms for high energy transport. These simulations are again illustrated for an additional oxide thickness of 15.4 Å and a detection energy of 8 eV. For each case, the left panel shows the detection probability and the right panel compares the resulting line shape (solid line) with the experimental data (dots). The multiplication factor refers to the simulated line. A factor of x1 means that the simulated intensity agrees with the experimental values. The dotted line shows the input distribution. The scattering rates for the various scattering processes have been fixed at the values determined

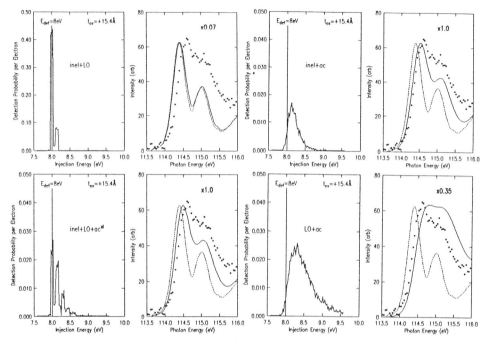

Fig. 8: Dependence of the calculated detection probability (left) and line shape (right) on the scattering processes included in the Monte Carlo transport simulation. The detection energy is 8 eV, the oxide thickness is 15.4Å and the absolute values for the rates are those shown in Fig. 10 The scattering processes included are deep inelastic scattering (inel) and longitudinal optical (LO) phonon scattering (top panels). Acoustic phonon scattering without phonon dispersion (ac^{el}) is included in addition in the bottom panels.

Fig. 9: Same as Fig. 8, but for different combination of scattering processes. Top panels: Deep inelastic (inel) and acoustic phonon scattering with phonon dispersion (ac). Bottom panel: LO-phonon and acoustic phonon scattering with dispersion.

above. In Fig. 8 (top two panels) we turned off the acoustic scattering. Transport through the oxide then has only a marginal effect on the line shape and intensity. The side peak in P_{det} corresponds to the first loss peak of the 153 meV LO-phonon. LO-phonon and inelastic scattering are very weak at 8 eV. An attempt to reproduce the line shape and intensity without acoustic scattering would require unreasonably high values for both the LO-phonon rate and the deep inelastic rate. Obviously, acoustic scattering plays a key role in the transport process. This role is twofold. First, quasi-isotropic acoustic scattering rapidly changes the direction of the electrons, leading to meandering paths in the overlayer, which are much larger than the sample thickness. Thus, the electron spends more time in the solid and emits more LO-phonons. This is illustrated in Fig 8 (bottom). P_{det} shows a whole series of phonon-replicas if pure elastic, isotropic scattering (acoustic scattering with hypothetical zero-energy phonons, at the same rate as determined for the acoustic phonons in the full simulation) is added to the two previous scattering processes. With this addition, the 2p-line is considerably broadened. Similarly, the path length

Invited Papers

amplification allows for additional deep inelastic events yielding the correct line intensity. Second, the acoustic phonons themselves are directly responsible for substantial carrier relaxation in spite of their smaller energies This can be illustrated by setting the LO phonon scattering to zero but including the acoustic scattering modes with their correct dispersion (Fig. 9 top). This shows, that direct energy loss to acoustic phonons cannot be neglected in the transport. Again, the line is correctly attenuated because of isotropic acoustic scattering. In the final example in Fig. 9 (bottom), only the deep inelastic rate is turned off. The agreement between simulation and experiment is very poor. Without deep inelastic scattering, long meandering paths with large numbers of emitted phonons are not eliminated. The line is much too broad and intense.

We have determined the scattering rates at all measured detection energies by applying the procedure illustrated above for 8 eV. In all cases, the reconstruction of the spectra is possible with comparable accuracy as for the 8 eV data. The scattering rates derived from these analyses are summarized in Fig 10. The acoustic phonon scattering rates (diamonds) are of the order of 4×10^{15} to 6×10^{15} at kinetic energies between 8 and 15 eV. The acoustic rate does not vary much with electron energy. In contrast, the deep inelastic rate rapidly increases with energy. The absolute values are in the range $2.4 \times 10^{14} < \tau_{inel} < 3.2 \times 10^{15}$. The dashed line shows the LO-phonon scattering rate as given by the Fröhlich formula.

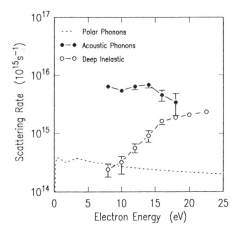

Fig. 10: Energy dependence of the scattering rates as derived from the evolution of the 2p-Si substrate core level. The LO-phonon scattering rate (dotted line) was assumed to be known in the analyses.

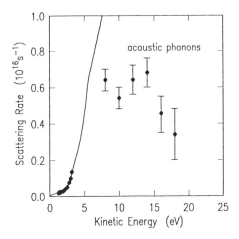

Fig. 11: Measured acoustic rates in comparison with theory. The solid line is calculated by eq (1) for a deformation potential of 6 eV.

IV Discussion and Conclusions

In Fig. 11, we compare the measured acoustic scattering rates with those (solid line) predicted for a constant deformation potential of 6 eV using eq (1). A value of 6 eV for the deformation potential was chosen, because it yields good agreement with high field transport experiments at lower electron energies.[14] A deformation potential of 6 eV is also in agreement with low energy internal photoemission results (corresponding measured rates are included in Fig. 11).[3] Eq. (1) yields considerably larger acoustic phonon scattering rates at high energies than those derived here. The calculated rate at 18 eV is about a factor of 10 higher than the value extracted from our experiments. Since $1/\tau_{ac}$ increases with C_{ac}^2, the deviation between experiment and theory can be accounted for with a variation of the deformation potential from 6 eV at low energies to less than 2 eV at high energies.

Equation (1) is derived assuming a single parabolic band. It is conceivable, that this might not be valid in the energy range studied. Inverse photo emission experiments[15] indicate a maximum in the conduction band density of states at about 6 eV above the bottom of the conduction band, and a multi-valley band structure might be more appropriate. Fischetti[16] et al have shown, that the scattering rates in semi-conductors are depend strongly on the conduction band density of states

We cannot at present give a full theoretical rationalization of the derived scattering rates. While these rates are unique, given our scattering model, different values might be obtained if scattering mechanisms we have neglected are important. Since there is little evidence for inter-valley scattering, we have not included this mechanism. An attempt to do so has been recently been made by Porod and Ferry. [17]

An other concern arises from the fact that we treat acoustic scattering as essentially isotropic at all energies. If acoustic scattering were becoming significantly forward directed, the rates we have derived would be incorrect, certainly too low for acoustic scattering. The energy above which forward scattering starts to play a role is poorly known. Fischetti has estimated that quasi-isotropic scattering might break down above about 25 eV.[18] Finally, TO-phonon scattering has not been included.

The resulting deep inelastic rates are shown in more detail in Fig 12. It is surprising that deep inelastic scattering sets in at such low energies. If electron-hole pair generation were the only deep inelastic scattering process, one would expect a threshold of about $E_{th} \simeq 13.5\,eV$.[19] However, E_{th} could in principle be as low as 9 eV. Since the impact ionization rate should increase rapidly with electron energy, we may assume that impact ionization is the major contribution to the measured deep inelastic rate at energies $E \gtrsim 2E_g$. The two curves in Fig. 12 are calculated with the Keldysh formula for impact ionization, [20] $1/\tau_{inel} = P \times (E - E_{th})/E_{th}^2$, for a threshold energy of 9 eV (dashed) and 12.5 eV (dotted) and the factor P adjusted to yield the rates at high energies. There are sustantial deviations with experiment at low energies.

These simple considerations suggest that impact ionization can not account for the inelastic rate observed in the vicinity of $E \simeq E_{gap}$.

Optical measurements in SiO_2 show preparation dependent absorption tails down to about 7 eV, and absorption is strong above band gap.[21] Such transitions might be excited by hot electrons and contribute to the observed inealstic rate.

In the experiment, the samples are irradiated by photons of the order of 100 eV, leading to high energy excitations not only in the Si, but also in the oxide overlayer. Strong permanent charge trapping in the additional oxide layer is unlikely due to

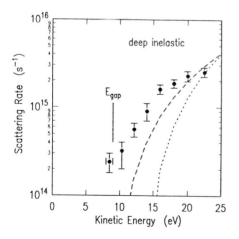

Fig. 12: Measured deep inelastic rate in comparison with theory. The lines show the energy dependence predicted by the Keldysh formula for impact ionization for a threshold of 9 eV (dashed) and 12.5 eV (dotted) normalized at 25 eV.

direct tunneling to the substrate (and we do indeed not observe significant oxide charging which would lead to shifts of the vacuum emission threshold). But there may still be charge trapping under experimental steady state conditions. Trap ionization of levels around midgap could thus contribute to the deep inelastic rate.

Our experiments are done on very thin oxides. Because of the proximity of the Si substrate and the Si/SiO_2 interface to the additional oxide overlayer, it is conceivable that hot electrons "within" the additional oxide layer might interact with the Si/SiO_2 interface and excite interface plasmons. The interface plasmon energy is 9.4 eV with a excitation line width of 4.1 eV.[22] This indicates that the threshold for surface/interface plasmon excitation could be considerably lower than that for impact ionization in the oxide.

The authors would like to thank Drs. M. V. Fischetti, D. Arnold and D.J. DiMaria for many stimulating discussions.

References

[1] S. Jones, P. Braunlich, R. Casper, X. Shen, and P. Kelly, Opt. Eng. **28**, 1039 (1989).
[2] D. DiMaria and J. Stasiak, J. Appl. Phys. **65**, 2342 (1989).
[3] E. Cartier and P. Pfluger, IEEE Trans. Electr. Insul. **EI-22**, 123 (1987).
[4] D. DiMaria, in *The Physics of Hot Electron Degradation in Si MOSFETS, Satalite Workshop, this conference* (1991).
[5] D. DiMaria, T. Theis, J. Kirtley, F. Pasavento, D. Dong, and S. Brorson, J. Appl. Phys. **57**, 1214 (1985).
[6] P. Pfluger, H. Zeller, and J. Bernasconiz, Phys. Rev. Lett. **53**, 94 (1984).
[7] E. Cartier and P. Pfluger, Phys. Rev. B **34**, 8822 (1986).
[8] E. Cartier, P. Pfluger, J. Pireaux, and M. R. Vilar, Appl. Phys. A **44**, 43 (1987).
[9] E. Cartier and P. Pfluger, Physisca Scripta **T23**, 235 (1988).
[10] F. McFeely, E. Cartier, L. Terminello, A. Santoni, and M. Fischetti, Phys. Rev. Lett. **65**, 1937 (1990).
[11] F. McFeely, E. Cartier, J. Yarmoff, and S. Joyce, Phys. Rev. B **42**, 5191 (1990).
[12] M. Fischetti, D. DiMaria, S. Brorson, T. Theis, and J. Kirtley, Phys. Rev. B **31**, 8124 (1985).
[13] W. Lynch, J. Appl. Phys. **43**, 3274 (1972).
[14] D. DiMaria and M. Fischetti, J. Appl. Phys. **64**, 4683 (1988).
[15] F. Himpsel and D. Straub, Surface Sci. **168**, 764 (1986).
[16] M. Fischetti and S. Laux, Phys. Rev. B **38**, 9721 (1988).
[17] W. Porod and D. Ferry, Phys. Rev. Lett. **54**, 1189 (1985).
[18] M. Fischetti and D. DiMaria, Phys. Rev. Lett. **55**, 2475 (1985).
[19] Z. Weinberg, M. Fischetti, and Y. Nissan-Cohen, J. Appl. Phys. **59**, 824 (1986).
[20] L. Keldysh, J. Sov. Phys.-J.E.T.P **37**, 509 (1960).
[21] Z. Weinberg, G. Robloff, and E. Bassous, Phys. Rev. B **19**, 3107 (1979).
[22] M. Fischetti, Phys. Rev. B **31**, 2099 (1985).

Paper presented at INFOS '91, Liverpool, April 1991
Invited Papers

Interface and oxide engineering for high quality SIMOX devices

Sorin Cristoloveanu

Laboratoire de Physique des Composants à Semiconducteurs (UA-CNRS)
Institut National Polytechnique, ENSERG, BP 257, 38016 Grenoble Cedex, France.

Abstract. The electrical properties of thin SIMOX films and the specific modes of operation of SIMOX transistors are discussed by emphasizing recent developments. SIMOX offers the possibility of engineering the Si film, buried oxide and interface quality and achieving a variety of structures. Typical phenomena such as the interface coupling, floating body, sidewall conduction and deep-depletion transients are described. The tolerance of the buried oxide to irradiation and hot carrier injection is investigated.

1. Introduction

Interest in silicon on insulator (SOI) has historically been generated by the need for radiation-hard devices. As additional virtues have been confirmed more recently (ease of the process, speed, reduced short-channel effects), the promise is to have SOI technologies promoted in the market of civilian CMOS circuits and eventually to overtake the conventional bulk Si process. This enthusiasm is however tempered by the fact that the frontiers/limits of bulk Si technology are being incessantly pushed forward and hence the wide application of SOI is more certain for deep-submicron MOSFET's ($\leq 0.25\mu m$). In addition, the rapid progress of SOI technologies has also revealed new difficulties (floating body and interface coupling effects) which were in the past obscured by the imperfect quality of SOI structures.

Although the SOI community has still to be patient, there is a reasonable optimism, well sustained by the outstanding development of SIMOX. The comparison of test CMOS circuits integrated on bulk Si and SIMOX with design rules of $1.2\mu m$, $1\mu m$, $0.8\mu m$ and $0.5\mu m$ has demonstrated that SIMOX is about one generation ahead (Auberton 1990). The success of SIMOX also comes from achievements in wafer availability and reproducibility. The wafer cost will continue to be significantly reduced.

This paper is aimed to address the properties of SIMOX material and devices which are inferred from electrical characterization. It is shown that the SIMOX structure can be engineered in order to obtain a wide range of properties for the film, buried oxide and interfaces, which have great impact on the performance of integrated devices. Most of the special characterization techniques (charge pumping,

© 1991 IOP Publishing Ltd

noise spectroscopy, PICTS, deep-depletion pulsing, spreading resistance, four point probing, ...) needed to develop an accurate electrical image of SIMOX have been described before (Cristoloveanu 1989). Details will be given only if improvements have occurred recently.

2. SIMOX synthesis

SIMOX is formed by deep implantation of oxygen doses in Si wafers and subsequent annealing. The implant energy is $150-200 keV$ and the oxygen dose ranges typically between 1.4×10^{18} and $2 \times 10^{18} cm^{-2}$. This results in a thin Si film ($0.1-0.3\mu m$) separated from the substrate by a buried oxide $0.35-0.45\mu m$ thick. Larger doses are intended for special applications requiring a thicker oxide. Thick film bipolar or CMOS devices are fabricated by completing the Si overlay by epitaxy. The trend is however to use, for CMOS circuits, ultra-thin films ($< 0.1\mu m$) which can be achieved by sacrificial oxidation, thinning or lowering the implant energy.

The wafer temperature is either imposed by the beam power or adjusted by an external source of heat. Increasing the temperature from $400°C$ to $600°C$ results in a clear improvement of the crystal and interfaces. Since temperatures in excess of $700°C$ are prohibited by an unacceptable surface sputtering, the optimum value is probably between $600-650°C$. High current implanters ($100mA$) have been designed especially for SIMOX in order to make it commercially available by drastically reducing the duration of the synthesis. The properties of these wafers are in many respects different from those fabricated earlier with currents three orders of magnitude lower. A yet more advanced machine is now being developed at IBIS.

The annealing stands as a key phase of the SIMOX process. Low temperature annealing ($1150-1250°C$) has been abandoned because it allows too many defects and SiO_2 precipitates to subsist in a relatively thick interface region. The advantage of annealing at $1300-1405°C$ is to dissolve the precipitates and to pump most of the residual oxygen atoms from the film. This results in a good quality homogeneous Si film and very sharp Si-SiO_2 interfaces. The annealing is still under optimization in terms of temperature ($1300-1350°C$), duration ($2-6$ hours) and ambient (argon or nitrogen). The typical problems arising from aggressive implantation and annealing (contamination and dislocations in the film, pipes and Si islands in the buried oxide) have been greatly alleviated. Most wafers are now implanted with $1.6 \times 10^{18} O/cm^2$ at 190 keV and $600°C$ and annealed at $1300°C$ in argon. The best quality SIMOX is so far fabricated by a sequence of multiple implants and anneals (three times $0.5 \times 10^{18} cm^{-2}$ and $1300°C$).

Various alternatives for SIMOX have been explored: (a) mixed implants (O+N, O+C, O+Ge) aimed to improve the interfaces and dielectric isolation, (b) two oxygen implantations at *different energies* in order to achieve a double-SIMOX structure with two separate oxides (useful for transducers and interconnections), (c) masked implantation through a patterned capping layer to form *totally-isolated* Si islands, low energy implants of deep "substoechiometric" doses ($1-5 \times 10^{17} cm^{-2}$) which surprisingly result, after high temperature annealing, in stoechiometric and continuous thin oxides. It is clear that the large number of process parameters offers the opportunity to engineer the geometry and quality of the SIMOX structure, in order to meet a wide spectrum of applications.

3. Properties of as-grown SIMOX

Hall effect measurements have been performed at various stages of SIMOX development. The best *low-current* implanted SIMOX we have probed, had a very low N-type residual doping ($10^{15} cm^{-3}$) and an excellent electron mobility ($1250 cm^2/Vs$) (Cristoloveanu et al 1987). Recent *high current* implanted SIMOX is in general N-type (although P-type films have also been reported), with a background doping of $5 - 7 \times 10^{15} cm^{-3}$ and a mobility of $700 - 850 cm^2/Vs$.

Accidental contamination of SIMOX can originate from: (i) implanter (metals, carbon), (ii) capping layer, (iii) anneal ambient (in particular if nitrogen is used) and (iv) subsequent CMOS process. The residual oxygen acts as an *intrinsic* source of contamination, via the formation of *thermal donors* (at $450 - 550°C$) and *new donors* (at $750°C$), which affect the intentional doping and device performance. The signature of oxygen donors, which is not yet clear in bulk Si, is even more puzzling in SIMOX because it depends on the implantation and annealing (Cristoloveanu 1990). A strong activation of thermal and new donors was found in SIMOX annealed below $1200°C$. After annealing above $1300°C$, the density of thermal donors is lowered to about $10^{15} cm^{-3}$ but many new donors are still being generated. Recent results performed on a matrix of device-grade SIMOX wafers tend to indicate that the oxygen activation can be reduced below the residual doping level.

The film-oxide interface can be probed *in situ* by taking advantage of the upside-down MOS structure which naturally exists in SOI: Si substrate (gate) – oxide – Si film. The pseudo-MOS transistor is operated by using low-pressure probes placed on the Si film to form source and drain point contacts and biasing the Si substrate as a gate (Fig.1(a)). Both inversion and accumulation channels can be activated on the same wafer giving insights into the electron and hole properties (Williams et al 1992). A full set of $I_D(V_G)$ characteristics is obtained and allows extracting the threshold voltage, subthreshold swing, breakdown voltage, leakage current and other related parameters. Shown in Fig.1(b) are typical transconductance curves which give the effective carrier mobilities.

4. Interface and sidewall engineering

The major parameters for interface engineering are the annealing temperature and implant temperature and dose. Increasing the annealing temperature from $1150°C$ to $1300°C$ results in a dramatic improvement of the back interface: the density of states drops from 10^{13} below $10^{12} cm^{-2} eV^{-1}$ and the electron mobility increases from $1 - 10$ to $400 - 600 cm^2 V^{-1} s^{-1}$. In early SIMOX annealed below $1250°C$,

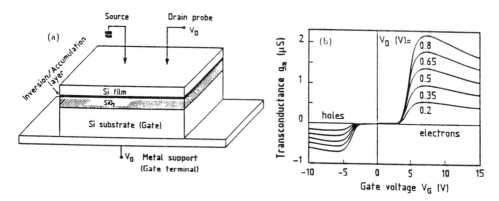

Figure 1 Configuration of the pseudo-MOS transistor (a) and transconductance curves (b) in a SIMOX wafer (after Williams et al 1991).

the increase of the implant temperature from 450 to 650°C or dose from 1.4 to 2.2 × $10^{18} cm^{-2}$ improved in both cases the quality of the buried interface (Mao et al 1987). The experiment has been repeated on device-grade SIMOX wafers implanted with a $100 mA$ beam and annealed at high temperatures ($\geq 1250°C$) in argon. No dose influence was observed on the front channel transistors which performed equally well. On the contrary, the back channel properties are surprisingly degraded by increasing the dose: higher threshold voltage, poorer subthreshold slope and reduced mobility. Varying the dose from 1.4 to 2.1 × $10^{18} cm^{-2}$, degrades the mobility by a factor of 4 and the density of states by a factor of 5 (Ioannou et al 1991). This result suggests that the balance between the beneficial aspects (sharper interfaces formed prior to any annealing) and detrimental aspects (increased number of implantation-induced defects) of high oxygen doses should be carefully considered. It is concluded that (a) increasing the dose results in better interfaces *only* for low-temperature-annealed SIMOX, (b) high temperature annealing yields very sharp interfaces whatever the dose and even for deep substoechiometric doses, and (c) very high doses just cause the implantation-induced defects to prevail.

The use of a capping oxide during the annealing is still questionable as detrimental effects have been observed. This is probably because the cap inhibits the surface reconstruction which normally occurs during the pumping of oxygen from the film towards the buried oxide. Top quality SIMOX is prepared by a sequence of multiple implants and high temperature anneals. No DLTS signal was obtained above the detection limit showing a very low density of traps in this material. The interfaces have a very good quality but this is also true for conventional single-implanted SIMOX. A major benefit is the improvement of the generation lifetime by one or two orders of magnitude (Ioannou et al 1991).

A typical problem in SOI devices is the parasitic conduction on the edges of the Si island. In LOCOS-isolated circuits, the edges can be deactivated by accumulating the back interface or increasing the back channel doping. Either solution cause in turn a subthreshold slope degradation. The edge properties have been investigated directly by performing charge pumping and noise measurements in the region where the sidewall conduction dominates (Elewa et al 1989). It was inferred that the density of defects is higher and *inhomogeneous* on the edges, decreasing from the back to the top interface. There is also evidence for a lateral inhomogeneity along the channel, with more defects close to the drain and source terminals.

In mesa-isolated MOSFET's, the threshold voltage is, in general, higher on the edges than at the front interface and the edge conduction is masked. In some cases, the shift of the threshold voltage on the edges (induced by accumulating the back interface) was surprisingly small. There are two possible explanations: (i) charge sharing between the front, back and sidewall gates do not allow the upper corners of the Si island to be deactivated by accumulating the back interface (Matloubian et al 1989) or (ii) if the defect density is very large on the edges, it causes the pinning of the Fermi level which in turn blocks the threshold voltage shift (Mazhari et al 1991).

5. Interface coupling effects

A fully-depleted structure is defined by the Si film being thinner than the depletion region controlled by either the front or the back gate. Total depletion enables a *strong interface coupling* which means that the front channel current can substantially be influenced by the back gate bias V_{G_2} and vice-versa. For instance the front gate threshold voltage V_{T_1} takes constant values if the back interface is accumulated or inverted, but decreases linearly with V_{G_2} when the back interface is depleted (Lim and Fossum 1983). Depletion at the back interface also causes the subthreshold swing to be a minimum. Excellent values ($69mV/decade$) have indeed been reported (Mazhari et al 1991) which are very close to the ideal swing ($60mV/decade$). In ultra-thin SOI films, the front and back interface defects have a comparable influence on the front gate subthreshold swing. The quality of the buried interface must therefore be optimized, otherwise the use of thick films or the accumulation of the back interface is preferable.

Experiments conducted on *long-channel* MOSFET's show that the shape of the front gate transconductance $g_{m1}(V_{G_1})$ depends dramatically on V_{G_2}. A deformation of the transconductance occurs when the back interface is inverted and a plateau emerges, indicating that the current flows at the *back interface* under control of the *front gate* (Fig.2). The carrier mobilities and threshold voltages at the front and back interfaces can be determined respectively from the transconductance peak and plateau. In *short-channel* MOSFET's the series resistances cause a degradation of the transconductance peak and plateau. Different parasitic resistances are associated with the front and back channel. Their influence on the front channel current depends on V_{G_2} and is a maximum when the back interface is inverted. Simple analytical expressions explain the lowering of the transconductance peak and plateau with increasing series resistances. The model is experimentally verified by associating *external* resistors to the transistor (Fig.2(a)) and applied to extract the intrinsic values of series resistances (Ouisse et al 1991).

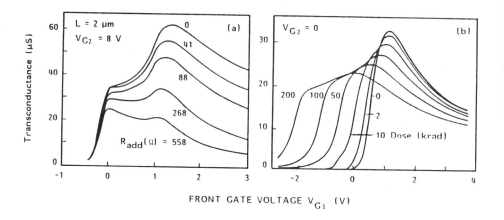

Figure 2 Transconductance curves for various (a) external series resistances and (b) irradiation doses in SIMOX–MOSFET's (after Ouisse et al 1991).

The modification of the transconductance curve under irradiation is very substantial (Fig.2(b)) because it combines several effects (changes in front and back channel threshold voltages, mobilities and series resistances). For doses above $10krad$, a plateau shows up in $g_{m1}(V_{G_1})$ and indicates that strong inversion occurs at the back interface. This confirms that the main physical mechanism is the gradual build-up of positive charges in the buried oxide which reduces V_{T_2} and is equivalent to an increase of V_{G_2}. The activation of the back channel illustrates an interesting situation of *radiation-induced interface coupling*. To explain the large degradation of the transconductance peak, we must assume the *presence of series resistances*. The irradiation also causes a large lateral shift of the transconductance (due to a decrease of V_{T_1}, via positive charge trapping in the gate oxide), a decrease of the back channel mobility and a variation of the series resistances.

More or less similar consequences of interface coupling can be observed for any type of electrical characterization of thin SOI films: charge pumping, dynamic transconductance, low-frequency noise spectroscopy, MOS capacitance, gated-diode current, DLTS. A more fundamental difference between bulk Si and thin SOI–MOS structures is related to the confinement of the electron gas. In Si, the potential bending at the surface induces a "triangular" potential well and typical 2-D quantum magneto-transport effects. In contrast, in thin SOI films, when both the front and back interfaces are inverted, the potential is almost flat. Although the potential drop in the middle of the film depends linearly on doping and quadratically on thickness, it remains very small ($\leq 5meV$ in films thinner than 100 nm and lightly doped). The minority carriers are no longer confined to a potential well

Invited Papers 59

but to a spatial (geometric) box which is delimited by the two interfaces. This *volume inversion* concept (Balestra et al 1987) has been confirmed by quantum mechanical calculations (Cristoloveanu and Ioannou 1990). The energy separation of the lowest subbands is extremely small in normal SOI films ($50 - 100nm$ thick) and would become significant only in $10 - 20nm$ films which are not yet manufacturable by conventional SOI technology. It follows that the density of states and the transport properties correspond to a 3-D system rather than to a 2-D gas. However, 2-D quantum effects could also be obtained in SOI by adjusting the back gate bias. Inversion at one interface and accumulation at the opposite interface results in a triangular well and quantization occurs in very thin films. Thus, a thin SOI–MOSFET is a very flexible device which gives the unique opportunity for studying the bias-induced transition from 3-D to 2-D systems.

In a practical point of view, the volume-inversion transistor has enhanced performance because carriers flowing in the middle of the film are less influenced by interface defects. This has been demonstrated either by biasing both gates ($V_{G_2} = 10V_{G_1}$, to account for the difference in oxide thicknesses) or by using a customized design. The *gate-all-around MOS transistor* is processed by (i) etching a cavity underneath the Si island, (ii) oxidizing and (iii) growing the gate around the Si bridge (Colinge et al 1990). Recently fabricated was a Δ-structure where the Si film is not only thin but also very narrow. Transconductance oscillations occur at low temperature due to 1-D quantum resonant transport (Takeda et al 1990).

6. Transient and floating body effects

The carrier generation properties which greatly influence the device performance are determined using current transient effects arising from the formation and relaxation of *deep-depletion* regions. Unlike the case of depletion-mode transistors where single gate operation is successful (Elewa et al 1988), the experiment with enhancement-mode MOSFET's requires a dual-gate control. The drain current variation at one interface is monitored while pulsing the opposite gate in accumulation. Majority carriers are immediately supplied by the ionization of acceptor impurities in the body which produces a deep-depletion region. The increase of the depletion charge is balanced by a decrease in the front interface inversion charge, so the drain current drops following the application of the pulse. For equilibrium to be reached, additional majority carriers are needed to compensate for the deep-depletion region and to adjust the back interface accumulation charge. According to the generation rates in the film and at the interfaces, the current increases with time more or less rapidly (Ioannou et al 1991).

In early SIMOX material annealed below $1200°C$, the generation lifetime was in the $10 - 100$ *nsec* range. The crystal improvement achieved by high temperature annealing is illustrated by the lifetime increasing to about 1 μsec and even to 100 μsec in multiple-SIMOX. This is explained by a drastic reduction of the density of dislocations which can be decorated with metal contaminants and therefore affect primarily the lifetime rather than the transport and interface properties. The generation velocity is very small at the front interface ($0.2 cm/s$) and much higher on the edges ($1 m/s$).

An undesirable consequence of high lifetimes is that SIMOX devices suffer from long transient effects. For instance, the shape of the "static" characteristics $I_D(V_{G_1}, V_{G_2})$ depends substantially on the hold time, delay time and direction of voltage scanning. A device may look as being totally-depleted after $0.1\mu s$ hold time, whereas after $500 s$ it turns out to be partially-depleted. Dramatic errors may arise in the evaluation of the transistor parameters (doping underestimation, mobility overestimation, inaccurate threshold voltage and subthreshold slope), unless very long relaxation times are allowed during the measurement.

Major parasitic effects are due to the build-up, in the Si film, of a majority carrier charge which is generated by impact ionization and does not recombine rapidly enough. The detrimental aspects of *floating body* are : (i) threshold voltage reduction, (ii) activation of the bipolar transistor leading to a kink in $I_D(V_D)$ characteristics, (iii) subthreshold behavior with nearly vertical slope and hysteresis, and (iv) loss of gate control (latch) allowing a high current in the off-state. Further details given in these proceedings (Ouisse et al 1991) demonstrate the correlation between the regions of negative conductance and transconductance and show that hysteresis effects in SIMOX may be treated as a phase transition. An optimum SIMOX material in terms of attenuated floating body effects would probably be a compromise between a high crystal quality and a reasonably low lifetime.

7. Reliability issues

A key argument for the commercial success of SIMOX circuits might be their improved tolerance to hot carrier induced degradation. Recent measurements confirm that thin film SOI transistors present a very good tolerance to aging which is explained by the reduction of the peak of the electric field in SOI as compared to bulk Si (Ouisse et al 1990). After more than 200 hours of stress, $1\mu m$ SIMOX-MOSFET's with optimized LDD configuration show quite acceptable rates of degradation of the threshold voltage ($\leq 10 mV$) and transconductance ($\leq 1\%$). No defects have been induced at the back interface under front channel stress.

Stressing the back channel transistor shows that the buried oxide is far less tolerant to hot carrier injection than the gate thermal oxide. The threshold voltage shift can exceed $15V$ after only 20 hours of stress. Although an increase of V_{G_2} or a decrease of the channel length accentuates the damage, the exponential time dependence law is not modified: $\Delta V_{th2} \sim t^{0.16}$. The aging of the sidewall transistor was monitored independently and found to be more rapid with time as many defects are generated in the bottom of the Si island edges. Figure 3(a) shows that instead of being reduced, the transconductance improves substantially (40%) after a few minutes of stress. This *transconductance overshoot* is due to the coupling of the defective and non-defective regions of the channel which have different threshold voltages and lengths. When the short damaged region is just entering in strong inversion, it dominates the total channel resistance (because the non-defective region is already in strong inversion) and imposes a higher transconductance. This two-region model allows extracting the density and location of generated defects. It was found that the aging kinetics consist of electron trapping followed by interface state formation (Ouisse et al 1990). After the stress, the kink effect occurs at a higher drain voltage and the excess current is reduced. The vulnerability of the buried oxide to hot carrier injection supports recent electron spin resonance data which suggests that the microstructures of the thermal and implanted oxides are different (Stesmans et al 1990). Hopefully, the aging of the buried oxide did not degrade the front channel properties.

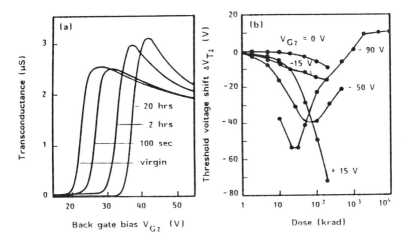

Figure 3 Back channel transconductance after various periods of hot carrier injection (a) and threshold voltage shift versus irradiation dose (b) in SIMOX–MOSFET's (after Ouisse et al 1990).

The evaluation of the radiation-induced damage is a basic test for SOI technologies. The influence of X-ray exposition on the back channel threshold voltage is shown in Fig.3(b). The initial decrease of V_{T_2} with dose is definitely governed by the trapping of holes which occur on the side of the oxide where the potential is lower. However, neither the field-induced shift of initially trapped holes nor the generation of interface states can account for the increase of V_{T_2} at higher doses. This rebound effect is due to the filling of *electron traps*. This scenario is supported by charge pumping measurements which confirm that the formation of back interface states is very limited ($\Delta N_{it2} \simeq 1.5 \times 10^{11} cm^{-2} eV^{-1}$).

The presence of numerous traps within the buried oxide seems therefore to govern the device degradation caused either by irradiation or hot carrier injection. An interesting point was to examine whether these traps can still be filled by hot electrons, even after irradiation at negative V_{G_2}. Several $1\mu m$ long n-channel MOSFET's, priorly irradiated with various doses, have been electrically stressed. The shift of V_{T_2} increases after irradiation which can be explained by (i) the recombination of hot electrons with trapped holes or/and (ii) a substantially enhanced electron trapping. It is inferred that during irradiation new electron traps are generated and only partially filled.

8. Conclusion

The potential of SIMOX and recent achievements in technology, characterization and modeling have been reviewed. Solutions exist to fabricate very good quality SIMOX with low densities of defects and contamination and high carrier mobility and lifetime. However, the interface coupling, floating body and transient phenomena become more severe in high standard SIMOX. The flexibility of the SIMOX process is a key advantage because a customized technological approach can be envisaged for each field of applications.

Acknowledgments

Thanks are due to my colleagues from Grenoble (T. Ouisse, T. Elewa, A. Chovet, S. Williams, M. Gri) and from G. Mason University in Virginia (D.E. Ioannou, X. Zhong, B. Mazhari) who have contributed to this work. Drs. A–J. Auberton-Hervé, J. Margail, C. Jaussaud, P. Hemment, J. Davis, H. Hughes and J–P. Colinge are also acknowledged for interest and support in this research.

References

Auberton-Hervé A-J 1990 Proc. *4th Int. Symp. on Silicon On Insulator Techn. and Devices* ed D Schmidt (Electrochemical Soc.) 90–6 pp 455–478
Balestra F, Cristoloveanu S, Benachir M, Brini J and Elewa T 1987 *IEEE Electron Device Lett.* **EDL-8** 410
Colinge J-P, Gao M H, Romano A, Maes H and Claeys C 1990 Proc. *IEEE SOS/SOI Technology Conf.* 137
Cristoloveanu S 1989 *Semiconductor Silicon* eds G. Harbeke and M.J. Schulz (Springer-Verlag: Berlin) pp 223–249
Cristoloveanu S 1991 *Vacuum* **42** 371
Cristoloveanu S, Gardner S, Jaussaud C, Margail J, Auberton-Hervé A-J and Bruel M 1987 *J. Appl. Phys.* **62** 2793
Cristoloveanu S and Ioannou D E 1990 *Superlattices and Microstructures* **8** 131
Elewa T, Haddara H and Cristoloveanu S 1988 *The Physics and Technology of Amorphous SiO_2*, ed R A B Devine (Plenum: New York) pp 553–9
Elewa T, Kleveland B, Cristoloveanu S, Boukriss B and Chovet A 1991 in press
Ioannou D E, Cristoloveanu S, Potamianos C N, Zhong X, McLarty P K and Hughes H L 1991 *IEEE Trans. Electron Dev.*
Lim H-K and Fossum J C 1983 *IEEE Transactions on Electron Devices* **ED-39** 1244
Mao B Y, Chang P H, Chen C E and Lam H W 1987 *J. Appl. Phys.* **62** 2308
Matloubian M, Sundaresan R and Lu H 1989 *IEEE Trans. Electron Dev.* **36** 938
Mazhari B, Cristoloveanu S, Ioannou D E and Caviglia A 1991 *IEEE Trans. Electron Dev.*
Ouisse T, Cristoloveanu S, Reimbold G and Borel G 1990 Proc. *ESSDERC'90* eds W Eccleston and P J Rosser (Adam Hilger: Bristol) pp 257–260

Ouisse T, Brini J, Ghibaudo G, Cristoloveanu S and Borel G 1991 Proc. *INFOS'91*
Stesmans A, Revesz A G and Hughes H L 1990 Proc. *IEEE SOS/SOI Technology Conf.* 162
Takeda E, Matsuoka H, Yoshimura T and Ichiguchi T 1990 Techn. Digest *IEDM'90* 387
Williams S and Cristoloveanu S, Ann. Conf. of the Condensed Matter Div.– Europ. Phys. Soc.

Paper presented at INFOS '91, Liverpool, April 1991
Workshop Papers

The relationship of trapping and trap creation in silicon dioxide films to hot carrier degradation of Si MOSFETs

D. J. DiMaria

IBM Thomas J. Watson Research Center, P.O. Box 218, Yorktown Heights, N.Y. 10598

ABSTRACT: Degradation mechanisms in Si MOSFETs are reviewed from the point of view of the basic physical mechanisms involving electron/hole transport, trapping, and defect generation. Alternative insulators to reduce these effects and improve device reliability are also discussed.

1. INTRODUCTION

Degradation of Si MOSFETs is caused after injection of either electrons, or holes, or both over the 3.1 eV interfacial, silicon-dioxide/silicon energy-barrier near the drain region of the device. These hot carriers are either injected locally from the Si channel or from the reversed-biased drain junction. For n-channel FETs, the dominant degradation effects are believed to occur when both electrons and holes are injected into the the oxide layer and produce interface states near the Si/oxide interface.[1] However for p-channel FETs, the degradation rate is not as rapid and is believed to be mostly due to electron injection and trapping.[2] The focus of this review will be to discuss studies using uniform-carrier-injection schemes which address the microscopic nature of the various degradation modes. Also, possible material modifications of the thermally-grown SiO_2 films which might inhibit degradation will be reviewed.

2. ELECTRON TRANSPORT and TRAPPING

Under electron injection conditions, the dominant effects are due to background trapping and trap/interface-state creation. Each will be discussed separately with more emphasis on trap creation due to its increased importance in current and future Si-based technologies. Hot-carrier bandgap ionization which would produce electron-hole pairs could be important for degradation at very-high electric fields ($\gtrsim$ 10 MV/cm) on thick oxides ($\gtrsim$ 250 Å), due to the holes produced. This ionization is not caused by the main portion of the hot-electron distribution, but only those few carriers in the high energy tail with energies exceeding $\approx$ 9 eV. Damage from gap ionization, if important, would come mostly from annihilation of trapped interfacial holes as discussed in section 3.

Background trapping occurs predominantly on bulk water-related impurities in the as-fabricated oxides.[3] Many of these sites can be removed with high temperature annealing in an inert ambient. For thinner oxides, these sites are less important, becoming negligible for films $\lesssim$ 50 Å in thickness. Other energetically-shallow background-sites are filled near the interfacial regions as the temperature is lowered. These shallow sites are believed to be due to oxygen vacancies near the Si/oxide interface which can be reduced with high-temperature annealing in oxygen.[4] Initial trapping probabilities on these background sites can vary from 1×10^{-7} to 1×10^{-4} trapped electrons per carrier depending on processing.

© 1991 IOP Publishing Ltd

Trap and interface-creation occurs in silicon dioxide due to carrier heating (particularly, when electron energies in the oxide conduction band exceed about 2 eV (Ref. 5) as will be discussed later). Electron heating in SiO_2 has been intensively studied since about 1982 and is caused by the energy gained from the applied electric field.[6,7] Figure 1 shows steady-state heating data from a variety of experiments where a threshold for rapid energy gain from the field is observed at $\approx$ 1.5 MV/cm. Stabilization at $\approx$ 4 MV/cm is caused by the dual action of large angle scattering of the hot electrons through their interaction with non-polar, acoustic-phonon modes and the subsequent energy loss in large chunks (0.153 and 0.063 eV) to the polar LO-phonon modes of this material.[6,7] Figures 2 and 3 demonstrate the trap creation phenomena in thick-oxide films for distributed electron traps and interface states, respectively. From these data, an onset for trap creation above the background is seen by the pronounced increase in trapped electrons (Fig. 1) and interface states (Fig. 2) at fields exceeding about 1.5 MV/cm which coincides with the observed electron heating threshold in Fig. 1 (Refs. 5, 8-12). At fields above $\approx$ 4 MV/cm, a slow down in the build-up of the hot-electron induced sites consistent with the electron heating dependence is also seen by comparing Figs. 1-3. For trapped electrons, field ionization of some charges influences the trap occupancy above 4 MV/cm (Ref. 8).

Insensitivity to changes in the oxide thickness and to the injection mode of the carriers from the cathode, imply that trap creation is produced by the release of a mobile species from near the anode/oxide interface.[5] This mobile species then moves to the cathode/oxide region where it reacts to produce the observed interface states and trapping sites which are distributed away from this surface. Figures 4 and 5 demonstrate this lack of a dependence on the bulk properties of the oxide layer for trap and interface-state generation, respectively. Consistent with these data on films thicker than $\approx$ 100 Å, it would be expected that trap generation should still occur on very thin films where no bulk oxide layer is left. This is demonstrated in Fig. 6 on a 50 Å film where electron energies are listed rather than the applied fields. On these very thin films, many of the carriers are traversing the oxide ballistically with nearly monoenergetic distributions.[13-15] Clearly from this figure, very little trap generation is observed below energies of 1.9 eV for injected-electron fluencies up to 50 Coul/cm^2. Figure 7 summarizes the data in Figs. 1-6 and shows that for poly-Si gated capacitors and FETs energies in excess of $\approx$ 2 eV are needed for electron trap creation in the oxide film independent of oxide thickness or injection mode. Similar conclusions can be made for interface-state generation.

Contrary to the observations previously discussed for background-trapping rates of sites in the as-fabricated oxides, the trap and interface-state generation rates are suppressed as the temperature is lowered.[5,16-19] This is shown in Fig. 8 where the generation rates increases strongly above $\approx$ 150 K. By fitting Arrhenius relationships to limited sections of these data, activation energies between 0.01 and 0.2 eV can be obtained. As observed in this figure, the insensitivity of these data to oxide thickness again suggest that the release of the mobile species from near the anode/oxide interface limits these observations.

Recent studies have shown that the trap and interface-state generation rates are sensitive to the presence of hydrogen in the oxide,[20-22] especially when it is incorporated into the oxide at high temperatures. This is demonstrated in Figs. 9 and 10 where trap and interface-state creation, respectively, are compared on forming-gas (10% H_2 - 90% N_2) annealed (1000°C for 60 min) and unannealed oxide-structures. Other studies using secondary-ion-mass-spectroscopy (SIMS) have shown directly the buildup of hydrogen near the cathode/oxide interface under electron heating conditions in the oxide.[21] Also, these studies and those of others implied that the hydrogen was released from near the anode/oxide interface by the energetic carriers.[21,22] Because of the hydrogen-sensitivity of

Workshop Papers

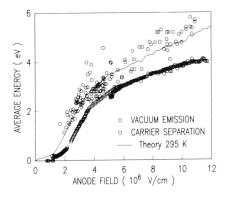

Fig. 1: Average energy of hot electrons (relative to the bottom of the oxide conduction band) as a function of oxide field at room temperature. The electron heating data was obtained from vacuum emission (open circles) and carrier separation measurements (open squares). The theoretical calculation was performed using the Monte Carlo approach where both polar and non-polar phonon scattering in the SiO_2 layer are included.

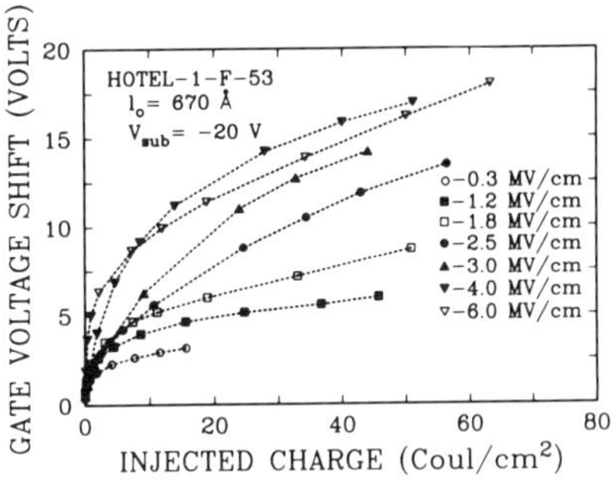

Fig. 2: Gate voltage shift as a function of injected charge for average oxide fields from 0.3 to 6 MV/cm on devices with a 670 Å gate oxide at room temperature. These voltage shifts are proportional to the electron trapping in the SiO_2 layer at distances $\gtrsim 13$ Å from the Si/SiO_2 interface.

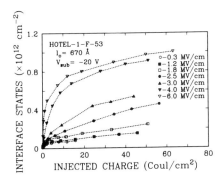
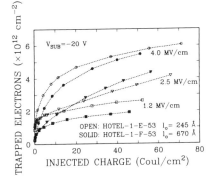

Fig. 3: Number of Si/SiO_2 interface states as a function of injected charge on the same devices as in Fig. 2 for fields from 0.3 to 6 MV/cm.

Fig. 4: Comparison of bulk-trapped-electron build-up for two oxide thickness's (245 and 670 Å) as a function of injected charge at room temperature for several different average electric fields (1.2, 2.5, and 4.0 MV/cm) through the electron heating threshold. These data show the relative insensitivity of trap creation to oxide thickness.

the generation rates and the direct observation of hydrogen evolution, this species is believed to be related to H, H⁺, or both.

In summary, these generation rates are sensitive to both processing and the electron energy that the carriers can gain from the applied field. These rates can vary from 1×10^{-9} to 1×10^{-4} generated sites per carrier and show no saturation up to densities of $\gtrsim 1\times10^{13}$ cm^{-2}.

3. HOLE TRANSPORT, TRAPPING, and ANNIHILATION

Hole transport and trapping may produce traps and interface-states, in a manner similar to that observed for electrons. However, as discussed below, it is the process of hole annihilation by electrons that is the most effective way to produce interface-states. For most device degradation processes, hot-holes are injected over large energy barriers at the SiO_2-anode interface. Some holes could possibly be produced at large fields due to hot-electron bandgap ionization. Radiation would also produce electron-hole pairs across the SiO_2 energy gap at energies exceeding ≈ 9 eV.

Hole injection into silicon dioxide is a very efficient way to produce positive-trapped-charge build-up. These holes are trapped near Si/oxide interfaces with an initial trapping probability that can range from about 0.01 to 0.3 trapped holes per carrier depending on applied-electric-field and processing temperatures.[23] These trapping sites are believed to be also related to oxygen vacancies. Usually few interface states are observed when the holes are initially trapped. At low temperatures ($\lesssim 150$ K) and low fields ($\lesssim 4$ MV/cm), holes can be trapped in the oxide bulk in energetically shallow sites which modulate their dispersive transport.[23]

Studies of trap creation induced by hole transport in the oxide valence band similar to those done for hot electrons have also been recently performed.[24] At oxide fields up to 5 MV/cm and injected-hole-fluencies up to 1×10^{-4} Coul/cm², little if any hole-trap or interface-state generation is observed above the background levels of the as-fabricated films. Since hole transport is highly dispersive and modulated by states in the oxide bandgap above the top of the valence band, this observation seems reasonable. However, it should be noted that these studies were done below the levels of injected charge at which the hot electron studies showed trap creation (for example, see Figs. 2, 3, and 9). Another recent report suggests that neutral bulk-traps (for electrons) can be produced by the passage of a large fluence of holes ($\gtrsim 1\times10^{-4}$ Coul/cm²), but no variations in the oxide field were studied.[25]

If electrons are also injected either during or after hole injection, electron-hole annihilation will occur. For hole annihilation near the Si/oxide surface, interface states are generated very efficiently as shown in Fig. 11. Low-field generation rates varying from 0.05 to 0.4 sites per annihilated-hole, depending on processing, have been reported.[12,26] Since the electron-hole annihilation rate is strongly dependent on the local electric fields, the interface-state generation rate should similarly track. These rates can drop orders of magnitude once the electron heating threshold is exceeded.[27,28] This is shown in Fig. 12 where the capture cross-section for Coulombic capture of a conduction band electron by a trapped hole is plotted as a function of the average field. The capture probability drops at a larger rate once the field for significant electron heating is exceeded. This is due to the net effect of both the onset for heating with dispersive transport above ≈ 1.5 MV/cm and the increasing tunneling probability with field of cascading electrons out of the excited states of these Coulombic trapping sites.[27] A few researchers have also reported induced

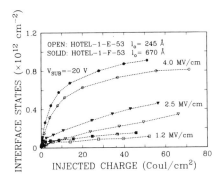

Fig. 5: Comparison of Si/SiO$_2$ interface-state build-up (for the same devices with 245 and 670 Å gate oxide thickness's used in Fig. 4) as a function of injected charge for different electric fields.

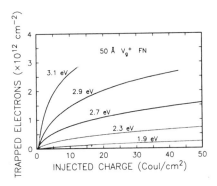

Fig. 6: Number of trapped electrons in bulk sites created by hot electrons in devices with a very thin SiO$_2$ gate-insulator (50 Å) as a function of injected charge at room temperature for various average energies of the oxide electron-distribution. The carriers were injected from the Si substrate and sensed by Fowler-Nordheim (FN) tunneling (at distances $\gtrsim$ than 20 Å from the cathode/oxide interface). This figure shows that trap creation starts to occur at hot-electron energies $\gtrsim$ 1.9 eV.

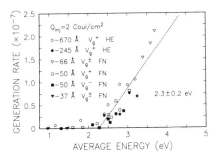

Fig. 7: Generation rate for creation of bulk electron-trapping-sites by hot carriers as a function of the average energy of the distribution for various oxide thickness's, voltage polarities, and means of injecting the electrons (either FN tunneling through or optically-induced hot-electron (HE) emission over the interfacial energy-barrier) after the injection of 2 Coul/cm^2 of charge at room temperature. This figure shows a threshold for trap creation at 2.3±0.2 eV from a linear fit of the data at energies $\gtrsim$ 2 eV. Carrier separation data like that in Fig. 1 were used to determine the average electron energy near the anode/SiO$_2$ interface from the value of the average electric field.

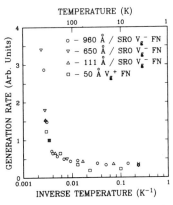

Fig. 8: Trap generation rate verses inverse temperature for injected electron fluencies $\gtrsim$ 20 Coul/cm^2. Device structures with oxide thickness varying from 50 to 960 Å with and without Si-rich oxide (SRO) electron injectors for both voltage polarities and varying gate area were used. A universal characteristic for the temperature dependence is observed. This temperature dependence of trap and interface-state creation is separable from dependence on electron energy observed in Fig. 7.

electron traps near the interfacial region under certain annihilation conditions.[29] From this discussion, FET operation under the condition of injection of both carriers at low internal fields would give the most rapid degradation of the device.

4. ALTERNATIVE INSULATORS

Recently, reduction in device degradation under hot carrier injection on n-channel devices has been observed on FETs with gate oxides that have undergone a nitridation/reoxidation sequence.[30] However, the opposite situation is observed for p-channel structures after the same processing.[31] These observations are believed to be due to increased electron trapping in background sites introduced by the incorporation of oxynitride layers near the interfacial regions and the decreased hole-trapping probability in energetically-deep interfacial sites in these same spatial regions.

Recent uniform injection studies on nitrided and reoxidized/nitrided structures tend to support these conclusions.[33,34] Increased electron capture in background sites in the oxynitride regions is observed to have initial trapping probabilities which can be at least two orders of magnitude larger than those for thermal oxides. This is shown in Fig. 9 where trapped electron buildup under high field injection conditions is shown as a function of injected charge. Decreases in trap and interface-state generation rates by hot electrons in the oxide are expected due to the blocking action of the denser oxynitride layers to any released mobile species. This is clearly consistent with the observations for interface-states in Figs. 10, but also for trap generation in Fig. 9 if detrapping effects are taken into account.[33,34] The reduction in steady-state occupancy of the trapping sites in the reoxidized-nitrided insulator compared to the nitrided insulator has been demonstrated to be mostly due to trap ionization at fields exceeding 8 MV/cm (Ref. 33). This latter observation could be important for FETs used in non-volatile memory applications where oxide layers are used under high fields to transfer charge to and from floating-gate-electrodes. To minimize degradation of these devices after many storage and discharge operations, trap creation and steady-state charge-buildup in background sites must be suppressed.

5. CONCLUSIONS

All Si-based memory-technologies (DRAM, Bipolar, and EEPROM) will suffer degradation effects if electrons, holes, or both enter the oxide over the blocking interfacial energy barriers. Electron-hole annihilation can be the most efficient for interface-state generation, but hot electrons in the oxide layers can also create traps in the oxide or at the interfaces. Although radiation effects have not been explicitly discussed, high energy photons or particles can produce damage similar to that discussed here.[23] The study of these degradation effects in the insulating layers of the various device technologies should continue to be important for both the commercial and military sectors now and in the future.

Acknowledgements: The author wishes to acknowledge the the contributions of M.V. Fischetti, D.A. Buchanan and J.H. Stathis to some of the work reviewed here.

References

1. P. Heremans, H.E. Maes, and N. Saks, IEEE Electron Dev. Lett. EDL-7, 428 (1986).
2. F. Matsuoka, H. Iwai, H. Hayashida, K. Hama, Y. Toyoshima, and K. Maeguchi, IEEE Trans. Electron Dev. ED-37, 1487 (1990).

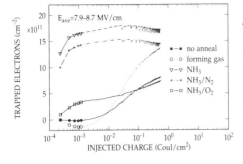

Fig. 9: Number of trapped electrons as a function of constant-current FN-injected charge at average fields varying from 7.9 to 8.7 MV/cm. The gate insulators were 517 Å of thermally-grown silicon-dioxide annealed in various gaseous ambients at 1000 °C sequentially for 60 min intervals where more than one gas was used.

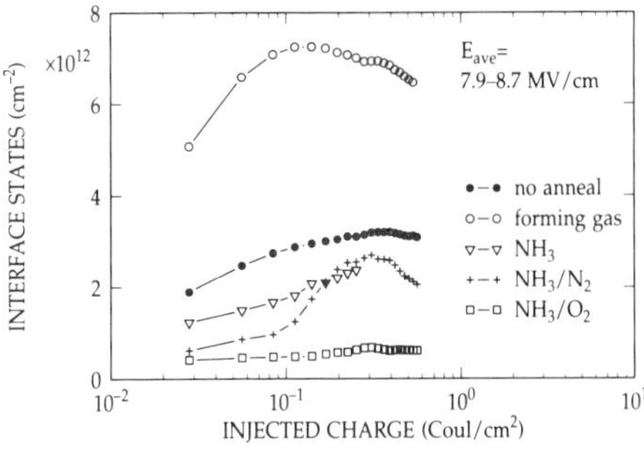

Fig. 10: Number of interface states as a function of FN-injected charge on the same devices and under the same conditions as in Fig. 9.

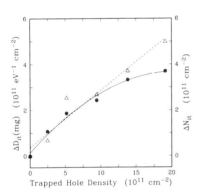

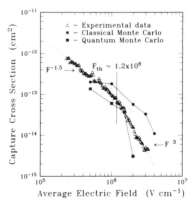

Fig. 11: Change in the midgap interface-state density ΔD_{it} (mg) indicated by solid circles and total integrated interface-state density ΔN_{it} indicated by open triangles as a function of trapped hole density after injection of an electron fluence of 2.5×10^{-4} Coul/cm² at very-low average-electric-fields ($\lesssim 0.7$ MV/cm). The electrons were injected to annihilate the trapped holes in SiO_2-Si interfacial sites.

Fig. 12: Coulombic capture-cross-section as a function of the average-electric-field for electron annihilation of trapped holes. A power-law fit for the low and high field regions gives exponents of -1.5 and -3, respectively, with a change in slope occurring at ≈ 1.2 MV/cm. Also shown are both classical and quantum Monte-Carlo simulations with the latter taking into account tunneling out of excited states of the potential wells formed by the positively-charged trapped-holes.

3. F.J. Feigl, D.R. Young, D.J. DiMaria, S. Lai, and J. Calise, J. Appl. Phys. 52, 5665 (1981).
4. M. Aslam, IEEE Trans. Electron Dev. ED-34, 2535 (1987).
5. D.J. DiMaria and J.W. Stasiak, J. Appl. Phys. 65, 2342 (1989).
6. D.J. DiMaria, T.N. Theis, J.R. Kirtley, F.L. Pesavento, D.W. Dong, and S.D. Brorson, J. Appl. Phys. 57, 1214 (1985).
7. M.V. Fischetti, D.J. DiMaria, S.D. Brorson, T.N. Theis, and J.R. Kirtley, Phys. Rev. B 31, 8124 (1985).
8. D.J. DiMaria, Appl. Phys. Lett. 51, 655 (1987).
9. S.N. Kuznetsov and V.A. Gurtov, in *Insulating Films on Semiconductors*, edited by G. DeClerck and R. DeKeersmaecker (North-Holland, Amsterdam, 1987), pp. 347-352.
10. C.C.H. Hsu, T. Nishida, and C.T. Sah, J. Appl. Phys. 63, 5882 (1988).
11. M.M. Heyns, D. Krishna Rao, and R.F. Dekeersmaecker, Appl. Surf. Sci. 39, 327 (1989).
12. D.A. Buchanan and D.J. DiMaria, J. Appl. Phys. 67, 7439 (1990).
13. G. Lewicki and J. Maserjian, J. Appl. Phys. 46, 3032 (1975).
14. D.J. DiMaria, M.V. Fischetti, J. Batey, L. Dori, E. Tierney, and J. Stasiak, Phys. Rev. Lett. 57, 3213 (1986).
15. M.V. Fischetti, D.J. DiMaria, L. Dori, J. Batey, E. Tierney, and J. Stasiak, Phys. Rev. B 35, 4404 (1987).
16. E. Harari, J. Appl. Phys. 49, 2478 (1978).
17. G.S. Gildenblat, C.L. Huang, and S.A. Grot, J. Appl. Phys. 64, 2150 (1988).
18. B.Balland, C.Plossu, and S.Bardy, Thin Solid Films 148, 149 (1987).
19. D.J. DiMaria, J. Appl. Phys. 68, 5234 (1990).
20. Y. Nissan-Cohen and T. Gorczyca, IEEE Electron Dev. Lett. EDL-9, 287 (1988).
21. R. Gale, F.J. Feigl, C.W. Magee, and D.R. Young, J. Appl. Phys. 54, 6938 (1983).
22. A.D. Marwick and D.R. Young, J. Appl. Phys. 63, 2291 (1988).
23. *Ionizing Radiation Effects in MOS Devices and Circuits*, edited by T.P. Ma and P.V. Dressendorfer (Wiley-Interscience, New York, 1989).
24. A.V. Schwerin, M.M. Heyns, and W. Weber, J. Appl. Phys. 67, 7595 (1990).
25. S. Ogawa, N. Shiono, and M. Shimaya, Appl. Phys. Lett. 56, 1329 (1990).
26. S.K. Lai, J. Appl. Phys. 54, 2540 (1983).
27. D.A. Buchanan, M.V. Fischetti, and D.J. DiMaria, Phys. Rev. B 43, 1471 (1991).
28. T.H. Ning, J. Appl. Phys. 47, 3203 (1976).
29. I.C. Chen, S. Holland, and C. Hu, J. Appl. Phys. 6, 4544 (1987).
30. G.J. Dunn and S.A. Scott, IEEE Trans. Electron Dev. 37, 1719 (1990).
31. G.J. Dunn and J.T. Krick, IEEE Trans. Electron Dev. 38, 901 (1991).
32. A. Yankova, L. Do Thanh, and P. Balk, Solid-State Electronics 30, 939 (1987).
33. D.J. DiMaria and J.H. Stathis, to be published in J. Appl. Phys.
34. D.J. DiMaria, J. Appl. Phys. 68, 5234 (1990).

Paper presented at INFOS '91, Liverpool, April 1991
Workshop Papers

Charge trapping and degradation of thin dielectric layers

M. M. Heyns and A. v. Schwerin[*]

Interuniversity Microelectronics Centre (IMEC), Kapeldreef 75, B-3001 Leuven, BELGIUM

> ABSTRACT : Charge trapping and degradation of thin dielectric layers are important reliability issues in small-geometry MOS transistors. This paper discusses the oxide field dependence of the defect generation during injection of electrons or holes in SiO_2 layers, the slow trapping instability, the degradation during high-field stressing and the characteristics of nitrided oxides.

1. INTRODUCTION

Thin thermal oxide layers are an important part of MOS-technologies because they are always related with the active parts of the device. When used as tunnel dielectric in memory applications they directly determine the overall reliability of the memory cell. When used as gate insulator in MOS-transistors they have a strong impact on the transistor characteristics and on the yield and reliability of MOS-circuits. The continuous scaling down of the minimum device dimensions, without the appropriate scaling of the supply voltage, has given rise to the presence of high fields in small geometry MOS-transistors. The injection of charges in the gate oxide resulting from this is a potential reliability problem due to the charge build-up in the SiO_2 layer and the degradation of the Si/SiO_2 interface which follows from it.

2. CHARGE INJECTION TECHNIQUES

Some of the characterization techniques which have been successfully used in the past to investigate the trapping properties of gate insulators are no longer applicable for very thin layers or do not control all relevant parameters. Most studies on the trapping properties of thermal oxide layers (DiMaria 1978, De Keersmaeker 1983, Balk 1984) have used the avalanche injection technique (Nicollian *et al* 1970) to introduce electrons or holes into the oxide layer. This technique works on simple capacitor structures. In order to assure a laterally uniform injection highly doped substrates are needed. The main drawback of the technique is that the oxide field (E_{ox}) during injection can not be controlled. This limitation can be overcome when homogeneous injection in MOS-transistors is used.

This injection technique is schematically illustrated in fig.1. Minority carriers generated in the Si-substrate by optical means or from an (underlying) diode (Verwey 1973, Ning *et al* 1974) are accelerated towards the Si/SiO_2 interface by a substrate bias while the source and drain of the transistor are grounded. Under these conditions the oxide field is determined by the gate voltage. Part of the carriers gain sufficient energy to overcome the barrier at the Si/SiO_2 interface and are injected into the oxide layer. The technique works relatively easy for electron injection but hole injection is more difficult. This is due to the shorter inelastic scattering length of holes in the Si-substrate and the higher energy barrier they have to overcome at the Si/SiO_2 interface. The hole injection efficiency can be increased by providing a large transverse field in the silicon (Schwerin *et al* 1990).

---

[*] now with: Siemens, Corporate Research and Development, Otto-Hahn-Ring 6, D-8000 München 83, Germany

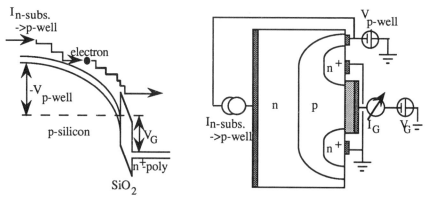

Fig. 1 : *Schematical representation of the electron injection process in the Si-SiO$_2$-poly-Si band diagram (left part) and of the measurement set-up (right part).*

Charge injection on capacitor structures can also be performed by applying fields large enough to cause Fowler-Norheim tunneling of electrons through the triangular barrier at the interface into the oxide conduction band (Lenzlinger et al 1969). Typically fields in excess of 7 MV/cm are needed before substantial currents begin to flow through a thermal oxide layer. The measurement of the injected tunneling current as a function of the applied field or of the voltage needed to sustain a fixed current was also used as a technique for sensing charges during or after stress (Solomon 1976). However, the validity of the interpretation of the shifts in these curves in terms of the charge build-up in the oxide layers can be questioned. Field-ionization of trapped charge is favoured by the high field and the injection mechanism was demonstrated to be affected by the high-field stress, even after very low fluences (Maserjian et al 1982), due to the high sensitivity of the injected current to small changes in the tunneling barrier. Moreover, the interpretation of the injection current in terms of oxide charge, taking into account the discreteness of the near-interface charge (Solomon 1976, Schmidlin 1966) is not straightforward. The technique is, however, very well suited to study degradation phenomena occurring at high fields and prior to breakdown and can provide valuable information on the characteristics of tunnel dielectrics operating under these conditions.

In this work the charge build-up and degradation during high-field stressing was investigated using an experimental procedure where stressing (and trap generation), trap filling and charge sensing were performed in consecutive steps of constant-current stress, avalanche injection and internal photoemission measurements (DiMaria 1976). This procedure also allows to make an unambiguous separation between bulk and interface charge.

3. EXPERIMENTAL CONDITIONS

The electron and hole injection experiments were performed on poly-Si gated transistors with a gate oxide thickness of 26 or 20 nm. The interface state density (D_{it}) is measured using the charge pumping technique (Groeseneken et al 1985), applied with constant pulses and a varying base level. The density of trapped oxide charge is obtained from the gate voltage shift for a fixed drain current level in deep subthreshold at a fixed low drain voltage. As the location of the charge is not known only an effective density of trapped charge is given with a centroid assumed to be located at the Si/SiO$_2$ interface. It was demonstrated (Schwerin et al 1990, Heyns et al 1989) that the effect of the interface state generation on this measurement can be neglected. More experimental details are given in Schwerin et al 1990 and Heyns et al 1989.

The capacitors used in the high-field stress experiments were fabricated on either p-type or n-type <100> Si wafers. Oxides were grown to thicknesses varying from 20 to 40 nm in dry O$_2$ at a typical temperature of 900°C. Either a thin transparent aluminum layer or a poly-Si layer was deposited as the electrode material. Capacitor structures were defined using standard lithography and processing techniques (Heyns et al 1986). All reported phenomena have been observed on a large number of wafers fabricated under a variety of processing conditions. The results are, therefore, thought to be at least qualitatively typical for high-quality SiO$_2$ layers.

4. DEFECT GENERATION DURING ELECTRON INJECTION

The D_{it} generation during electron injection was found to strongly increase with increasing oxide field (Heyns *et al* 1989), as shown in fig.2. In this linear plot two 'threshold' fields, at 1.5 and 4 MV/cm respectively, can be observed, which could suggest the existence of two generation mechanisms. A threshold of 1.5 MV/cm was already reported for the D_{it} generation during photoinjection of electrons (Zekeryia *et al* 1983). This threshold may be correlated with the mechanism by which the electrons loose their energy ("thermalize") when injected in the oxide. Below this threshold the scattering mechanism is dominated by LO-phonons while above this field acoustical phonon scattering (non-polar scattering) becomes important (Fischetti *et al* 1985). The gate current density and the energy of the incoming electrons were observed to have no effect on the D_{it} generation rate during electron injection. The interface state distribution as a function of energy in the Si-bandgap is independent of the oxide field and shows a large density in the upper half of the bandgap (Schwerin *et al* 1991a).

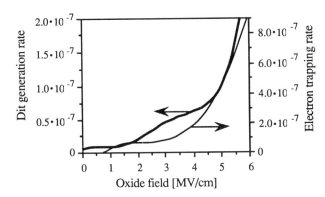

Fig. 2 : *Oxide field dependence of the interface state generation (left scale) and the electron trap generation (right scale) during electron injection.*

In agreement with previously reported results the generation efficiency of electron traps during electron injection is also observed to increase with increasing oxide field (Heyns *et al* 1989). The generation rate of the charged trap centres was calculated from the linear part of the trapped charge versus injected charge curves with the charge trapping at the lowest measured field subtracted from the measurements as background trapping in pre-existing electron traps. This generation rate is plotted as a function of the oxide field in fig.2. A 'threshold' around 4 MV/cm is observed on this linear plot. This result is in contrast with the 1.5 MV/cm threshold reported earlier (DiMaria 1987). This threshold was inferred from plots of the total trapped charge as a function of the oxide field for various amounts of injected charge and the probability of trap generation was not taken into account. Much larger fluences were used and measurements were performed up to saturation of the trap generation. In contrast with this the curves presented in fig.2 show the generation rate at the start of the experiment for low fluences and far removed from saturation. Beside this the importance of this threshold (and its existence) can be questioned as the change in the generation rate is not very steep.

On some series of samples a decrease of the trap occupation during the experiment with increasing oxide field was observed (Schwerin *et al* 1991a). This is due to field detrapping which occurs at the higher fields. Under these conditions the number of generated electron traps is much larger than the amount of negative charge measured after the stressing and evidence for the field enhancement of the trap generation process at higher fields is found only when a short low-field injection is carried out after the stressing to charge the generated electron traps. When not taken into account this effect can lead to erroneous conclusions on the field dependence of the electron trap generation rate.

The energy of the electrons at the moment of injection, as controlled by the p-well bias, does not seem to affect the electron trapping when the current flow through the oxide is kept fixed. This most probably indicates the fast thermalization of the carriers in the SiO_2 layer. On the other hand for oxide fields below approximately 4 MV/cm the trap generation efficiency depends on the current density during injection as illustrated in fig 3. This effect disappears for oxide fields larger than approximately 4 MV/cm, both before and after trap filling at a lower field. This gate current dependence was found both during experiments with an underlying pn-junction and with optically stimulated substrate hot electron injection. The dependence of the electron trapping on the oxide current density can not be explained with a simple trapping-detrapping model as it was demonstrated that the total integrated time that the field is applied has no major effect on the electron trapping (Schwerin et al 1991a). This is an important result because when no special precautions are taken the injected gate current depends on the substrate bias as well as on the oxide field. Therefore, this gate current dependence could eventually be misinterpreted as an oxide field dependence at low oxide fields.

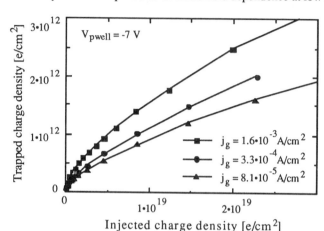

Fig. 3 : *Trapped electron density as a function of the density of electrons injected at an oxide field of 3.5 MV/cm, a p-well bias of -7 V and different gate current densities (jg).*

The field dependence of the D_{it} generation and the charge trapping, shown in fig.2, suggests a common origin for both generation mechanisms at fields larger than 4 MV/cm. A correlation between electron trapping and D_{it} generation was already suggested (Do Thanh et al 1986) and could occur by the release of hydrogen from water-related electron traps when an electron is captured at this defect. The hydrogen can diffuse towards the Si/SiO$_2$ interface where it may generate a dangling silicon bond, acting as an interface state. Another possible origin for the generation of both the interface states and the electron traps is the injection of holes from the anode (Fischetti 1985). The importance of holes in the D_{it} generation has already been clearly demonstrated (Lai 1983, Wang et al 1988) while the generation of electron traps by the recombination of electrons and trapped holes in SiO$_2$ has also been reported (Chen et al 1987). Within this model the D_{it} generation and electron trap generation are not directly correlated but have the same origin, i.e. the injection of holes from the anode. The average electron energy (above the oxide conduction band) at an oxide field of 4 MV/cm is approximately 2.5 to 3 eV (Fischetti et al 1985). This is in good agreement with the threshold energy for damage generation which was reported to be 2.3 eV (DiMaria et al 1989). Summed with the conduction band offset between the SiO$_2$ layer and the poly-Si electrode (taken as 3.1 eV), this leads to a threshold energy for the injection of holes between 5.6 and 6.1 eV. This energy is larger than the energy barrier for holes at the poly-Si electrode/SiO$_2$ layer interface but smaller than the suggested 7.5 eV threshold for the generation of surface plasmons (Fischetti 1985).

5. HOLE TRAPPING AND INTERFACE STATE GENERATION

Using homogenous injection of holes in transistor structures it was demonstrated (Schwerin et al 1990) that the trapping of holes does not depend on the substrate bias (and therefore on the energy of the injected holes) during injection. In fig.4 the hole trapping curves for different average oxide fields are compared. In all cases a very high trapping efficiency is found. In contrast with the results for electron injection there is apparently no strong field dependence for the hole trapping. The small trend towards a decreased trapping at higher fields can be due to either a field dependent capture cross section or to detrapping from shallow hole traps at higher fields (Schwerin et al 1990). The possibility of electron injection from the cathode at the higher fields, which recombine with the trapped holes, can be excluded (Schwerin et al 1990). When the injection is prolonged to larger injected holes densities a clear saturation level is observed for all oxide fields. The saturation level is independent of the oxide field during injection and no evidence could be found for the generation of hole traps during hole injection.

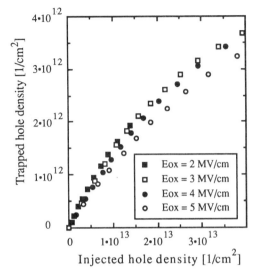

Fig. 4 : *Effective number of trapped holes as a function of the number of injected holes for various oxide fields ranging from 2 to 5 MV/cm.*

The D_{it} generation during irradiation, high-field stressing or hot electron injection is still a point of controversy in literature (Do Thanh *et al* 1986, Lai 1983, Wang *et al* 1988, Lyon 1989, Heremans *et al* 1987, Krishna Rao *et al* 1988 and Hofmann *et al* 1985). One of the suggested models assumes the D_{it} generation to be a two-step process (Lai 1983). The first step is the trapping of holes without any D_{it} generation. In the second step electrons recombine with the trapped holes causing the creation of interface states. On the other hand, experiments on hot carrier injection in transistor structures observed the maximum D_{it} generation under conditions where electrons and holes are simultaneously injected in the gate oxide layer (Saks *et al* 1986) and no evidence could be found for a two-step mechanism (Heremans *et al* 1987, Krishna Rao *et al* 1988). However, in most experiments on transistor structures it is very difficult to investigate the effects of hole injection while completely avoiding the injection of (a small number of) electrons. As the capture efficiency for a trapped hole to capture an electron is very high (DiMaria *et al* 1977) neutralization of the trapped holes occurs very efficiently and the second step of the two-step process may pass unnoticed. This will lead to the generation of an interface state and a re-structured hole trap which can act as a slow state (Heyns *et al* 1988a). The positive charge in these states gives rise to the positive charge observed after the hole injection which can be neutralized without the generation of interface states (Heremans *et al* 1987, Krishna Rao *et al* 1988).

The direct generation of interface states during hole injection was observed to decrease with increasing oxide field (Schwerin *et al* 1991b). The amount of generated interface states is only a small fraction of the number of trapped holes. This clearly demonstrate that the efficiency for D_{it} generation when only holes are injected is very low compared to the hole trapping efficiency. Normalizing the number of generated interface states to the number of injected holes results in a rough estimate of about $5 \cdot 10^{-3}$ for the D_{it} generation efficiency during hole injection (dependent on the oxide field during injection). As a comparison the D_{it} generation rate during electron injection is of the order of $5 \cdot 10^{-6}$ (also depending on the oxide field during injection) (Heyns *et al* 1989). This shows that hole injection, even without a two step process, is still about 1000 times more efficient in generating interface states than electron injection, in good agreement with results obtained with gate-controlled diodes (Heremans *et al* 1988).

When the D_{it} values were re-measured one week after the hole injection (with the devices stored at room temperature with the gate floating) a strong increase was observed together with a decrease in the trapped hole density (Schwerin *et al* 1990). A more detailed study of the time and field dependence of the two phenomena is shown in fig.5 (Schwerin et al 1991b). No one-to-one correlation of hole detrapping and interface state build-up after hole injection is found. The time dependence of the trapped charge decrease after the hole injection is qualitatively consistent with a simple model of direct tunneling of the charge carriers between the SiO_2 and the silicon substrate. The hole detrapping depends on both the magnitude and the polarity of the applied gate voltage. In contrast with this the simultaneous build-up of interface states is apparently only dependent on the field polarity. These observations can not be easily explained in the simple two-step model (Lai 1983) where trapped holes are transformed into interface states.

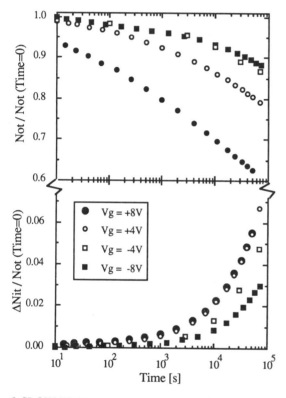

Fig. 5 : *Change in the density of trapped holes (upper part of the figure) and the interface state density (lower part of the figure) as a function of time after the hole injection. The values are normalized to the effective density of trapped holes measured directly (at time 0) after the hole injection (Not (Time=0)). A constant voltage (Vg) was applied to the gate after hole injection. Note the different scales for detrapping and interface state build-up.*

6. SLOW TRAPPING INSTABILITIES

The generation of slow trapping instabilities (or slow states) has been observed in the form of anomalous positive charge near the Si/SiO_2 interface during electron avalanche injection (Young et al 1979, Lai et al 1981, Heyns et al 1981, Feigl et al 1981, Fischetti et al 1982a, Fischetti et al 1982b, Sah et al 1983) and high-field stressing (Hillen et al 1983, Hofmann et al 1981). This charge responds to changes in the silicon surface potential (and/or the internal field) with time constants typically ranging from seconds to several hours, depending on temperature and on the history of the sample (Heyns et al 1981). A variety of physical models have been invoked to explain the build-up of this positive charge: the diffusion of hydrogen (Feigl et al 1981) or excitons (Weinberg et al 1978) to the Si/SiO_2 interface, hot hole injection from the anode (Fischetti 1985), electron-hole pair generation via band-to-band (DiStefano et al 1974, Shatzkes et al 1976) or trap-to-band impact ionization (Nissan-Cohen et al 1985) and field-stripping of electrons from valence band orbitals (Olivo et al 1983). Using a precise quantification technique (Heyns et al 1981) based on the charging/discharging characteristics of these slow states their generation kinetics were studied during avalanche injection of electrons or holes and during high-field stressing (Heyns et al 1986, Heyns et al 1988). The results indicated that there are only a limited and fixed number of sites in the oxide which can be converted into slow states, independent of the stress mode responsible for generating the slow states. It was demonstrated (Heyns et al 1988) that the positive charge due to slow states is located in initially present hole traps and no new hole traps are generated during electrical stressing. The hole trapping and slow state generation is not a fully reversible process. After the capture and detrapping of a hole the capture cross section of the centre for re-capturing a hole is strongly increased, indicating a change in the local structure of the hole trap. From this a model was proposed (Heyns et al 1988) according to which slow states originate from hole traps upon sequential trapping of a hole and an electron.

7. CHARGE AND DEFECT CREATION DURING HIGH-FIELD STRESSING

The degradation and charge build-up in the SiO_2 layer during high-field injection provides information on the wearout and breakdown mechanisms of these layers. The total charge-to-breakdown during these tests is found to depend on the electrode material, the injected current density and the stress conditions (Haywood et al 1985). The midgap voltage shift (ΔV_{MG}) during high-field stressing often indicates the generation of positive charge. This charge was demonstrated (Hillen et al 1983) to be located near the Si/SiO_2 interface in slow states. Under similar stress conditions these slow states form to a much smaller extent in poly-Si than in Al-gate structures. The exact shape of the ΔV_{MG}-vs-time curve during stressing is the net result of various charge components in the SiO_2-layer and at the Si/SiO_2 interface and is qualitatively dominated by the charge state of the slow traps during the high-field stress (Heyns et al 1988b). ΔV_{MG} can, therefore, not be directly interpreted in terms of the charge build-up in the oxide layer, as is often done. From more detailed measurements it was concluded (Heyns et al 1984) that no charge is built up in the bulk of the oxide layer during high-field stressing, the reason being that the high applied field (9-11 MV/cm) causes any trapped charge to be detrapped so that no bulk oxide charge remains.

Using a combination of avalanche injection and internal photoemission experiments it was demonstrated that during negative high-field stressing on Al-gate capacitors electron traps are generated close to the Si/SiO_2 interface (Heyns et al 1988). The generation rate was found to display a sublinear regime (Heyns et al 1988b), in contrast with other findings assuming a linear generation rate up to the occurrence of breakdown (Liang et al 1981, Liang et al 1984, Chen et al 1986). A positive high-field stress on Al-gate capacitors generates a large density of positive charge under the Al-electrode (Heyns et al 1985) while no electron trap generation close to the Si/SiO_2 interface was observed after this stress. Experiments conducted on poly-Si gate capacitors showed that during negative high-field stressing electron traps are generated close to the substrate-Si/SiO_2 interface (Heyns et al 1985), confirming the results obtained on Al-gate MOS-capacitors. Under positive high-field stress conditions electron traps are generated close to the poly-Si electrode (Heyns et al 1985). From comparing the results on Al-gate and poly-Si gate capacitors it follows that the positive charge present under the Al-gate after positive stress is directly associated with the Al-gate, because it is not found on poly-Si gate capacitors. Furthermore, it can be concluded that during high-field stressing of MOS-structures electron traps are generated in the vicinity of the non-injecting interface. While the generation of electron traps during high-field stressing was clearly demonstrated, no evidence could be found for the generation of hole traps during high-field stressing (Heyns et al 1988b). The results on the charge build-up and degradation of SiO_2 layers during high-field stressing are summarized in fig.6. It is important to notice that all the degradation and charge build-up phenomena encountered during the high-field stressing are occurring at the interfaces and no electrically measurable effects have been observed in the bulk of the oxide in the thickness range investigated here (20-40 nm). This suggests that the quality of the SiO_2 layer, as far as wearout and breakdown are concerned, is mainly determined by the quality of these interfaces. Furthermore, all charge build-up and defect generation phenomena were found to display regimes with strongly decreased generation rates before breakdown occurs. It is, therefore, not possible to simply point to one of the charge build-up and defect generation mechanisms as the primary cause for breakdown, for then a continuous increase of such degradation (and eventually run-away) would be expected. Most probably a complex interplay of the different observed phenomena will locally generate a critical condition leading to destructive breakdown or 'local' degradation phenomena, which are not detected by the measurement techniques used in this investigation, are dominating the breakdown behaviour.

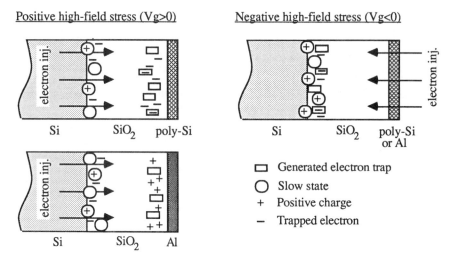

Fig. 6 : *Schematic illustration of the charge and defect distribution after negative and positive constant current stressing of MOS-capacitors with Al and poly-Si electrodes.*

8. NITRIDED OXIDES

Nitrided oxides and re-oxidized nitrided oxides have been proposed as alternatives to thermal SiO_2 layers for very thin gate insulators. The characteristics of these layers are a complicated function of the oxide thickness, the nitridation conditions and the re-oxidation and annealing conditions. A complete discussion of the characteristics of these layers is, therefore, beyond the scope of this paper. In general good breakdown properties, low interface state densities and fixed oxide charge densities have been reported for these layers but usually an increase in the trapping characteristics as compared to thermal SiO_2 is observed.

A short rapid thermal nitridation (RTN) step was found to generate a large density of water-related electron traps (with capture cross-sections of $\sim 10^{-17}$ and 10^{-18} cm^2 (Hartstein *et al* 1981)) in 15 to 30 nm oxide layers (Dooms *et al* 1989a, Dooms *et al* 1989b). More severe nitridation results in the generation of a nitrogen-related 10^{-16} cm^2 trap (Dooms *et al* 1989a, Dooms *et al* 1989b), which is also observed after furnace nitridation (Severi *et al* 1987a, Severi *et al* 1987b). Other traps with a variety of capture cross sections (from 10^{-14} cm^2 to 10^{-17} cm^2) have been reported (Chang *et al* 1984, Ferry *et al* 1985, Yankova *et al* 1987 and Lai *et al* 1982). Re-oxidation can be used to lower the electron trap density (Dooms *et al* 1989b, Yang *et al* 1988), most probably because hydrogen can be removed from the films (Hori *et al* 1989). A strongly increased hole trapping is observed after a short RTN-step, while prolonged nitridation and re-oxidation decreases the hole trapping again (Dooms *et al* 1989a, Dooms *et al* 1989b). These results are consistent with investigations on furnace nitrided oxide layers (Severi *et al* 1988, Dunn 1989) where it was reported that nitridation at relatively low temperatures and for short times increases the density of hole traps. More severe furnace nitridation conditions were found to result in a hole trap reduction (Yankova *et al* 1987, Severi *et al* 1988). Only limited information is available up to now on the field dependence of the electron trap generation in nitrided oxides (DiMaria *et al* 1990). This is an important subject which deserves a more detailed investigation as it was already demonstrated (DiMaria *et al* 1990) that although reoxidized-nitrided oxides can be superior under high-field injection conditions compared to thermally grown oxides, the situation can be reversed at lower fields.

9. CONCLUSIONS

The field dependence of the trapping and interface state generation during electron and hole injection was investigated using homogeneous injection in transistors. Holes are much more efficient in generating interface states than electrons but the exact generation mechanism is still unknown. When the oxide field is increased above the Fowler-Nordheim injection threshold other degradation and charge build-up mechanisms start to become important. These results have important consequences in the study of the hot carrier degradation of MOS-transistors. When the results on oxide degradation obtained from investigations on capacitor structures or homogeneous injection on transistor structures are extrapolated towards hot-carrier stress of MOS transistors, it must be realized that the oxide field can strongly differ. Also the validity of accelerated lifetime tests on capacitor structures must be re-evaluated within this framework because the degradation mechanisms during normal operating conditions are different from the mechanisms during accelerated tests, rendering the extrapolation between the two conditions invalid. Nitrided (re-oxidized) oxides are a possible alternative to thermal oxide layers but, in general, they exhibit larger trapping properties.

REFERENCES

Balk P 1984 *Solid State Devices 1983* ed E H Rhoderick *The Institute of Physics Conf. Ser. No. 69* pp 63
Chang S, Johnson N M and Lyon S A 1984 *Appl. Phys. Lett.* **44** 316
Chen C F and Wu C Y 1986 *J. Appl. Phys.* **60** 3926
Chen I C, Holland S and Hu C 1987 *J. Appl. Phys.* **61** 4544
De Keersmaecker R F 1983 *Insulating Films On Semiconductors* eds J F Verweij and D R Wolters (North-Holland: Amsterdam) pp 85
DiMaria D J 1976 *J. Appl. Phys.* **47** 1082
DiMaria D J 1978 *The Physics of SiO2 and its interfaces* ed S T Pantelides (Pergamon: New York) pp 160
DiMaria D J 1987 *Appl. Phys. Lett.* **51** 655
DiMaria D J and Stasiak J W 1989 *J. Appl. Phys.* **65** 2342
DiMaria D J and Stathis J H 1990 *Trapping and trap creation studies on nitrided and reoxidized-nitrided silicon dioxide films on silicon* Presented at the 1990 SISC, San Diego, California, Dec. 1990
DiMaria D J, Weinberg Z A and Aitken J M 1977*J. Appl. Phys.* **48** 898
DiStefano T H and Shatzkes M 1974 *Appl. Phys. Lett.* **25** 685
Dooms E E, Heyns M M and De Keersmaecker R F 1989a *Appl. Surf. Sci.* **39** 227
Dooms E E, Heyns M M and De Keersmaecker R F 1989b *internal IMEC-report*
Do Thanh L, Aslam M and Balk P 1986 *Solid State Electron.* **29** 829
Dunn G J 1989 *J. Appl. Phys.* **65** 4879
Feigl F J, Young D R, DiMaria D J, Lai S K and Calise J A 1981 *J. Appl. Phys.* **52** 5665
Ferry F L, Wyatt P W, Naiman M L, Mathur B P, Kirk C T and Senturia S D 1985 *J. Appl. Phys.* **57** 2036
Fischetti M V 1985*Phys. Rev. B* **31** 2099
Fischetti M V, DiMaria D J, Brorson S D, Theis T N and Kirtley J R 1985 *Phys. Rev. B* **31** 8124
Fischetti M V, Gastaldi R, Maggioni F and Modelli A 1982a *J. Appl. Phys.* **53** 3129
Fischetti M V, Gastaldi R, Maggioni F and Modelli A 1982b *J. Appl. Phys.* **53** 3136
Groeseneken G, Maes H E, Beltran N and De Keersmaecker R F 1985 *IEEE Trans. Electron. Dev.* **32** 375
Harari E 1978 *J. Appl. Phys.* **49** 2478
Hartstein A and Young D R 1981 *J. Appl. Phys.* **50** 6321
Haywood S K, Heyns M M and De Keersmaecker R F 1985 *paper presented at the SISC Conf. 85, Fort Lauderdale, Florida, Dec. 5-7*
Heremans P, Bellens R, Groeseneken G and Maes H 1988 *IEEE Trans. Electron. Devices* **35** 2194
Heremans P, Groeseneken G and Maes H E 1987 *paper presented at IEE "Colloquium on hot carrier degradation in short channel MOS* London, Jan. 1987
Heyns M M and De Keersmaecker R F 1981*paper presented at the 6th Solid State Device Technol. Symp. (ESSDERC), Toulouse, France, Sept. 1981*
Heyns M M and De Keersmaecker R F 1985 *J. Appl. Phys.* **58** 3936

Heyns M M and De Keersmaecker R F 1986 *Dielectric layers in semiconductors : novel technologies and devices 1986* eds G Bentini, E Fogarassy and A Golanski (Les Editions de Physique: Les Ulis Cedex, France) pp 303
Heyns M M and De Keersmaecker R F 1988a *The Physics and Technology of Amorphous SiO_2* ed Devine R A (Plenum: New York) pp 411
Heyns M M and De Keersmaecker R F 1988b *Mat. Res. Symp. Proc.* **Vol.105**, pp 205
Heyns M M, De Keersmaecker R F and Hillen M W 1984 *Appl. Phys. Lett.* **44** 202
Heyns M M, Krishna Rao D and DeKeersmaecker R F 1989 *Appl.Surf.Sci.* **39** 327
Hillen M W, De Keersmaecker R F, Heyns M M, Haywood S K and Daraktchiev I S 1983 *Insulating Films On Semiconductors* eds Verweij J F and Wolters D R (North-Holland: Amsterdam) pp 274
Hofmann K R and Dorda G 1981 *Insulating Films On Semiconductors* ed Schulz M (Springer: Berlin) pp 122
Hofmann K R, Werner C, Weber W and Dorda G 1985 *IEEE Trans. Electron Devices* **32** 691
Hori T and Iwasaki H 1989 *J. Appl. Phys.* **65** 629
Jenq C S, Ranganath T R, Huang C H, Jones H S and Chang T T L 1981 *IEEE Int'l Electron Devices Meeting 1981, Technical Digest* pp 388
Krishna Rao D, Heyns M M and De Keersmaecker R F 1988 *Proc. of ESSDERC 88* eds Nougier J P and Gasquet D, (Les Editions de Physique: France) pp 669
Lai S K 1983 *J. Appl. Phys.* **54** 2540
Lai S K, Dong W D and Hartstein A 1982 *J. Electrochem. Soc.* **129** 2042
Lai S K and Young D R 1981 *J. Appl. Phys.* **52** 6321
Lenzlinger M and Snow E H 1969 *J. Appl. Phys.* **40** 278
Liang M S, Choi J C, Ko P K and Hu C 1984 *IEEE Int'l Electron Devices Meeting 1984, Technical Digest* pp 152
Liang M S and Hu C 1981 *IEEE Int'l Electron Devices Meeting 1981, Technical Digest* pp 396
Lyon S A 1989 *Appl. Surf. Sci* **39** 552
Maserjian J and Zamani N 1982 *J. Appl. Phys.* **53** 559
Nicollian E H and Berglund C N 1970 *J. Appl. Phys.* **41** pp 3052
Ning T H and Yu H N 1974 *J. Appl. Phys.* **45** 5373
Nissan-Cohen Y, Shappir J and Frohman-Bentchkowsky D 1985 *J. Appl. Phys.* **58** 2252
Olivo P, Ricco B and Sangiorgi E 1983 *J. Appl. Phys.* **54** 5267
Sah C T, Sun J Y and Tzou J J 1983 *J. Appl. Phys.* **54** 944
Saks N S, Heremans P L, Van den hove L, Maes H E, De Keersmaecker R F and Declerck G J 1986 *IEEE Trans. Electron. Dev.* **33** 1529
Schmidlin F W 1966 *J. Appl. Phys.* **37** 2823
Schwerin A v and Heyns M M 1991a *Oxide field dependence of bulk and interface trap generation in SiO_2 due to electron injection* to be presented at INFOS 91, April 1991
Schwerin A v and Heyns M M 1991b *Homogeneous hole injection into gate oxide layers of MOSFET's : injection efficiency, hole trapping and Si/SiO_2 interface state generation* to be presented at INFOS 91, April 1991
Schwerin A v, Heyns M M and Weber W 1990 *J. Appl. Phys.* **67** 7595
Severi M and Impronta M 1987a *Appl. Phys. Lett.* **51** 1702
Severi M, Impronta M and Bianconi M 1987b *Proc. 17th European Solid State Device Research Conf. ESSDERC 87* pp 845
Severi M, Impronta M, Dori L and Guerri S 1988 *Proc. 18th European Solid State Device Research Conf. ESSDERC 88* pp.417
Shatzkes M and Av-Ron M 1976 *J. Appl. Phys.* **47** 3192
Solomon P 1976 *J. Appl. Phys.* **47** 2089
Verwey J F 1973 *J. Appl. Phys.* **44** 2681
Wang S J, Sung J M and Lyon S A 1988 *The Physics and Technology of Amorphous SiO_2* ed Devine R A (Plenum: New York) pp 465
Weinberg Z A and Rubloff G W 1978 *Appl. Phys. Lett.* **32** 184
Yang W, Jayaraman R and Sodini C G 1988 *IEEE Trans. Electron. Dev.* **35** 935
Yankova A, Do Thanh L and Balk P 1987 *Solid State Electron.* **30** 939
Young D R, Irene E A, DiMaria D J, De Keersmaecker R F and Massoud H Z 1979 *J. Appl. Phys.* **50** 6366
Zekeryia V and Ma T P 1983 *Appl. Phys. Lett.* **43** 95

Paper presented at INFOS '91, Liverpool, April 1991
Workshop Papers

Hot carrier-induced degradation modes in thin-gate insulator dual-gate MISFETs

Hiroshi Iwai

ULSI Research Center, Toshiba Corporation
1, Komukai-Toshiba-cho, Saiwai-ku, Kawasaki, 210, Japan,
Tel: +81-44-549-2183, Fax: +81-44-549-2266

ABSTRACT: Hot carrier induced degradation modes are investigated for deep sub-micron CMOS (or CMIS) devices. In particular, the effects of scaling-down and new process (or material) technologies on degradation are explained. These are the effects of ultra-thin gate oxide films, p^+ poly gate electrodes, and nitrided oxide gate films. In general, as scaling-down and the introduction of new technologies proceed, hot-carrier reliability improves more than predicted. Other hot carrier issues are also discussed.

1. INTRODUCTION

Over the last 20 years, progress in MOS LSI development has been made mainly by scaling down the feature size as seen in case of the logic device shown in Fig.1, where the design rule shrinks twice every two or three years. With this scaling, the performance or speed of devices continues to grow as seen in the propagation delay time of CMOS inverters shown in Fig.2. In addition to performance, reliability issues such as hot carrier-induced degradation of MOSFETs, time-dependent dielectric breakdown (TDDB), and long-term low-field reliability (bias temperature stress) are always great concerns. In this paper, the results of investigations into hot carrier-induced degradation with deep sub-micron CMOS devices in mind are reported.

Changes in MOSFET characteristics during operation, induced by hot carrier injection into the gate oxide, were the major issue in CMOS LSI reliability in the 1980s. The fundamental mechanisms of hot carrier induced degradation have been studied intensively, and degradation of scaled MOSFETs can now be predicted accurately (Hu et al 1985, Takeda et al 1985). Degradation defined by changes in drain current or threshold voltage can be predicted using the following expression:

$$\Delta I_D / I_D \text{ or } \Delta V_{TH} / V_{TH} \cong A_1 t^n \cong A_2 I_{SUB}^m t^n \cong A_3 \exp(-\alpha/L) \exp(\beta/V) t^n$$

where t is stress time, L is channel length, and V is stress voltage.

In the late 1980s, new process and material technologies, such as ultra-thin gate-oxide films, p^+ poly Si gate electrodes, nitrided oxide gate films, and salicide structures were investigated as technology moves towards deep submicron CMOS devices. Figure 3 shows an example of advanced CMOS (or CMIS) structures for deep sub-micron devices. At that time, several new phenomena accompanying the technologies were found. Do these new phenomena improve hot carrier-induced degradation or not? How would the above expression predict the degradation? In this paper, the new hot-carrier phenomena arising from the new process technologies, as well as from the scaling-down of size, are reviewed and discussed.

© 1991 IOP Publishing Ltd

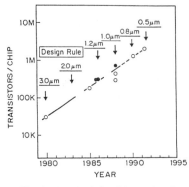

Fig 1. Trend of CMOS Logic LSI   **Fig 2.** Propagation delay time of CMOS inverter

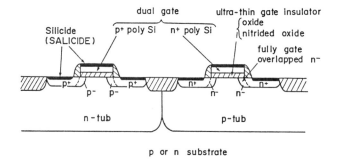

Fig 3. Advanced CMOS structure towards deep sub-micron

2. EXPERIMENTAL CONDITIONS

Table 1 shows the sample fabrication conditions used in these experiments. In most cases, the dual-gate CMOS structure was studied, which means n^+ poly Si gate for n-MOSFETs and p^+ poly Si gate for p-MOSFETs. Single drain structures or non-LDD structures were used in order to simplify the situation and to clarify the effect of each new technology separately. Silicide was not used for the same reason. Table 2 shows the hot carrier test conditions used in the experiments. Only the dc stress case was investigated – ac stress was not studied.

Table 1   Sample Conditions

- Substrate
  P(100) Si wafer
- Gate electrode
  n^+/p^+ poly si, non-silicide
- Gate insulator
  Pure oxide/nitrided-oxide
  10~3nm
- Source/drain
  Single, non-LDD, $X_{jn} \approx 0.15 \mu m$, $X_{jp} \approx 0.2\sim 0.3 \mu m$
- Channel Length, Width
  $L = 0.5\sim 10 \mu m$, (typically $0.8\sim 1.2 \mu m$), $W = 10 \mu m$
- Passivation
  under $A\ell$: BPSG on $A\ell$: PSG, non-nitride

Table 2   Test Conditions

Stress
- Applied drain voltage
  6~3V (typically 6~5V)
- Bias conditions
  n-MOS : typically ISUB MAX Condition
  p-MOS : typically IG MAX Condition
- Stress time
  typically 1000sec

Measurement
- V_{TH} : $V_D = 0.05\sim 0.1V$, $I_D = 1nA$
- N_{IT} : Charge pumping

3. EFFECTS OF ULTRA-THIN GATE OXIDE FILM

Hot carrier degradations in MOSFETs with ultra-thin gate oxide films are explained first. Even in MOSFETs with gate insulator less than 2.5 nm, some transistor action was observed. However, significant gate tunneling leakage current was seen in such devices (Morimoto et al 1990a, 1991), which reduces drivability of the transistor. Thus, the practical limit on gate insulator thickness for conventional MOSFET operation is 2.5 nm. In this section, the effects of gate oxide thickness on hot carrier degradation are discussed down to the 5 nm case, a thickness expected to be used in 0.25 - 0.15 μm CMOS devices.

3.1. n-MOSFET case

In the case of n-MOSFETs, the reduction in drain current or mobility caused by interface state generation is the major degradation mode, as shown in Fig.4, and the degradation is defined by the drain current change as shown in Fig.5. Figure 6 (Toyoshima et al 1990) shows the dependence of drain current degradation on gate oxide thickness. It should be noted that the drain current degradation falls off as the gate oxide becomes thinner, even if the peak substrate current increases.

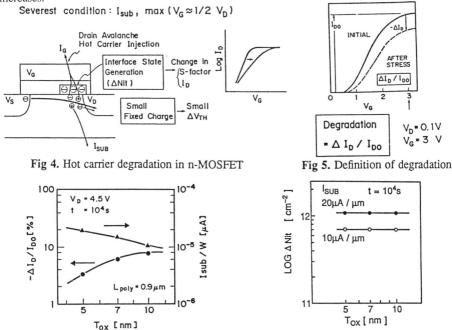

Fig 4. Hot carrier degradation in n-MOSFET

Fig 5. Definition of degradation

Fig 6. Dependence of $\Delta I_D/I_{DO}$ on t_{OX}

Fig 7. Dependence of ΔN_{IT} on t_{OX}

Three possible mechanisms for this phenomenon can be proposed. They are, that in thinner gate oxide MOSFETs, i)N_{IT} generation is smaller, ii)qN_{IT}/Q_{INV} becomes smaller because Q_{INV} increases, and iii)the loss of mobility due to N_{IT} becomes small. Cause i) was disproved by direct measurement of ΔN_{IT} using the charge-pumping method (Groeseneken et al 1984) as shown in Fig.7. The other two causes are considered further by calculating the degradation with a simple two-series-conductances (in non-degraded and degraded regions) model as shown in Fig.8. Here, ΔN_{IT} and ΔL (the length of the degraded region) were again measured by the charge-pumping method. The drain current degradation is calculated from the change in series conductance. First, the effect of the smaller qN_{IT}/Q_{INV} of the thinner gate oxide device is considered without the effect of mobility degradation due to N_{IT}. The results, shown in Fig.9, explain only one third of the degradation. The dependence of mobility on N_{IT} and E_{EFF} (vertical effective electric field in the channel) is obtained in other experiments (Toyoshima et al 1990) as

$$\frac{1}{\mu_{eff}} = 1100 + 1.5 \times 10^{-4} N_{it} + 1.7 \times 10^{-5} E_{eff}^{1/3} + 2.45 \times 10^{-15} E_{eff}^2 \quad [cm^{-2} Vsec]$$

and the degradation is calculated again incorporating this effect, as shown in Fig.10. The calculation agrees with measurements very well. Thus, the smaller mobility degradation in the thinner gate oxide device (mechanism iii)) is found to be the primary cause. The reason for the smaller mobility degradation is that the effect on mobility of electron scattering by N_{IT} becomes less important compared with the scattering due to surface roughness as the vertical electric field (E_{EFF}) becomes higher in a device with thinner oxide. This is clear from the above expression and also in Fig.11. Usually supply voltage scaling does not follow dimension scaling.

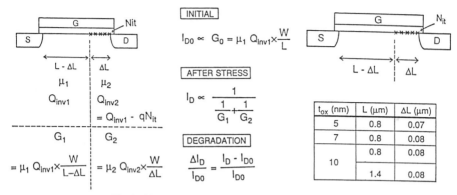

Fig 8. Simple model for locally degraded MOSFET

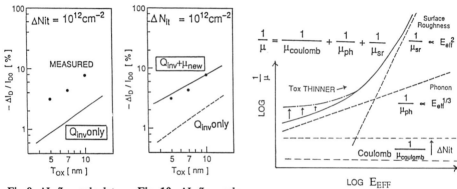

Fig 9. $\Delta I_D/I_{DO}$ calculated by simple model. Only mechanism ii) is considered.

Fig 10. $\Delta I_D/I_{DO}$ calculated by simple model. Mechanism ii) and iii) are considered.

Fig 11. Dependence of mobility on E_{EFF}

3.2. P-MOSFET case

In the case of p-MOSFETs, a shift in threshold voltage due to electron trapping is the major degradation mode, as shown in Fig.12. In the p-MOSFET case, too, the degradation is less severe for a thinner gate oxide (Hiruta et al 1989a), as shown in Fig.13, in spite of the high gate injection current during stress. In this case, the gate current is a better measure of degradation than substrate current. The smaller ΔV_{TH} is due to simultaneous tunneling of the trapped charge during the stress. With oxide layers less than 6nm thick, direct tunneling of the trapped charge occurs either to the upper poly-Si electrode or to the lower Si substrate. It should be noted that, in the case of p-MOSFETs with gate oxide less than 6nm, the problem of hot carrier degradation almost disappears because of the tunneling effect.

Meanwhile, ΔN_{IT} is higher in a thinner gate oxide device, as shown in Fig.14. Fortunately, however, the value of ΔN_{IT} is not large enough to change the device characteristics, as shown in Fig.15.

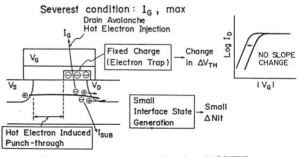

Fig 12. Hot carrier degradation in p-MOSFET

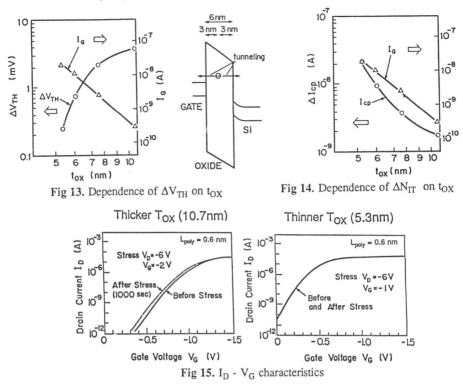

Fig 13. Dependence of ΔV_{TH} on t_{OX}

Fig 14. Dependence of ΔN_{IT} on t_{OX}

Fig 15. I_D - V_G characteristics

4. EFFECT OF P⁺ POLY GATE ELECTRODE

Conventionally, n⁺ poly Si is used for both n- and p-MOSFETs. Because gate oxide usually has positive charges at the interface, p-MOSFET V_{TH} is always too negative. Thus, in the p-MOSFET case, counter doping with boron and the formation of a p⁻ layer in the substrate is necessary to reduce the over-negative V_{TH}. Due to the buried p⁻ layer in the substrate, the hole current flows in the deep region, leading to a buried channel. The buried channel current is relatively difficult to control using gate bias, because the channel is so distant from the surface. Thus, the conventional n⁺ poly p-MOSFET has a significant short channel effect on V_{TH} as shown in Fig.16. Instead of counter doping the substrate, p⁺ poly gate also reduces over-negative V_{TH} because of 1.1eV work function difference between n⁺- and p⁺ poly Si. In this case, the channel remains at the surface and control of V_{TH} is easy. Thus, the p⁺ poly gate is regarded as necessary in deep sub-micron CMOS devices. Here, it should be noted that p⁺ poly Si should not be used in n-MOSFET, because of long-term low-field reliability of p⁺ poly n-MOSFETs was found to be extremely bad due to hole injection from the p⁺ poly (Hiruta et al 1989b, Iwai et al 1989). Thus, in deep sub-micron CMOS devices, a dual-gate structure should be used where an n⁺ poly electrode is used for n-MOSFETs and a p⁺ poly electrode for p-MOSFETs.

Hot carrier degradation in p⁺ poly gate p-MOSFETs is compared with that in conventional n⁺ poly gate p-MOSFETs. Is the degradation more severe, in n⁺ poly p-MOSFETs or in p⁺ p-MOSFETs? Figure 17 shows the results (Matsuoka et al 1990). The hot carrier reliability of p⁺ poly gate p-MOSFETs is better than that of conventional n⁺ poly gate p-MOSFETs for the following reasons: i)injected I_G is

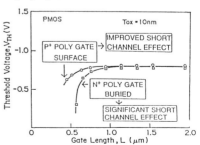

Fig 16. Short channel effect

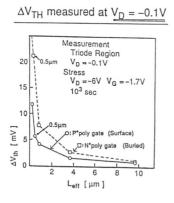 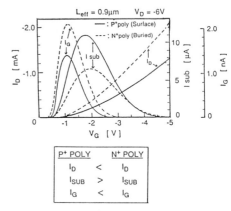

Fig 17. Dependence of ΔV_{TH} on L_{EFF}

Fig 18. I_D, I_{SUB}, and I_G during stress

smaller, ii)ΔV_{TH} for the trapped electrons is smaller, and iii)there is a compensation resulting from both hole and electron injection.

Figure 18 compares I_D, I_{SUB}, and I_G for a p^+ poly p-MOSFET with the values for an n^+ poly. In p^+ poly p-MOSFETs, I_D is smaller because holes flow at the surface, and thus the mobility is lower. In spite of the smaller I_D, I_{SUB} is larger because the hole current flows near the peak of the lateral electric field which is close to the surface as shown in Fig.19. In spite of the larger I_{SUB}, it is interesting to note that I_G is smaller. This is because of the difference in work function between p^+ and n^+ poly gates. In the p^+ poly gate case, the electric field at the drain edge is higher, which results in a lower barrier for electron injection into the gate as shown in Fig.20.

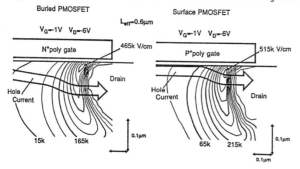

This smaller injected I_G corresponds to i) above. Reason ii) is obvious (see Matsuoka et al 1990). In a p^+ poly gate p-MOSFET, it is easy to control the drain current with gate bias. Thus, ΔV_{TH} is affected less by the trapped charge.

Fig 19. Lateral electric field distribution under stress

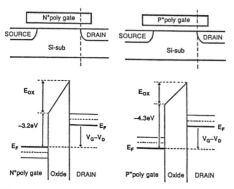 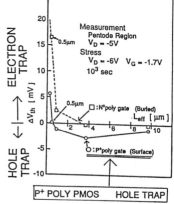

Fig 20. Vertical electric field at drain edge

Fig 21. Dependence of ΔV_{TH} on L_{EFF}

It is interesting to investigate iii) above. Figure 21 shows ΔV_{TH} measured at a high drain voltage. It should be noted that hole trapping is observed in the p$^+$ poly p-MOSFET case. The dependence of ΔV_{TH} on measurement V_D shown in Fig.22 suggests both hole and electron injections as shown in the figure. Simulated vertical electric field in Fig.23 also supports hole injection in the p$^+$ poly gate case.

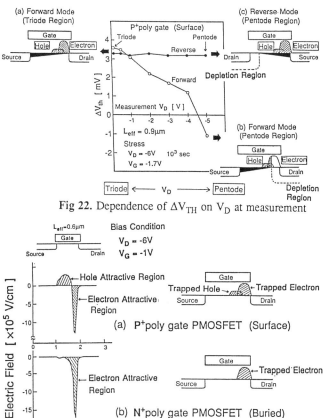

Fig 22. Dependence of ΔV_{TH} on V_D at measurement

Fig 23. Vertical electric field distribution under stress

5. EFFECT OF NITRIDED OXIDE GATE FILM

Recently, nitride oxide gate films have been investigated for use in deep sub-micron CMOS structures because of their high reliability in TDDB (Hori et al 1989) and their ability to suppress boron diffusion from p$^+$ poly Si into the silicon substrate (Morimoto et al 1990b, c). Among the electrical characteristics of nitrided oxide gate MOSFETs, several interesting phenomena such as mobility (Momose et al 1989b, c, 1990c, Wu et al 1989, Iwai et al 1990a) and short channel effect (Momose et al 1990a, b) have been reported.

Figure 24 shows a typical nitrogen profile in the gate oxide. Figure 25 explains the nitridation process. Most of the nitrided films were re-oxidized in order to reduce interface states. The nitridation was accomplished by RTN (Rapid Thermal Nitridation) in NH$_3$ gas. The nitrogen concentration in the oxide is 0 - 10 %, depending on the nitriding conditions, and the profile is relatively uniform. In most cases, the final film thickness as evaluated by C-V measurements is unchanged from the initial oxide thickness.

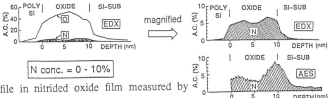

Fig 24. Nitrogen profile in nitrided oxide film measured by EDX and AES

Now, the effects of nitrided oxide gates on hot carrier phenomena are discussed (Momose et al 1989a). Figure 26 shows I_{SUB} during stress. I_{SUB} for the nitrided oxide ("NO") gate samples is smaller than for pure oxide ("PO") gate samples. Figure 27 shows I_G during stress. It is interesting that in the n-MOSFET case (hole injection case), I_G in the "NO" sample is larger than that in the "PO" sample, while to the contrary, in the p-MOSFET case (electron injection case), I_G in the "NO" sample is smaller than that in the "PO" sample. This interesting phenomenon can be explained by assuming there are trapped electrons in "NO" films, which act to increase electric field in the oxide during hot hole injection in the n-MOSFET case, and act to decrease electric field during hot electron injection in the p-MOSFET case, as shown in Fig.27.

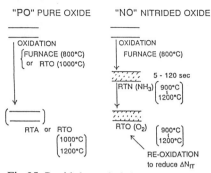

Fig 25. Rapid thermal nitridation process

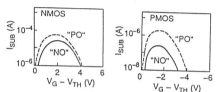

Fig 26. Substrate current under stress

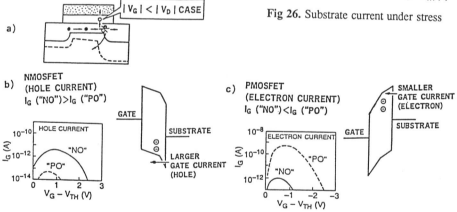

Fig 27. Gate current under stress

Interface state generation during hot carrier stress was compared for "NO" and "PO" n-MOSFETs and the results are shown in Fig.28. The nitrided oxide gate MOSFETs have greater reliability than those with pure gate oxide, because the nitrided oxide gate film is more resistant to interface state generation (Yankova et al 1987).

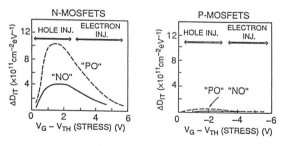

Fig 28. Interface state generation

Workshop Papers

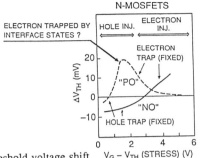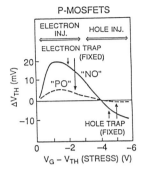

Fig 29. Threshold voltage shift

Threshold voltage shifts were also compared as shown in Fig.29. In the p-MOSFET case, the "NO" film traps more injected carriers than the "PO" film. Fortunately, however, the threshold voltage shift in the "NO" p-MOSFET is the same level as that in the "PO" n-MOSFET. Usually, threshold voltage shift in "PO" n-MOSFETs is not a problem. (Usually, interface state generation is a problem.) In the n-MOSFET case, the "NO" film traps injected carriers. Fortunately again, the absolute value of the threshold voltage shift in the "NO" sample is smaller than that in the "PO" sample. Note that in the case of the "PO" film, opposite type (electrons) to the carriers (holes) injected into the gate electrode are trapped. These trapped electrons are probably those trapped in the interface state generated by hot carriers. Finally it should be noted that the "NO" film property to easily trap injected carriers strongly depends on the nitrogen concentration in the film, and thus the concentration should be optimized.

6. OTHER EFFECTS

Other technologies which should be considered for deep sub-micron devices are simply listed in this section. They are Bi-CMOS (Momose et al 1989a), other gate insulator gate materials (LPCVD-SiO_2, -Si_3N_4 (Iwai, 1990b, 1990c), -Ta_2O_5), silicide, salicide, and drain engineering (LDD, GDD, etc). The effects of residual mechanical stress in the film, detrapping under high temperature, very short channel length, and very low power supply voltage should be also considered.

7. SUMMARY AND CONCLUSION

The effects of scaling-down and new process (and material) technologies leading towards deep sub-micron CMOS devices on hot carrier-induced degradation are discussed and their mechanisms explained. In particular, the effects of ultra-thin oxide gate films, p^+ poly gate electrodes, and nitrided oxide gate films are investigated.

In general, with the introduction of these new technologies, hot carrier reliability improves more than predicted.

It should be noted that hot carrier degradation falls in ultra-thin gate MOSFETs. In the n-MOSFET case, mobility degradation due to ΔN_{IT} falls because the surface roughness effect becomes more important. Larger Q_{INV} also reduces the effects of N_{IT} on carrier reduction in the channel. In the p-MOSFET case, simultaneous tunneling of the trapped charges in the ultra-thin gate film results in very small ΔV_{TH}.

Continuous efforts to find new technologies and to examine them is very important if high hot-carrier reliability is to be achieved in deep sub-micron devices.

8. ACKNOWLEDGEMENT

The author gratefully acknowledges H. S. Momose, T. Morimoto, Y. Toyoshima, F. Matsuoka, Y. Hiruta, H. Oyamatsu, H. Hayashida, and K. Hama, who have been studying hot carrier phenomena in the course of developing sub-micron logic CMOS devices, for the experimental data. He would like to thank N. Goto, S. Mimura, H. Hara, and A. Kasami for encouragement to this work.

9. REFERENCES

Groeseneken G, Maes H E, Beltran N and Keersmaecker R F 1984 *IEEE Trans. Electron Devices*, vol. ED-31, no.1, pp 42-53

Hu C, Tam S C, Hsu F-C, Ko P-K, Chan T-Y and Terrill K W 1985 *IEEE J. Solid-State Circuits*, vol. SC-20, no.1, pp 295-305

Hamada A, Furusawa T and Takeda E 1990 *Dig. of Tech. Papers VLSI Symp. on Tech. (Honolulu)* pp 113-114

Hiruta Y, Oyamatsu H, Momose H S, Iwai H and Maeguchi K 1989a *ESSDERC (Berlin)* pp 732-735

Hiruta Y, Iwai H, Matsuoka F, Mama K, Maeguchi K and Kanzaki K 1989 *IEEE Trans. Electron Devices*, vol. ED-36, no.9, pp 1732-1739

Hori T and Iwasaki H 1989 *IEDM Tech. Dig. (Washington D. C.)* pp 459-462

Iwai H, Matsuoka F, Oyamatsu H, Momose H S, Hama K, Toyoshima Y and Hayashida H 1989 *IEEE Semiconductor Interface Specialists Conference, (Ft. Lauderdale, FL)* pp II.5

Iwai H, Momose H S, Takagi S, Morimoto T, Kitagawa S, Kambayashi S, Yamabe K and Onga S 1990a *Dig. of Tech. Papers VLSI Symp. on Tech. (Honolulu)* pp 287-290

Iwai H, Momose H S, Morimoto T, Takagi S and Yamabe K 1990b *ESSDERC (Nottingham)* pp 149-152

Iwai H, Momose H S, Morimoto T, Ozawa Y and Yamabe K 1990c *IEDM Tech. Dig. (San Francisco)* pp 235-238

Takeda E 1985 *Dig. of Tech. Papers VLSI Symp. on Tech. (Kobe)* pp 2-5

Toyoshima Y, Iwai H, Matsuoka F, Hayashida H, Maeguchi K and Kanzaki K 1990 *IEEE Trans. Electron Devices*, vol.ED-37, no.6, pp 1496-1503

Matsuoka F, Iwai H, Hayashida H, Hama K, Toyoshima Y and Maeguchi K 1990 *IEEE Trans. Electron Devices*, vol.ED-37, no.6, pp 1487-1495

Momose H S, Niitsu Y, Iwai H and Maeguchi K 1989a *IEEE Bipolar Circuit and Technology Meeting (Mineapolis)* pp 98-101

Momose H S, Kitagawa S, Yamabe K and Iwai H 1989b *IEDM Tech. Dig. (Washington D.C.)* pp 267-270

Momose H S, Takagi S, Kitagawa S, Yamabe K, and Iwai H 1989c *IEEE Semiconductor Interface Specialists Conference, (Ft. Lauderdale, FL)* pp I.5

Momose H S, Morimoto T, Takagi S, Yamabe K, Onga S and Iwai H 1990a *International Conference on Solid State Device and Materials (Sendai)*, pp 279-283

Momose H S, Morimoto T, Takagi S, Yamabe K, Onga S and Iwai H 1990b *ESSDERC (Nottingham)* pp 149-152

Momose H S, Morimoto T, Yamabe K and Iwai H 1990c *IEDM Tech. Dig. (San Francisco)* pp 65-68

Morimoto T, Momose H S, Yamabe K and Iwai H, 1990a *International Conference on Solid State Device and Materials (Sendai)*, pp 361-364

Morimoto T, Momose H S, Yamabe K and Iwai H, 1990b *ESSDERC (Nottingham)* pp 73-76

Morimoto T, Momose H S, Ozawa Y, Yamabe K and Iwai H, 1990c *IEDM Tech. Dig. (San Francisco)* pp 429-432

Morimoto T, Momose H S, Ozawa Y, Yamabe K and Iwai H, 1991 *submitted to International Conference on Solid State Device and Materials (Yokohama)*

Wu A T, Chan T Y, Murali V, Lee S W, Nulman J and Garner M 1989 *IEDM Tech Dig. (Washington D.C.)* pp 271-274

Yankova A, Thanh L D and Balk P 1987 *Solid-State Electron.*, vol.30, no.9, pp 939-946

Paper presented at INFOS '91, Liverpool, April 1991
Workshop Papers

The impact of hot carrier degradation on scaling of sub-μm CMOS processes

H.-M. Mühlhoff, M. Steimle and J. Dietl
Siemens AG, Semiconductor Division
Munich, Germany

Introduction: Recent advances in silicon process technology such as i-line lithography using phase shift masks, RTP, low temperature reactive ion etching and reduction of defect density have made CMOS processes with device dimensions approaching 0.5µm and below manufacturable. The full performance potential of modern CMOS can thus be realized. Shrinking device dimensions improves speed and packing density. However, both technological problems and device physics aspects have to be addressed in the scaling scenario. The main scaling problems associated with device physics are: 1) Integrity of thin gate oxide, 2) junction breakdown due to high concentration gradient and gated diode effect, 3) isolation (surface leakage and punchthrough), 4) hot carrier induced degradation. After giving a brief summary of the first 3 issues, this paper will focus on the hot carrier problem.

The quality of conventional furnace grown oxide usually poses no problem for thicknesses above 10nm. Below 10nm, stacked films are being studied which have been proven in DRAMs to exhibit high breakdown fields, low defect density and low leakage currents. Application in MOS transistors puts much more stringent requirements on shift of flatband voltage from carrier trapping and change of surface states from hot carrier degradation. Therefore stacked films in MOS transistors require low density of traps and surface states. 3-4nm LPCVD nitride films on thermal oxide have been shown to satisfy these requirements /1,2/. Also nitrided SiO_2 is being investigated because it shows reduced interface state density and higher resistance to boron penetration when using boron doped polysilicon gates. Avoiding hydrogen by using N_2O instead of NH_3 for the nitridation process improves hot carrier reliability because electron traps are reduced /3,4/.

Reducing device dimensions requires higher doping concentrations and shallower and more abrupt p-n junctions in order to scale the size and lateral extension of space charge regions. This causes the electric field in reverse biased junctions to increase and lowers avalanche breakdown voltages in source/drain diodes. High field conditions exist both at the transistor gate edge and at the edge facing the isolation region. Fig.1 shows a typical result for breakdown at the LOCOS edge in a process which is used in 4Mbit and 16Mbit DRAMs. Increasing the field doping improves the isolation properties of the LOCOS isolation but at the expense of lower junction breakdown voltages. However, if boosted voltages in on chip circuits are avoided there is enough margin left to continue to use planar isolation (LOCOS type) for future chip

© 1991 IOP Publishing Ltd

generations /5,6/. The other high field region is formed at the gate edge when the transistor is biased at 0V for turn-off. The gate voltage pins down the surface potential at the gate edge causing a large band bending between the gate edge and the diffusion (s. Fig.2). For thin gate oxides this band bending can be large enough for tunneling of electrons through the bandgap to occur. The resulting leakage current puts a lower limit on oxide thickness and an upper limit on the voltage between gate and drain in turned-off transistors /7,8/. Fig.2 shows simulated tunneling currents and their dependence on voltage and gate oxide thickness. From these calculations it can be concluded that 7nm thick gate oxide can be used in sub-micron devices operating at 3.3V.

Keeping devices safely turned off has become an increasingly difficult task in sub-µm CMOS. Surface channel transistors pose less of a problem than buried channel devices. Thinner gate oxide provides a straightforward route to smaller device dimensions (Fig.3). The scaling potential has been plotted in Figs.4 and 5. For NMOS surface channel transistors, LDD type drain structures are normally used. Optimization of LDD-implantation dose leads to a variation of about 0.1µm in the minimum usable gate length. Due to velocity saturation in n-channels, a direct dependence of output current on gate length is absent /9,10,11,12/ and can only be felt by Vt-lowering in the short channel regime. The maximum driving current an NMOS transistor can deliver is therefore only dependent on gate oxide thickness and threshold voltage. Experimental data can best be described by the following expression:

$$I_{DS} = A \, (1-\exp(-Bt_{ox}))/t_{ox} \, V_{GS} - C \, V_T \qquad (1)$$

with A = 2.3, B = 0.115 and C = 0.25. Fig.5 shows the range of minimum gate lengths and maximum saturation currents and their dependence on operating voltage and gate oxide thickness. Reducing operating voltage from 5V to 3.3V permits the use of about 0.1µm shorter gate lengths. At 10nm gate oxide thickness, minimum transistor size ranges from 0.4 to 0.5µm for 3.3V and from 0.5 to 0.6µm for 5V.

Turn off characteristics in buried channel devices is much more difficult to control than in surface channel transistors /13,14/. The coupling of gate potential to the potential barrier isolating source and drain is much weaker than in surface channel devices. Subthreshold leakage is controlled by drain voltage and the subthreshold slope varies with gate length (Fig.6). DIBL causes punchthrough at small gate lengths. Minimum gate length is therefore more restricted than in surface channel devices and strongly dependent on operating voltage. For buried channel PMOS devices it has to be set at about 0.1µm more than for surface channel NMOS devices. Scaling of gate length with gate oxide thickness for buried channel PMOS transistors is shown in Fig.7. At 10nm gate oxide thickness, gate length cannot decrease below 0.7µm for 5V and 0.6µm for 3.3V. Further complications arise from high temperature processing. Outdiffusion of boron source/drain diffusion during high temperature silicidation steps and reflow of intermediate dielectrics in interconnect layers causes channel shortening with respect to short channel effect and punchthrough but no gain in current drive due to L_{eff} reduction /15/. Considerably better PMOS performance can be achieved by changing over to surface channel devices. This requires the use p^+ doped poly silicon gates. But penetration of boron atoms through the gate insulator has been a technological problem for many years. Recently however, progress has been reported in manufacturing high quality gate films which resist boron penetration /3,4/. But when the added process complexity of integrating both n- and p-doped poly silicon as gate material is considered, using surface channel PMOSFETs in production is still a subject of discussion and a switch over is unlikely to occur before all means of optimization have been exhausted.

Of all devices problems associated with scaling, hot carrier degradation constitutes the most difficult to quantify. Degradation of devices and the circuits which contain them is slow and gradual and in most cases not observable in a laboratory situation unless accelerated stress conditions are used. Obtaining meaningful lifetimes from stress tests is difficult because the criteria for device and circuit failure are vague at best. Circuits respond to hot carrier stress with lower switching speeds and higher leakage currents but rarely fail completely. Since hot carrier degradation does pose a limiting factor in microelectronics the technologist is forced to resort to estimated guesses to evaluate the safety of a new or scaled process. The remainder of this paper is devoted to a) justify lifetime criteria used for process evaluation b) derive minimum geometries based on these criteria and c) show how processing parameters can affect hot carrier reliability.

Device Fabrication: Experimental data presented was taken from a 16Mbit DRAM technology. This technology is driven by ease of processing and high yield requiring highly manufacturable transistor structures. Only n-poly gates are used and RTP is restricted to tempering of the contact barrier metal. PolySi-gate lengths range from 0.5µm to 1µm and gate oxide thicknesses from 10nm to 18nm. LDD transistors were used for NMOS and offset drain for PMOS (S/D implant after sidewall spacer formation) /16/. The process employs a silicide interconnection layer requiring a very high temperature budget which makes the optimization of PMOS transistors especially difficult.

The Effect of Hot Carrier Degradation on Circuit Speed: The simplest circuits for measuring switching speeds are ring oscillators. In Fig.8 delay times measured in a 33 stage ring oscillator are compared with NMOS and PMOS currents of single transistors. Calculated delay times have also been plotted. These calculated times were obtained from the total gate capacitance and the saturation currents neglecting all parasitic RC delays. Agreement within 10% between measured data and calculated numbers has been obtained. This demonstrates that besides gate capacitance the saturation current is the key parameter for switching speed. A 10% decrease in saturation current will slow down circuit speed by 10%. However, degradation of LDD NMOS transistors occurs at the drain side gate edge close to or inside the spacer with little impact on threshold voltage and saturation current /17,18/. The main degrading parameter is the series resistance which affects mostly the triode region. But degradation in the triode region has a much weaker effect on circuit speed than degradation in the saturation region /19/. Nevertheless a 10% decrease in linear current or gm change is usually accepted as lifetime criterion for NMOS transistors. This criterion gives ample safety margins against process tolerances and errors in extrapolating from accelerated to operating conditions.

Hot carrier degradation in PMOS transistors takes place mainly by electron trapping in the gate oxide on the drain side of the channel. This effect lowers the channel length and gives rise to an increase of saturation current /20,21/. Degradation in p-channel transistors has the opposite effect on circuit speed as in n-channel transistors. PMOS reliability is therefore of no concern as far as circuit speed is concerned except perhaps in critical circuits where timing accuracy is important.

The Effect of Hot Carrier Stress on Leakage: Leakage current between source and drain of turned off transistors is the main contribution to standby current in complex circuits. For example a 16Mbit DRAM contains transistors with a total width of about 5m in its peripheral circuits. In order to deliver high saturation currents, threshold voltages and leakage currents in these transistors are higher than in the 16 million cell array transistors. Their contribution to standby current is therefore estimated at about 70%, assuming 0.1nA leakage per 1μm transistor W. NMOS-LDD transistors do not suffer any leakage current increase after hot carrier stress. PMOS transistors, however, exhibit a reduction in channel length which manifests itself in reduced Vt and higher leakage currents. Fig.6 shows subthreshold characteristics of PMOSFETs before and after stress. In very short channels, leakage current is increased due to both threshold lowering and deterioration of subthreshold slope. Therefore hot carrier induced leakage poses a serious reliability issue in submicron PMOSFETs. Leakage currents before and after stress for devices with different gate lengths have been plotted in Fig.9. The dependence of leakage current on Vt is maintained after stress, permitting the use of a Vt based degradation criterion. For 5V operating voltage and a minimum Vt before stress of 0.7V, a 100mV reduction of Vt over the device lifetime appears tolerable if leakage currents in excess of 0.1nA/μm are to be avoided. This criterion is used to evaluate PMOS lifetime.

Estimation of Lifetimes: Using 10% degradation of current in the triode region and 100mV change of Vt for NMOS and PMOS transistors respectively, lifetimes can be estimated. To obtain meaningful results in a limited amount of time, accelerated stress conditions have to be used. V_{DS} during accelerated stress is about 3 to 4 V higher than the actual operating voltage. Stress is carried out at maximum substrate current for NMOS and at maximum gate current for PMOS devices. But even then degradation may be less than the failure criterion. This means that already at accelerated stress conditions lifetimes can only be obtained through extrapolation. This extrapolation is performed using a power law relationship /22/ and is demonstrated in Fig.10. Estimation of lifetimes under operating conditions is based on fitting stress data at different voltages with an exponential function between lifetime and the reciprocal of the stress voltage. The results are shown in Fig.11. PMOS slopes are lower than NMOS slopes. Using this diagram, acceleration factors can be obtained and lifetime dependence on gate length can be estimated (Fig.12). It is evident from Fig.12 that PMOS degradation, which poses no problem at gate lengths above 0.7µm is more critical than NMOS in the half micron regime. Decrease of lifetime due to gate length reduction is actually quite low in NMOS-LDD transistors and has been found to be less than a factor of 4 per 0.1µm. From a hot electron point of view NMOS-LDD transistors can therefore still be used at 5V with a gate length of 0.5µm. This is however not the case for buried channel PMOSFETs, which require 3.3V at this scaling level.

The lifetimes of Fig.12 are underestimated if thinner gate oxide is used. Thinner gate-oxide has been found to improve lifetime after hot carrier degradation in both NMOS and PMOS transistors /23,24,25/. There are several theories explaining the observed improvement including less effect of trapped charge on channel inversion layer in thin oxide transistors, less mobility degradation and more restricted lateral extension of damage-region in PMOSFETs.

Impact of Processing on Hot Electron Lifetime: Though degradation after hot electron stress does not appear to be an insurmountable obstacle in scaling sub-µm CMOS processes, nevertheless its effect has to be carefully monitored and devices have to be optimized for maximum reliability. Processing can have a significant impact on device lifetime and minor technology changes can mean large improvement or deterioration of reliability. Some of the effects studied in recent years are a) dose and energy of LDD implant, b) shadowing of LDD implant, c) reoxidation of PolSi gate after patterning, d) mechanical stress, e) spacer technology. They will be discussed in the following paragraphs.

Over the last few years many optimizations on the familiar LDD structure have been carried out to improve current drive and reliability of transistors. The main findings of these studies are that overlap between gate and n⁻ region has to be maximized /26,27,28/. Satisfying this requirement, however, one is faced with a trade off involving short channel effect and punchthrough. Fig.13 demonstrates this problem. It compares both the improvement obtained through higher dose or higher energy LDD implantation and the channel length shortening that goes with it. It is evident from this data that hot carrier reliability and short channel effect cannot be optimized independently.

Though LDD drain structures have also been studied in PMOS transistors /20/ they appear to be of less advantage than in NMOS transistors. Doping profiles of BF_2 doped conventional drain structures after high temperature reflow steps are comparable to phosphorus doped NMOS-LDD regions. Additional reduction of the electric field peak is therefore unlikely. Especially in DRAM processes with their high temperature budget, PMOS source/drain regions diffuse far enough under the gate to create an LDD like situation /15/.

Lack of overlap between LDD region and transistor gate can also result from shadowing effects of the LDD implantation angle. Depending on the location of the source and drain with respect to the implantation angle, large asymmetries in lifetime after HE-stress can result. Up to six orders of magnitude difference in lifetime between best and worst devices were found /29,30/. Rotation of wafers during LDD implantation all but eliminates this asymmetry.

It has been found necessary in CMOS technology to oxidize the PolySi gate after patterning. There are two reasons for this. One concerns the quality of the gate oxide at the gate edge where extra damage due to plasma etching is likely. A reoxidation creates a small bird's beak and prevents oxide breakdown at the gate edge /31,32/. The second reason involves removing poly silicon stringers after etching. In DRAMs the topology underneath the transistor gate is already so high that dry etching leaves behind small spacers, which can only be removed after a short oxidation. The gate bird's beak aggravates HE degradation because overlap between drain and gate is reduced and charge trapped in the bird's beak or the surface states below has a stronger effect on the inversion layer than in the thin oxide region /33,30/. The following processing steps have been found to minimize the effect on hot carrier lifetime. a) LDD implantation has to be performed before reoxidation to guarantee maximum overlap between gate and drain. b) Reoxidation thickness should be as small as possible and in diluted atmosphere. c) Reoxidation should be in dry ambient because in wet ambient, oxidizing species diffuse faster under the gate edge creating a longer bird's beak.

In modern CMOS processes, many layers of different materials are usually deposited. These different layers can cause mechanical stress with negative consequences for the electrical performance of circuits. For example in order to relieve alignment tolerances in the cell array of a DRAM, a self aligned contact technology has been developed /34/. This technology requires a silicon nitride layer as etch stop and oxidation barrier. Then the nitride layer has to be removed in contact windows. Process architecture leads to large chip areas which are totally covered by this nitride layer. The resulting mechanical stress generates leaky junctions and lower HE lifetimes as is evident from Fig.14. Minimizing nitride thickness, however, proved to remedy the situation. Silicon nitride layers used for passivation have been found to cause similar stress related HE problems especially if the passivation film contains large amounts of SiH which may release hydrogen during subsequent processing steps /35,36/. Recently however it has been demonstrated, that tensile stress, which is responsible for higher HE degradation, is reduced by lowering the gate length. Future generation MOS devices may therefore be less affected by stress related HE degradation /37/.

Spacers in LDD transistors are usually made from CVD oxide. In modern DRAMs, these oxide spacers also provide isolation between bitlines and wordlines. Using other spacer materials such as poly silicon or silicon nitride have also been evaluated. Si_3N_4 appears to be a promising candidate because selectivity of nitride against oxide etching can be made high enough to prevent noticeable field oxide thinning during spacer formation. Si_3N_4 spacers have also been found to reduce HE degradation because of their higher dielectric constant. The higher gate fringing pushes the high drain potential edge into the n^+ region lowering the electrical field as a consequence /38,39/.

Conclusion and Outlook: Hot carrier degradation continues to be an issue which has to be dealt with when scaling sub-µm CMOS processes. In n-poly gate processes at gatelengths below 0.7µm, buried channel PMOS transistors are more affected than surface channel NMOS. Though 5V operation at these gatelengths and below may be possible, gate insulator breakdown, band-to-band tunneling and off state isolation require a reduced operating voltage. On the other hand, smaller device dimensions are possible at 3.3V. Minimum gatelength and inverter delay times for 5V and 3.3V are shown in Fig.15. An increase in speed can be realized when reducing operating voltage from 5V to 3.3V.

Some effects of processing on HE lifetime have been described. High quality gate insulators with low interface and trap density, good overlap between source/drain regions and gate, minimum gate bird's beak and suppression of mechanical stress are essential to achieve high reliability. Monitoring HE lifetime during each phase of process development is necessary for the understanding and elimination of unwanted effects.

References:

1. L.Dori, J.Sun, M.Arienzo, S.Basavaiah, Y.Taur, D.Zichermann, "Very thin nitride/oxide composite gate insulator for VLSI CMOS", Digest of Technical Papers, VLSI Symposium on Technology, Karuizawa, 25 (1987)

2. H.Iwai, H.Sasaki, T.Morimoto, Y.Ozawa and K.Yamabe, "Stacked-nitride oxide gate MISFET with high hot-carrier-immunity", IEEE Techn. Dig. IEDM, 235 (1990)

3. H.Hwang, W.Ting, D.L.Kwong and J.Lee, "Electrical and reliability characteristics of ultrathin oxynitride gate dielectric prepared by rapid thermal processing in N2O", IEEE Techn. Dig. IEDM, 421 (1990)

4. A.Uchiyama, H.Fukuda, T.Hayashi, T.Iwabuchi and S.Ohno, "High performance dual-gate sub-halfmicron CMOSFETs with 6nm-thick nitrided SiO2 films in an N2O ambient", IEEE Techn. Dig. IEDM, 425 (1990)

5. T.Nishihara, M.Inuishi, T.Ogawa, H.Miyatake, K.Tsukamoto and K.Kobayashi, "A 0.5µm isolation technology using advanced poly silicon pad LOCOS (APPL)", IEEE Techn. Dig. IEDM, 100 (1988)

6. H.-M.Mühlhoff, J.Dietl, P.Küpper and R.Lemme, "Field isolation and active devices for 16MBit DRAMs", 20th European Solid State Device Research Conference (ESSDERC), Nottingham, 531 (1990)

7. C.Chan and J.Lien, "Corner Induced Drain Leakage in Thin Oxide MOSFETs", IEEE Techn. Dig. IEDM, 714 (1987)

8. T.Y.Chan, J.Chen, P.K.Ko and C.Hu, "The Impact of Gate-Induced Drain Leakage Current on MOSFET Scaling", IEEE Techn. Dig. IEDM, 718 (1987)

9. B.Hoeneisen and C.A.Mead, IEEE Trans. Electron Devices, ED-19, 382 (1972)

10. Y.El-Mansy, "MOS Device and Technology Constraints in VLSI", IEEE Trans. Electron Devices, ED-29, 567 (1982)

11. K.M.Cham et al., "Computer-Aided Design and VLSI Device Development", 2nd ed. Kluwer Academic Publishers, 271 ff (1988)

12. W.Müller and I.Eisele, "Velocity Saturation in Short Channel Field Effect Transistors", Solid State Communications, Vol.34, 447 (1980)

13. F.M.Klaasen and W.Hes, "Compensated MOSFET Devices", Solid State Electron., Vol.28, 359 (1985)

14. K.M.Cham and S.Y.Chiang, "Device Design for the Submicrometer p-Channel FET with n^+ Polysilicon Gate", IEEE Trans. Electron Devices, ED-31, 964 (1984)

15. J.Dietl, L.DoThanh, K.H.Küsters, L.Kusztelan, H.-M.Mühlhoff, W.Müller and F.X.Stelz, "Buried stacked capacitor cells for 16M and 64M DRAMs", 20th European Solid State Device Research Conference (ESSDERC), Nottingham, 465 (1990)

16. S.Odanaka, M.Fukumoto, G.Fuse, M.Sasago, T.Yabu and T.Ohzone, "A new half-micron p-channel MOSFET with efficient punchthrough stops", IEEE Trans. Electron Devices, ED-33, 317 (1986)

17. F.-C.Hsu and H.G.Grinolds, "Structure-Enhanced MOSFET Degradation Due to Hot-Electron Injection", IEEE Electron Dev. Let., EDL-5, 71 (1984)

18. Y.Toyoshima, H.Nihira, M.Wada and K.Kanzaki, "Mechanism of Hot Electron Induced Degradation in LDD NMOS-FET", IEEE Techn. Dig. IEDM, 786 (1984)

19. J.Winnerl, A.Lill, D.Schmitt-Landsiedel, M.Orlowski and F.Neppl, "Influence of transistor degradation on CMOS performance and impact on life time criterion", IEEE Techn. Dig. IEDM, 204 (1988)

20. M.Koyanagi, A.G.Lewis, J.Zhu, R.A.Martin, T.Y.Huang and J.Y.Chen, "Investigation and Reduction of Hot Electron Induced Punchthrough Effect in Submicron PMOSFETs", IEEE Techn. Dig. IEDM, 722 (1986)

21. H.-M.Mühlhoff, P.Murkin, M.Orlowski, W.Weber, K.H.Küsters, W.Müller, C.M.Rogers and H.Wendt, "Sub Micron P-MOSFETs Under Static and Swap Stress", Symposium on VLSI Technology, Karuizawa, 57 (1987)

22. C.Hu, S.C.Tam, F.C. Hsu, P.K.Ko, T.Y.Chan and K.W.Terrill, "Hot-Electron-Induced MOSFET Degradation - Model, Monitor and Improvement", IEEE Journ. of Solid-State Circ., SC-20, 295 (1985)

23. M.Yoshida, D.Tohyama, K.Maeguchi and K.Kanzaki, "Increase of Resistance to Hot Carriers in Thin Oxide MOSFETs", IEEE Techn. Dig. IEDM, 254 (1985)

24. Y.Toyoshima, H.Iwai, F.Matsuoka, H.Hayashida, K.Maeguchi and K.Kanzaki, "Analysis on gate oxide thickness dependence of hot carrier induced degradation in thin gate oxide nMOSFETs", IEEE Trans. Electron Devices, ED-37, 1496 (1990)

25. S.Odanaka and A.Hiroki, "Gate oxide thickness dependence of hot-carrier-induced degradation in buried p-MOSFETs", IEEE Techn. Dig. IEDM, 565 (1990)

26. R.Izawa, T.Kure, S.Iijima and E.Takeda, "The Impact of Gate-Drain Overlapped LDD (GOLD) for Deep Submicron VLSIs", IEEE Techn. Dig. IEDM, 38 (1987)

27. H.M.Mühlhoff, K.H.Küsters and H.Melzner, "Transistor Optimization for 0.6μm CMOS processes", Proc. 2nd Int. Symp. ULSI Science and Technology, (1989)

28. T.Hori, "1/4μm LATID (Large-Tilt-angle-Implanted Drain) Technology) for 3.3V operation", IEEE Techn. Dig. IEDM, 777 (1989)

29. T.Mizuno, Y.Matsumoto, S.Sawada, S.Shinozaki and O.Ozawa, "A New Degradation Mechanism of Current Drivability and Reliability of Asymmetrical LDD-MOSFETs", IEEE Techn. Dig. IEDM, 250 (1985)

30. H.M.Mühlhoff, P.Murkin, K.H.Küsters, M.Orlowski and W.Müller, "Analysis and Optimization of Submicron MOSFETs with Gate Bird's Beaks", Proc. 1st Int. Symp. ULSI Science and Technology, 632 (1987)

31. C.M.Osburn, A.Cramer, A.M.Schweighart and M.R.Wordemann, "Edge Breakdown of PolySi Gates over Thin Oxides During Ion Implantation", Proc. ECS Symposium on VLSI Science and Technology, ed. C.J.Dell'Oca and W.M.Bullis, 354 (1982)

32. C.Y.Wong, T.N.Nguyen, Y.Taur, D.S.Zichermann, D.Quinlan and D.Moy, "Process Induced Degradation of Thin Oxides", Proc. 1st Int. Symp. ULSI Science and Technology, 155 (1987)

33. P.K.Ko, S.Tam and C.Hu, "Enhancement of Hot-Electron Currents in Graded-Gate-Oxide MOSFETs", IEEE Techn. Dig. IEDM, 88 (1984)

34. K.H.Küsters et al, "A High Density 4Mbit dRAM Using a Fully Overlapping Bitline Contact (FOBIC) Trench Cell", Symposium on VLSI Technology, Karuizawa, 93 (1987)

35. J.Mitsuhashi, S.Nakao and T.Matsukawa, "Mechanical stress and hydrogen effects on hot carrier injection", IEEE Techn. Dig. IEDM, 386 (1986)

36. W.H.Stinebaugh, A.Harrus and W.R.Knolle, "Correlation of gm degradation of submicrometer MOSFETs with refractive index and mechanical stress of encapsulation materials", IEEE Trans. Electron Devices, ED-36, 542 (1989)

37. A.Hamada, T.Furusawa, E.Takeda, "A new aspect on mechanical stress effects in scaled MOS devices", Symposium on VLSI Technology, Honolulu, (1990)

38. T.Mizuno, S.Sawada, Y.Saitoh and S.Shinozaki, "Si3N4/SIO2 spacer induced high reliability in LDDMOSFET

and its simple degradation model", IEEE Techn. Dig. IEDM, 234 (1988)

39. T.Mizuno, T.Kobori, Y.Saitoh, S.Sawada and T.Tanaka, "High dielectric LDD spacer technology for high performance MOSFET using gate-fringing field effects", IEEE Techn. Dig. IEDM, 613 (1989)

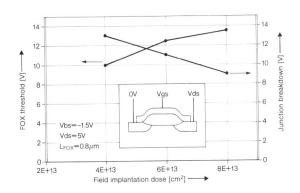

Fig.1. Tradeoff between junction breakdown and parasitic field transistor turn on in LOCOS isolation.

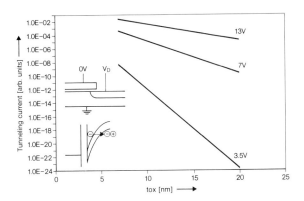

Fig.2. Leakage current in gated diode structure due to band-to-band tunneling.

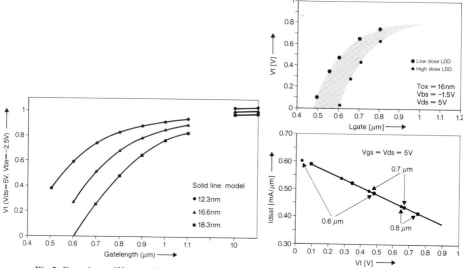

Fig.3. Dependence of Vt on gate length and gate oxide thickness for surface channel NMOS-LDD transistors. Drain structures are identical for all three types of devices.

Fig.4. Vt-roll off and its dependence of LDD-structure. Saturation drain current is only dependent on threshold voltage.

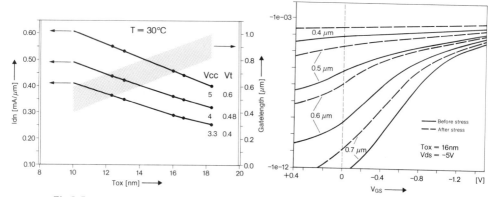

Fig.5. Range of gate lengths and saturation drain currents as function of gate oxide thickness in surface channel NMOS-FETs.

Fig.6. Buried channel PMOSFET subthreshold characteristics before and after hot carrier stress.

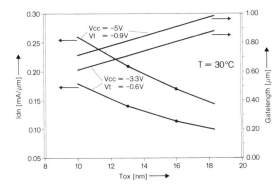

Fig. 7. Effect of gate oxide scaling on PMOS performance. Minimum gate length is strongly dependent on operating voltage due to punchthrough.

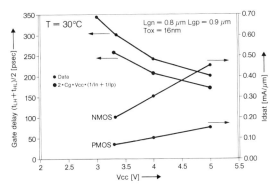

Fig. 8. Comparison between measured and calculated gate delays and NMOS and PMOS saturation currents in 33 stage ring oscillator.

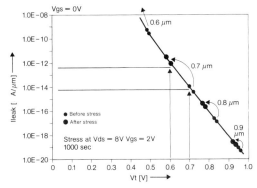

Fig. 9. Relationship between threshold voltage and leakage current in PMOSFET and its dependence on hot carrier stress.

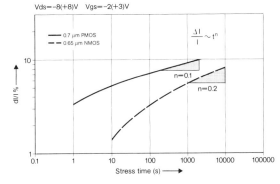

Fig. 10. Power law relationship on time for NMOS and PMOS degradation.

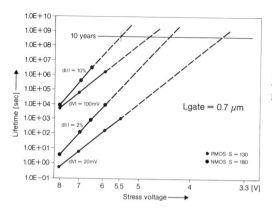

Fig.11. Dependence of lifetime on stress voltage and failure criterion in n- and p-channel transistors.

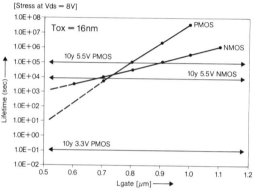

Fig.12. Dependence of lifetime on gate length. In short channel devices, degradation of p-channel is more critical.

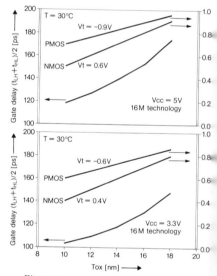

Fig.15. Comparison between minimum gatelengths and delay times in CMOS inverters at 5V and 3.3V.

Fig.13. Trade off between higher lifetime after HE stress and channel shortening when increasing LDD dose.

Fig.14. Stress related device problems resulting from silicon nitride layer. The nitride layer is used as etch stop in self aligned contact technology.

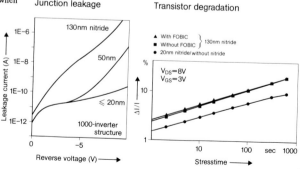

Paper presented at INFOS '91, Liverpool, April 1991
Workshop Papers

SOI CMOS devices

M.Haond and O.Le Néel*

Centre National d'Etudes des Télécommunications, BP 98, 38243 Meylan Cedex, FRANCE

*On leave from Matra-MHS, Route de Gachet, 44087 Nantes Cedex 03, FRANCE.

Abstract: In this paper, we first discuss the main advantages and specific features of Silicon-On-Insulator MOS devices compared to bulk silicon devices. We further discuss the potential advantages of using thin-film fully depleted SOI MOS and show that some of them are reduced by parasitic effects which become very drastic in the submicron regime where the VLSI technology has entered nowadays. We thereby indicate some open fields for further research in the SOI device area, if it is to be applied to ULSI circuits.

1. INTRODUCTION

Silicon-On-Insulator (SOI) has long been considered a good candidate for replacing bulk silicon in VLSI CMOS devices owing to its intrinsic isolation properties. It has known in the last decade a new attention from VLSI MOS researchers thanks to the important progress in the material quality. This has promoted intensive work on SOI devices, namely for submicron dimensions. The era of thin fully depleted transistors has started aiming at fast circuit applications. These advances have brought out SOI-specific technological and electrical problems, which have again delayed the introduction of SOI into the VLSI circuit manufacturing. In this paper, we compare the SOI thin-film device, looking at its specific features, to bulk silicon MOS. We discuss the actual limitations of SOI devices and the problems to be overcome if these are to be used for VLSI CMOS circuits.

2. BULK-SILICON vs SOI

2.1 Radiation hardness

SOI has long (and mainly) been used in military applications, mainly the Silicon-On-Sapphire (SOS) for its intrinsic high radiation hardness, as compared to bulk. This is due to the small collection volume accessible to the incoming ionising particles (Figure 1), which is important for the transient dose and SEU (soft error) hardening.
Furthermore, the quasi-semiinfinite thickness of the underlying insulator avoid field induced migration of charges created in it by the radiation, resulting in a high cumulative dose hardness. This is somewhat different in nowadays SOI substrates such as SIMOX where the buried oxide is in the order of 400 nm. This reduced oxide thickness might allow the separation of holes and electrons by the electric field corresponding to the voltage drop between the SOI film and the underlying Si-substrate. These charges, when they reach the interface can induce a back leakage channel. This effect has usually been suppressed by using a deep implant at the interface (boron in an NMOS). Intensive work is carried out to identify the suitable SOI substrate and buried oxide thicknesses together with an appropriate technological process, for increasing the radiation hardness /1/.

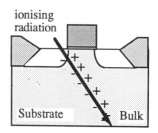
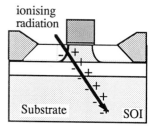

Figure 1: This schematics shows the difference in the silicon volume accessible to the incoming ionisation radiation in the active parts of a MOS transistor in bulk silicon and SOI.

2.2 Capacitance reduction

Another important advantage of SOI relies on its intrinsic isolation properties. It is possible to diffuse the drain/source regions down to the buried oxide. This provides an important reduction of the drain/source junction capacitances as indicated in Table 1.

Capacitance	SOI (SIMOX)*	Bulk Silicon*	Gain
Gate	1.3	1.3	1
Drain (Source) junction	0.05	0.2...0.35	4...7
Poly gate/substrate	0.04	0.1	2.5
Metal1/substrate	0.027	0.05	1.85
Metal2/substrate	0.018	0.02	1.11

(*) the values are in $fF/\mu m^2$

Table 1: Comparison of the capacitance values for a 1 μm LOCOS-isolated CMOS process in SOI and bulk silicon /2/.

Moreover, the interconnection-to substrate capacitances, corresponding either to polysilicon or to metal lines, are generally reduced since the buried oxide thickness is added to the other dielectrics present in the field regions. This is indeed the case when a LOCOS isolation is used, instead of mesa etching. Table 1 compares the capacitance values for a bulk silicon and aSOI 1μm LOCOS-isolated CMOS process /2/. The corresponding expected gain in speed ranges from 30 to 50%. This is indeed obtained on ring oscillators as visible in Figure 2 which exhibits the propagation delay time of ring oscillators with different design rules, both in bulk silicon and SOI.

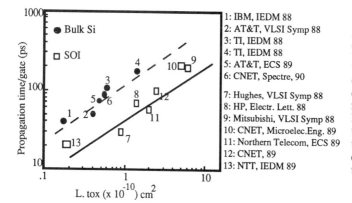

Figure 2: Diagram of the propagation delay time per gate for ring oscillators corresponding to various design rules. The gate length has been normalized to the gate oxide in order to be able to compare different generations of technologies.

It is, however, worth mentioning that with the reduction of the dimensions, all features are scaled down, namely the pitch of the metal lines. This results in a dramatic increase of the fringing capacitance between metal lines, and it has been shown /3/ that these become preponderant in submicron circuits with double (or triple) level metallisations. This will also be the case in SOI circuits, and the total capacitance will tend to be close to the bulk equivalent circuit value. The gain in speed in SOI (if the otherwise device current densities are similar), will then mainly rely on a higher integration density, and the associated interconnection line length (and resistance) reduction.

2.3 Lateral isolation and latch-up

This is probably the most interesting feature of SOI regarding its potential application to VLSI circuits. In bulk silicon, the isolation techniques are becoming more and more complex in order to satisfy the scaling of the dimensions of the active and field regions. Moreover new wells designs and implant procedures (such as retrograde wells) must be used to avoid the field transistor action and the latch-up triggering: Figure 3a) and b).

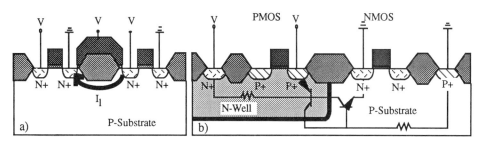

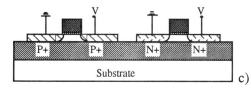

Figure 3: Schematics of the isolation problems in bulk: a) the field parasitic NMOS transistor and b) the latch-up situation. c) presents the "naturally" isolated SOI structure free of both features.

It is clear that its "natural" isolation feature provides SOI the advantage of an easier scaling of the isolation. Moreover, since the complementary wells are completely isolated in SOI, it is no more necessary to specify any N+/P+ distance rule, and all active regions can be treated uniquely, without any latch-up triggering: Figure 3c). It remains true, however, that the isolation procedure in SOI, must be well engineered, both to avoid the presence of silicon filaments between active regions and to avoid transistor width reduction, in order to satisfy the design rules. The lateral isolation is however also concerned by intradevice problems, such as edge and back-interface leakage, which can be avoided by specific boron implants (for the NMOS): sidewall and deep implants for the edge and back-interface parasitic channels respectively /4,5/.

2.4 Floating body

This is the main electrical difference between a bulk and a SOI MOS transistor: in bulk silicon, the substrate region below the channel is always connected to the source potential, whereas it is floating in SOI. Therefore, the neutral region below the depletion region under the channel has a floating and hence fluctuating potential. In the conduction mode, i.e. above threshold, in a NMOS, the electrons are attracted at the upper interface and the holes rejected down to the buried interface. At the drain end, the electrons accelerated by the longitudinal electric field create electron-hole pairs by impact ionisation. Most of the electrons are collected by the drain and the holes diffuse back to the source. They however accumulate in the low potential region of the substrate. This leads to a positive biasing of the floating substrate, which in turn reduces the threshold voltage, and thereby increases the channel current: it results in a positive feedback. A "kink" appears in the saturation part of the output characteristics (I_D-V_D) of a SOI

NMOS transistor: Figure 4. This effect is mainly visible in the NMOS since the electron ionisation rate is higher than its hole counterpart. The positive feedback action is active until the substrate potential is fixed by the substrate-source junction forward biasing. This effect results in non-linearities in the output characteristics, to an anomalous subthreshold slope at high V_D and to transient overshoots in the drain current /6,7/. At short gate length (below 1μm), this will trigger a single latch of the lateral parasitic bipolar transistor. This is visible in Figure 4 at VG=0 & 1V and will be discussed below.

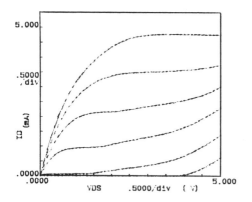

Figure 4: I_D-V_D plot of a SOI NMOS fabricated in a 150 nm thick film, with a mean doping level of about 1.5 10^{17} cm^{-3}, which does not provide a full depletion of the SOI film and allows the "kink" to appear (see below in thin-films).

W=20μm, Leff=0.8μm.

VG is ranged from 0 to 5V by steps of 1V.

It is possible to suppress this effect by using body contacts connected to the source potential: Figure 5a). However, this reduces the circuit density, which is supposed to be a great advantage of SOI, and it is not efficient for large transistors, where the access resistance to the body increases with its width. It is possible to introduce P+ diffusions in the N+ source (split-source) of the NMOS (N+ in a PMOS): Figure 5b). Again there is a loss in density because it is necessary to increase the transistor width to take into account the P+ diffusions. Moreover in any case, there is some increase in the drain/source junction capacitances, which reduces the gain indicated in Table 1. We will show in the next paragraph, that the use of fully depleted SOI films reduces the floating volume and its effects.

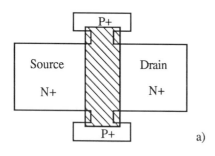 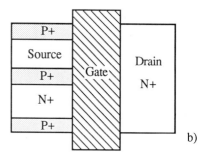

Figure 5: Schematics showing 2 different device structures which can be used to avoid a floating body below the gate: a) a lateral contact and b) a split-source device.

3. THIN-FILM SOI

This name is given to SOI films where the silicon under the gate is completely depleted at V_G=Vt, i.e. T_{Si} < W_{Dmax} where T_{Si} is the film thickness and W_{Dmax} the maximum calculated depletion width in (bulk) silicon. The use of fully depleted films has a number of consequences which we will discuss below: suppression of the "kink", Vt lowering, subthreshold slope increase, short channel and punchthrough improvements.

3.1 Front- and back-gate coupling

In fully depleted transistors, a strong coupling exists between the front gate and the back substrate which acts as a back gate. As a consequence, the "kink" effect is suppressed, since the region below the channel is depleted, avoiding the neutral region /8/. The potential of the depleted region is controlled by the front and somewhat by the back gate potential. It is therefore necessary to fix the back gate potential (at VDD or VSS), in order to avoid that its potential fluctuates.

At low V_D, the thin-film SOI MOS transistor can be roughly described by the coupling equations obtained by writing the potential continuity and the electric field at each interface in a 1-D model and exchanging the front and back gates /9/. We can use these equations to derive further parameters, such as the threshold voltage dependance on doping, silicon film thickness, back and front interface states and oxide fixed charges. We have shown /10/ that they could be further expanded to derive technological parameters such as the actual silicon film thickness under the gate and the buried oxide thickness. We can also derive the interface states density and/or the fixed oxide charge /11/.

3.2 Vt lowering and subthreshold slope and low-field mobility improvements

An important consequence of the full depletion in thin-film SOI is the lowering of the threshold voltage (Vt) of the MOS transistors as compared to the equivalent thick film (or bulk silicon) devices. It can be explained by using a schematical 1-D diagram of the band curvatures in bulk and SOI as shown in Figure 6: in thin-film SOI, hen the gate potential is increased, the band curvature at the upper surface increases as soon as the depletion region reaches the buried interface, and the inversion is reached rapidly. There will be a compromise between the film thickness and the doping level necessary to maintain a given threshold voltage value imposed by the circuit designers' specifications. The choice of this value depends on the subthreshold slope factor (S), which is improved in fully-depleted MOS transistors, since the depletion charge is bounded by the silicon film thickness: $Q_D = q\, N_a\, T_{Si}$ (where Q_D is the depletion charge and N_a the doping level).

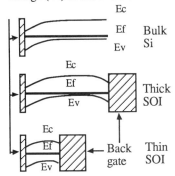

Figure 6: Schematics of the band curvature at the surface in bulk silicon and SOI. In thin SOI, the coupling between front and back gates leads to a surface reduction of both Ec-Ef and electric field intensity.

We can understand the improvement in the subthreshold slope as follows: let us write: $S = Ln10\, dV_G/dLnI_D$ /12/. Since the inversion is rapidly reached when the depletion reaches the buried oxide, the transition from accumulation to inversion is fast and the gate control over the charge in the conduction channel is more effective. A high subthreshold slope (low S factor) allows a reduction of Vt, and this provides high saturation currents (I_D is roughly proportionnal to $(V_G-Vt)^2$ for long transistors and to (V_G-Vt) for short devices). This is one of the main parameters impinging on the switching speed of the logic circuits. Values close to the theoretical value of 60 mV/decade are reported in SOI /13/.

We could use almost the same explanation for the increase in the surface low-field mobility in very thin SOI films /14/, since the surface electric field is lower in this case than in bulk transistors at the same doping level (this can be deduced from Figure 6, by drawing the 1-D potential distribution across the film below the gate). This increase in the surface mobility results in an increase of the maximum transconductance value.

However, it is important to keep in mind that this is obtained at low Vt, which is not always compatible with the designer's specifications. If we assume a realistic subthreshold slope of 63 mV/decade which is close to the minimum theoretical value (kT/q Ln10 = 60), a minimal OFF-state current of 1pA/μm of channel width and a drain current of 0.1 mA/μm at V_G=Vt & V_D=5V (which is realistic for a 0.7 μm CMOS process), then the minimal value for Vt is 0.5V (8 decades x 63 mV/dec). Taking into account the unavoidable process fluctuations in a real process, it will be difficult to allow a worst case value for Vt below 0.6V. If the gate material is kept similar to bulk CMOS (N+ gate on an NMOS), this value will be obtained in fully depleted NMOS transistors by increasing the doping level under the gate, which will degrade the S value, the mobility (see Figure 6 in Ref.15) and the transconductance. A rough calculation (verified on processed devices) gives the following process values for an NMOS device with Vt = 0.6 V: for Toxf = 15 nm and Toxb = 400 nm, for full depletion, T_{Si} < 80 nm and Na > 2 10^{17} cm^{-3}.

3.3 Short-channel effects and punchthrough

Let us consider another important parameter in submicron CMOS devices: the short channel effect and the punchthrough susceptibility. Figure 7 presents schematics of the extension of the various depletion regions in a bulk and an SOI NMOS transistor. It shows that the use of thin-film SOI limits the extension of the depletion region from the drain into the region below the gate. There is a two-dimensional limitation to this extension. This should provide a good short-channel behaviour and the absence of punchthrough. Some results have been reported concerning the short channel behaviour /16/. However these reports have only considered the evolution of Vt at low V_D which is not of much concern for a transistor in a circuit. Unfortunately, in the NMOS, this short channel behaviour remains theoretical since at high V_D, and for small gate lengths, the punchthrough is completely hidden by the triggering of the parasitic lateral bipolar transistor, which we will discuss below. In the PMOS, as long as an N+-gate is used, we have an accumulation mode transistor and a buried channel associated to a buried junction, as long as the SOI film is not thin enough. The punchthrough is favoured in the region situated between the upper interface and the buried junction /17, 18, 19/. It is well-known in bulk silicon that in order to suppress punchthrough in this region, it is necessary to reduce this junction depth. In SOI, it has been claimed that the best solution was the use of thin films, i.e. with a thickness of the order of what would be the buried junction depth in bulk silicon at the same design rules, and to fabricate depleted accumulation mode PMOS transistors. It must be compatible with the process possibilities: etch stops, homogeneity of process,.../5/ It is a process limitation for deep submicron devices.

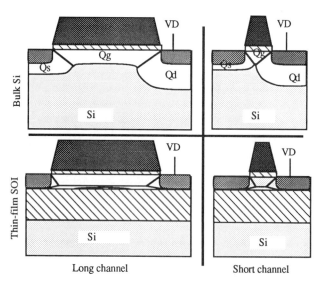

Figure 7: Schematics comparing the extension of the depletion zones in long and short channel bulk (top) and thin-film SOI (bottom) MOS transistors. In the last case, the extension is limited by the gate controlled depleted region and by the film thickness.

4. THIN-FILM SOI DEVICE PROBLEMS

4.1 Thermal effects

If we look at a body-tied SOI NMOS transistor (since in transistors without body contact, the effect can be hidden by the "kink"), a negative resistance is visible in the output characteristics /20, 21, 22/. This effect was also observed in PMOS transistors /20/. This is attributed to a thermal effect. We have conducted many experiments in order to confirm this hypothesis. The thermal behaviour is much different from bulk silicon, since the thermal conductivity K_{SiO2} of silicon dioxide is 100 times lower than that of silicon K_{Si}. A temperature rise results in a reduced mobility and a reduced transconductance. We have shown by transient pulse measurements that the time constant of this self-heating effect is of the order of the microsecond /20/. Different methods can be used to measure the temperature rise: IR thermography /21/, micro-Raman /23/, liquid crystals /22,24/. Figure 8a) presents the self-heating temperature rise in a ZMR SOI transistor as measured by checking the phase transition of liquid crystals. A temperature increase of about 20K was measured at a power of 5mW, which corresponds to a thermal resistance of 4000 K/W.

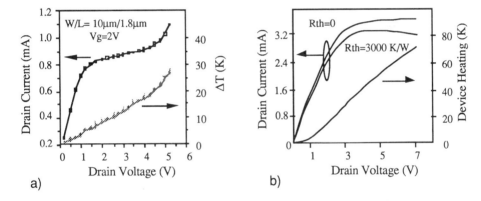

Figure 8: a) ID-VD plot of a ZMR-SOI NMOS and the corresponding temperature rise as measured by using the phase transition of a liquid crystal/23/; b) simulated ID-VD characteristic with the following parameters /24/: W/L=20/1.6, Vt=0.45V, V_G=7V, μ_0=500cm^2/V.s. The simulation accounts for the temperature dependance of both the mobility and threshold voltage.

Figure 8b) presents the results of simulated output characteristics for a SOI material with a thermal resistance of 3000 K/W /24/. The parameters used for the simulation are given in the legend. This thermal effect will of course be reduced in proportion with the buried oxide thickness and SIMOX material will be better than ZMR material for power applications. We have estimated that we had a 15 % saturation reduction for an NMOS transistor with 1µm design rules in SIMOX SOI as compared to bulk silicon..

4.2 Parasitic lateral bipolar transistor

This is probably the main problem to be solved rapidly if SOI wants to have its chances of being used for submicron CMOS applications. This phenomenon is a direct consequence of the floating substrate. In a SOI NMOS, all electrons travelling under the gate (namely below Vt) are accelerated by the drain electric field and create electron-hole pairs at the drain end. Most of the electrons are collected by the drain and the holes drift along the minimum potential lines down to the source region where they accumulate until the barrier is lowered (by their accumulation). In a very thin-film, the potential in the depletion region below the gate might be high enough to suppress the barrier height for the holes. In that case, in the lateral bipolar constituted by the drain (collector), the depleted region (floating base) and the source (emitter), the emitter-base junction is forward biased and the bipolar transistor is triggered: see Figure 9.

If the current gain (ß) of this bipolar multiplied by the avalanche multiplication factor (M) is greater than 1, there is a positive feedback and the MOS transistor is triggered into snapback. This explains why in open base transistors, the collector breakdown voltage is lower than in base-grounded devices /25/ and why the drain breakdown voltage of MOS transistors is lower in SOI than in bulk silicon, where the substrate is connected to the source potential. It has therefore been proposed to use body ties in SOI to increase the drain breakdown voltage in submicron devices. However, in addition to the drawbacks indicated previously for this solution, the access resistance to the body is very high in fully depleted transistors and it reduces the efficiency of such a solution.

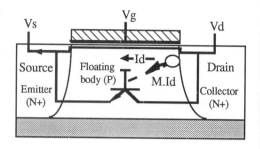

Figure 9: Schematic representation of the SOI NMOS transistor and the associated parasitic bipolar transistor with its floating base (floating body).

There has not been so far any completely satisfactory solution to this problem. Different technological approaches have been proposed which might allow to go down to 0.5 µm devices, where the bias voltage will be 3.3V. The most widely used approaches are dealing with the drain/source engineering in order to find the appropriate LDD structure which would allow the drain to sustain VDD+20% at ambiant temperature. We have fabricated 4KSRAM's on 100 nm thick SOI SIMOX wafers in a double level metal CMOS process, with 1 µm design rules. Figure 10 exhibits the power consumption of these devices versus the NMOS LDD dose, for different applied bias voltages. We can notice that the power consumption is comparable to equivalent bulk values for LDD doses below 10^{13} at/cm^3. We believe that above this value, the effective bias applied to the effective channel is higher and the lateral bipolar transistor is triggered earlier. It has been proposed to use gate-overlapped low dose tilted implants /26/. It seems possible to find a working window for 0.5 µm devices. Another approach is to use self-aligned TiSi2 /27/ which is supposed to kill the emitter efficiency (source) by increased electron-hole recombinations, which reduces the ß of the bipolar. These solutions seem to allow a narrow working window for one device generation but do not give a real answer to the problem of the floating base parasitic lateral bipolar in SOI.

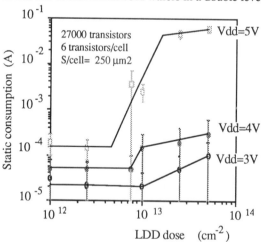

Figure 10: Plot of the static power consumed in a 4KSRAM made in 100 nm SIMOX-SOI at various Vdd, as a function of the NMOS LDD dose. This illustrates the effect of the parasitic bipolar on a circuit consumption and how an adapted drain/source engineering can reduce it..

4.3 Drain saturation current

It has been reported /28/ that the drain saturation current should be increased in thin-film fully-depleted SOI transistors thanks to a mobility and transconductance increase. In our experiments, even in films down to 80 nm, we have never measured drain saturation currents higher in SOI transistors than in bulk transistors, for devices where the threshold voltage was adjusted to a value higher or equal to that indicated above (Vt = 0.6V). We believe that this results from the addition of a few thin-film SOI parasitic effects. We have already indicated the thermal effect leading to current reduction. We have also indicated that in order to have a threshold voltage compatible with the electric design rules, an increased doping level was necessary in thin-film SOI transistors, and this reduces the mobility by increasing both the impurity scattering and the surface scattering as a result of an increase in the vertical electric field, and this reduces the transconductance and the saturation current.

Last, but not least, in thin-film fully depleted SOI transistors an important access resistance is present /29,30/. The reason is twofold: first and quite obviously, for a given doping level in the drain/source, if the film is thinned, the sheet resistance will be reduced; second, the increase of the bipolar triggering voltage is obtained by decreasing the LDD doping (see Figure 10), and this will increase the drain/source serial resistance.

4.4 Back interface leakage, hot electrons and reliability

It must be pointed out that in fully depleted devices, the quality and the behaviour of the back interface is of much concern, since we have seen that because of the coupling, it has a strong influence on the device parameters. Among the most important parameters influenced by the back gate, let us mention the threshold voltage and the leakage current. The back-side leakage problem is particularly relevant for the PMOS, since in a CMOS inverter, the back substrate is biased at 0V and the source of the PMOS at 5V. Therefore the back V_{GS} is equal to -5V, and if negative charges are to be trapped in the oxide, the back threshold voltage might fall in that range leading to a back channel. The leakage current can result from various technological problems during the process steps, which may create charges inside the oxides: plasma etching, dielectrics deposited with or without plasma enhancement,... The charges in the oxide may also appear while the device is working. It can result from the impact of ionising radiations or from the injection of hot carriers. A fully-depleted MOS transistor will never be as radiation-hard as a partially-depleted transistor, since in a fully-depleted device, the trapped charges induce a leakage channel, whereas in a partially-depleted one they are screened by the neutral region (as long as their density is low enough).

In that respect, the hot electron behaviour is also of important concern in deep submicron devices. In that dimension range, the bulk devices need an appropriate drain/source engineering in order to support an extrapolated device lifetime of about ten years. If this also works for partially-depleted SOI transistors /31/, it seems to be difficult to obtain the same lifetime extrapolation for fully-depleted SOI deep submicron devices /32/. This might result from the fact that in SOI fully-depleted devices, the energetic electrons created by impact ionisation at the drain end in the pinch-off region, have additionnal chances to be trapped in an oxide, since in addition to the gate oxide, there is a buried oxide at the same distance, and its thickness is higher. Furthermore, we have seen above that the avalanching impact ionisation was increased by the floating body, because of the parasitic bipolar action. This explains why the thin-film SOI NMOS transistors exhibit very rapid Vt shifts compared to bulk devices /32/. However, Vt then saturates, probably because trapped electrons both in the buried oxide and the front oxide spacers deplete the LDD region inducing a highly resistive region. This field of problems needs probably to be further investigated in order to study these phenomenons and find out some SOI specific solutions.

5. CONCLUSION

We have presented a rapid review of the main features of thin-film fully depleted transistors in comparison to bulk and thick-film SOI devices. Besides the complexity of the technological process of thin-film SOI, we have shown that some of the potential electrical advantages were applicable to VLSI manufacturing, namely the isolation features, but that some of them had not been confirmed for further use, since they were annihilated by parasitic effects, namely the transconductance reduced by thermal effects and serial resistance or the good short-channel behaviour annihilated by the parasitic bipolar effect. However, the field remains open for further research, keeping in mind both the applications and the complexity of the technology. It has been proposed for instance to use accumulation mode NMOS transistors (depleted/2, 18/ or not /17/) with a P+ gate symmetrically to the PMOS with its N+-gate (and controversely to the bulk trend of N+/P+ gates on N- and PMOS respectively). New developments might also be opened by the use of selective epitaxial growth for making elevated drain/source, which would reduce the transistor serial resistance and modify the electric field at the drain end and improve the avalanching problem. The most important effort of research and imagination is however probably to be put on the suppression of the parasitic lateral bipolar and the SOI single latch.

REFERENCES

/1/ J-L.Leray et al. 1989 IEEE SOS/SOI Technology Conference, pp 114-115, 1989.
/2/ A-J.Auberton-Hervé, Proc. of the Fourth Int. Symp.on SOI Techn. and Dev., pp.455-478, ECS Montreal, 1990.
/3/ Y.Ushiku et al., IEDM Tech.Dig., pp.340-343, 1988.
/4/ M.Haond, Proc. of 1990 IEEE SOS/SOI Technology Conference, pp.117-118, 1990.
/5/ J.Alderman, these Proceedings, 1991.
/6/ J.R.Davis et al. IEEE Electron.Dev.Lett., vol.EDL-7, pp. 570-572, 1986.
/7/ M.Haond et al., Microcirc. Eng.n°8, pp. 201-218, 1988.
/8/ J-P.Colinge, IEEE Electron Device Letters, vol. EDL-9, pp. 97-99, 1988.
/9/ H.Lim and J.Fossum, IEEE Trans.Electron.Dev., vol.ED-30, pp. 1244-1251, 1986.
/10/ M.Haond and M.Tack, IEEE Trans. Electron.Dev., vol.ED-38, pp. , 1991.
/11/ M.Haond, unpublished results, 1990.
/12/ S.M.Sze, in Physics of Semiconductor Devices, J.Wiley Eds, p.447, 1936.
/13/ J-P.Colinge, IEEE Electron Dev.Lett., vol.EDL-7, pp 244-246, 1986.
/14/ M.Yoshimi et al., Electron.Lett., vol.24, pp.1078-1079, 1988.
/15/ A.Yoshino, Proc. of the Fourth Int. Symp.on SOI Techn. and Dev., pp.544-553, ECS Montreal, 1990.
/16/ S.D.S.Malhi et al., IEDM Tech.Dig., pp. 107-110, 1982.
/17/ M.Haond, Proc. of ESSDERC 89, pp 881-884, 1989.
/18/ J.R.Davis et al., IEEE Trans.Electron.Dev., vol.ED-38, pp. 32-38, 1991.
/19/ T.Skotnicki et al., IEEE Trans.Electron.Dev., vol.ED-36, pp. 2546-2556, 1989.
/20/ O.Le Néel and M.Haond, Electron.Lett., vol.26, pp.73-74, 1990.
/21/ L.McDaid et al., Electron.Lett., vol.25, pp.827-828, 1990.
/22/ H.Lifka and P.H.Woerlee, Proc. of ESSDERC 90, pp. 453-456, 1990.
/23/ R.Ostermeier et al., Proc. of ESSDERC90, pp.591-594, 1990.
/24/ O.Le Néel, unpublished results, 1989.
/25/ M.Haond and J-P.Colinge, Electron. Lett., vol.25, pp.1640-1641, 1989.
/26/ Y. Yamagushi et al., Proc. of IEEE 1990 SOS/SOI Technology Conference, pp 23-24, 1990.
/27/ F.J.Lai, L.K.Wang, Y.Taur, J.Y-C.Sun, K.E.Petrillo, S.K.Chicotka, E.J.Petrillo, M.R.Polcari, T.J.Bucelot, and D.S.Zicherman, IEEE Trans. Electron. Dev., vol.ED-33, pp.1308-1319, 1986.
/28/ J.Sturm, K. Tokunaga and J-P.Colinge, IEEE Electron Device Letters, VOL.9, N°9, 1988, pp 460-463
/29/ H.Miki et al., IEEE Trans. Electron.Dev., vol.ED-38, pp. 373-377, 1991.
/29/ P.H.Woerlee et al., IEDM Tech.Dig., pp. 821-824, 1989.
/31/ T.Ouisse et al., Proc. of ESSDERC 90, pp. 257-260, 1990.
/32/ P.H.Woerlee et al., IEDM Tech.Dig., pp 583-586, 1990.

Poly-Si thin film transistors

S D Brotherton

Philips Research Laboratories, Redhill, Surrey RH1 5HA

> The different low temperature poly-Si technologies of columnar deposition, thermal crystallisation and laser crystallisation are discussed in this paper in the context of the different final grain structures. Features of device behaviour which are specifically correlated with the particular grain structure are presented as well as features common to all types, such as field enhanced leakage currents and output characteristic degradation.

1. INTRODUCTION

The present interest in poly-Si thin film transistors results from the developing technology of large area electronics on glass. The applications of this technology are in displays, sensors and printers, of which flat panel displays is one of the most industrially advanced. These displays are addressed by a matrix of active devices (TFTs or diodes) with at least one device per picture point. The dominant, and most wide spread, technology at the moment makes use of amorphous silicon thin film transistors. However, one limitation in the fabrication and assembly of these display panels is the requirement to make an external connection to each row and column of the display with peripherally mounted integrated circuits. For a full resolution display this amounts to approximately 1300 connections. If the scanning and addressing functions of the circuits could be fabricated on the same substrate as the display devices, the number of external connections would be reduced by one to two orders of magnitude. The prime requirement for these circuits is that the TFT has a sufficiently large mobility to allow the row and column shift registers to work at $\sim$30kHz and $\sim$11MHz, respectively, and for the output driver transistors to be able to charge the appropriate row or column capacitances. In simple terms, this requires the field effect mobility to be in excess of $\sim$1cm^2/Vs for row drivers and >25cm^2/Vs for column drivers with a limited degree of matrixing. The electron field effect mobility of amorphous silicon is <1cm^2/Vs, being typically within the range 0.3-0.7cm^2/Vs, whereas poly-Si TFTs exhibit mobilities over the range 5->100cm^2/Vs. The large spread in quoted mobilities for poly-Si is, in part, due to the different technologies being used to produce this material. The differences in grain structure resulting from these technologies are discussed below together with the common features, as well as the differences in device behaviour which can be related to the grain structure.

2. POLY-SI TECHNOLOGIES

© 1991 IOP Publishing Ltd

The technologies to be summarised involve low temperature processing which is compatible with the use of alkali free alumino-borosilicate glass substrates with strain points of ~640°C. The three major approaches are:

i) columnar poly-Si, in which the material is deposited in a fine grain (<0.1µm) form by low pressure CVD at temperatures of 600-620°C. TFTs made from this material typically have mobilities of 5-10cm^2/Vs[1]

ii) thermally recrystallised α-Si, in which the pre-cursor material is amorphous and can be obtained from LPCVD layers deposited at ~540°C[2], PECVD material deposited at ~250°C[1] or implant amorphised poly-Si[3,4]. In all cases, the amorphous starting material converts to large grain (~1µm) poly-Si as a result of random nucleation at anneal temperatures of ~600°C. Mobilities quoted for this material range from ~20-80cm^2/Vs

iii) finally, poly-Si may be obtained by laser crystallisation of amorphous starting material[5-7]; depending upon the detailed conditions, the mobility can range from 10-175cm^2/Vs. One feature of this approach that distinguishes it from the others is the fact that the use of a laser permits selective local recrystallisation of material, so that poly-Si drivers could be fabricated on an α-Si:H TFT plate[5]. The other two technologies would yield monolithic, rather than hybrid, display and driver TFTs.

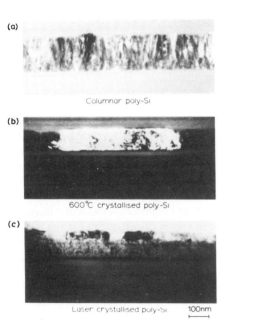

FIG.1 Cross section TEM micrographs   FIG.2 Poly-Si TFT transfer characteristics

These three different approaches result in significantly different grain structure as illustrated by the cross-sectional TEM micrographs in figure 1. As might be expected, the different grain structure can result in detailed differences in device behaviour, although with appropriate processing similar device characteristics may be obtained with all three technologies as illustrated in figure 2. The major difference in these characteristics is the lower on-current in the columnar device due to the lower carrier mobility. The device structures fabricated with these technologies are shown in figure 3; the auto-registered structure was used with the columnar and thermally recrystallised material and the channel etched, coplanar device was used for the laser annealed structures. The laser recrystallisation was accomplished with a KrF excimer laser with the sample mechanically scanned through the laser beam. The beam size was approximately 0.6x2.0cm^2 and the samples were scanned in the 'long' direction yielding 0.6cm wide stripes. The laser wavelength was 248nm and it was operated at a low duty cycle of a few hertz with a pulse width of 20ns.

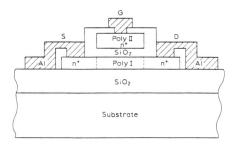

Auto-registered TFT

3. LEAKAGE CURRENT

A common feature of all three poly-Si technologies is the way in which the device leakage current increases with negative gate bias. As previously discussed[1], this behaviour, for columnar material, is a strong function of trap passivation with the characteristics shown in figure 2 being observed in fully passivated structures.

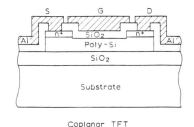

Coplanar TFT

FIG.3 Schematic cross sections of TFT structures

The leakage current characteristic shown in figure 2 after full hydrogenation has been reported by most workers on poly-Si TFTs and the mechanism for this field enhanced generation has been variously attributed to direct tunnelling[8], Poole-Frenkel emission[9] and phonon-assisted tunnelling[10] from traps in the drain space charge region. As shown below, the interpretation of our data is only consistent with the final mechanism. The leakage current characteristics shown in figure 2 were found to be independent of source-drain separation over the measured range of 4-20μm, confirming that the effect was associated with generation at the drain junction, rather than being a hole channel effect. The leakage current activation energy was strongly dependent upon both gate and drain bias as shown by the results in figure 4 and the leakage current at a given value of V_G rose exponentially with drain bias as shown in figure 5. The thermal activation of the leakage current would rule out a simple tunnelling process from traps[8], but the exponential rise of drain current with V_D is consistent with the approximate expression[11] for carrier emission due to phonon associated tunnelling. For the Poole-

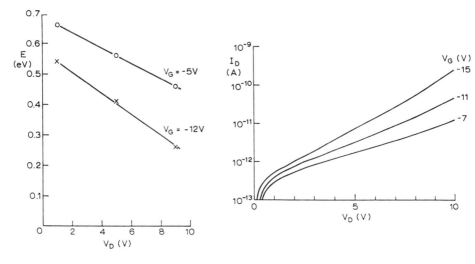

FIG.4. Leakage current activation energy as a function of V_D and V_G

FIG.5 I_D-V_D plots of fully hydrogenated TFT in off-state

Frenkel mechanism the emission rate increases exponentially with $\sqrt{E}$, where E is the field. Assuming that the field is developed in a space charge region near the drain, the emission rate should vary with $\exp V_D^{1/4}$. No samples have shown a linear relationship between $\ln I_D$ and $V_D^{1/4}$; this would argue against the Poole-Frenkel mechanism. This conclusion was further supported by the activation energy variation with V_D in figure 4. Illustrative curves are shown in figure 6 of the calculated dependence of the activation energy of the carrier emission rate versus drain bias for the Poole-Frenkel and the phonon-assisted tunnelling mechanisms. The functional dependence of the Poole-Frenkel effect upon drain bias is inconsistent with all our measured data. However, the Poole-Frenkel effect will dominate at lower fields and it is interesting to note that the results of Madan and Antoniadis[9], reporting the Poole-Frenkel effect in their TFTs, were obtained at lower drain biasses then used in our measurements. Hence the interpretation of the two data sets may not be inconsistent.

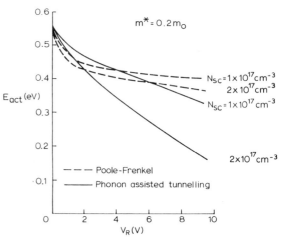

FIG.6 Calculated dependence of leakage current activation energy on reverse bias

4. LARGE GRAIN DEVICES

As shown in figure 1, large grain poly-Si material is obtained following the thermal crystallisation of amorphous silicon, as reported by many workers[2-4]. The primary interest in this material is due to the enhanced carrier mobility which increases with grain size[1-4]. However, the large grain material can also have a deleterious effect upon leakage current if the film is too thick. This is demonstrated by the results in figure 7 comparing TFTs made on 1500Å of fine grain columnar and thermally crystallised large grain poly-Si. In the latter devices, the leakage current scaled inversely with source-drain separation and the I_D-V_D curves were linear. Both features are consistent with an electron bulk channel being present due to partially depleted grains in the large grain material. The effect of grain size on layer resistivity is readily calculated for 1-D bulk conduction and calculations are shown in figure 8 for a background shallow donor density of 2×10^{15}cm^{-3}. This dopant incorporation is assumed to occur unintentionally during layer deposition. The source material for the results in figure 7 was PECVD α-Si, but

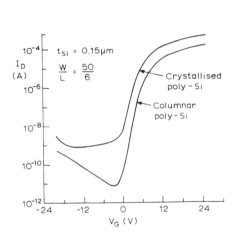

FIG.7 Comparison of transfer characteristics for fine grain columnar and large grain recrystallised poly-Si TFTs

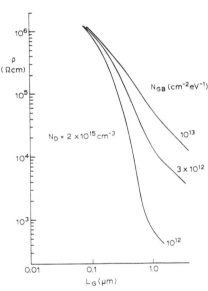

FIG.8 Calculated poly-Si film resistivity versus grain size

similar results have also been seen with α-Si layers obtained from LPCVD and MBE deposition systems. The calculations in figure 8 are for a uniform distribution of donor and acceptor grain boundary trapping states in the lower and upper halves of the band gap, respectively. As will be seen, with fine grain material (<0.1μm) the films are fully depleted of free carriers and are intrinsic, whereas in films with 1μm grains the resistivity is a sensitive function of the grain boundary trap density and can be orders lower. The leakage current activation energy for the large grain device in figure 7 was 0.36eV and independent of a wide range of gate and drain biases. A comparable activation energy is calculated for 1μm grain material assuming a donor incorporation density of 2×10^{15}cm^{-3} a grain boundary trap density of 1×10^{12}cm^{-2}eV^{-1}. To date we have no independent chemical measurement of unintentional donor incorporation below the SIMS detection level of $\sim 5\times 10^{16} \text{cm}^{-3}$ for phosphorus. Therefore the calculations in figure 8 should be regarded as illustrative rather than accurate modelling. For this reason nothing more complex than a uniform distribution of trapping states was put into the model.

Low leakage currents in these large grain films, as shown by results already published[1] and those in figure 2, were obtained by reducing film thickness to $<0.1\mu\text{m}$ so that the bulk electron population could be fully depleted at negative gate bias. Reference to the literature will indicate that most workers are also using films in this thickness range.

5. LASER ANNEALED DEVICES

In the preceding section, effects were discussed which could be directly correlated with the particular grain structure of the large grain material. Similarly, in this section results will be presented which are particular to the grain structure resulting from excimer laser annealing. The cross sectional TEM micrograph of a laser annealed specimen in figure 1b, shows the film to be stratified into a larger grain ($\sim 800\text{Å}$) near surface layer and an underlying fine grain (0-250Å) region. The former region is believed to be the result of surface melting by the strongly absorbed 248nm radiation (absorption depth $\sim 60\text{Å}$), whilst the lower layer has undergone solid phase recrystallisation. The thickness of the large grain surface layer is a strong function of incident laser energy as shown in figure 9. The data points were obtained both from direct TEM measurements and also inferred from SIMS measurements in which the movement of a 200Å thick boron doped marker layer, on the surface of a 1500Å undoped α-Si:H layer, was taken to be indicative of the melt depth. The implicit assumption being that the boron moved rapidly in the molten phase but had limited solid state diffusion. The agreement between the TEM and the SIMS results supports this assumption. The results show that the threshold for melting was of the order of 100mJ/cm^2/pulse and that the molten layer attained a limiting maximum thickness of 600-700Å. The pulse energy values should only be regarded as approximate since the energy measurement averaged over the total beam area and did not monitor the spatial energy profile within the beam. For this reason it is inappropriate

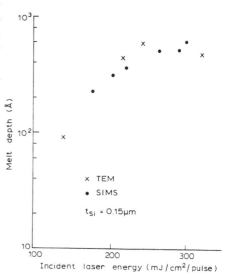

FIG.9 Poly-Si melt depth versus incident excimer laser power

to make a detailed comparison with other published values of the threshold energy density for melting[6,12], except to note that they are comparable. The limiting maximum melt depth with incident laser energy has also been observed by other workers[12] and is a function of the shallow absorption depth of the radiation and the detailed heat loss and conduction mechanisms within the particular structure.

The grains within the large grain surface region are of much greater crystallographic perfection than the large dendritic grained material obtained by thermal recrystallisation at 600°C. This is possibly why the laser annealed TFTs do not require exposure to a hydrogen plasma to passivate trapping states. One can readily anticipate that a sufficient depth of large grain material will be required for high quality TFT characteristics and this is illustrated in figure 10. The particular feature of interest in these curves is the progressive degradation in the sub-threshold slope near the threshold voltage as the laser energy is reduced. These results are interpreted in terms of the large grain layer thickness near the melt threshold. For gate biasses just above zero it will be easy to bend the bands at the surface whilst the band bending is contained within the large grain material, but once this material is depleted and the band bending starts to extend into the underlying fine grain material, the Fermi level is likely to be pinned by the high trapping state density. This will cause a corresponding degradation in sub-threshold slope. The magnitude of this effect will diminish with increasing thickness of the large grain layer. As will be seen from the results, the degradation progressively reduced with increasing power over the range 90-163mJ/cm^2/pulse. From this it can be concluded that at 163mJ/cm^2/pulse the depletion layer at inversion was substantially contained with the large grain material. Changes in mobility with power beyond this energy, as illustrated in figure 11, are assumed to be due to the increase in mean grain width. Only limited TEM results are available, but mean grain width has been found to increase from 600Å to 900Å for recrystallisation energies of 230 and 350mJ/cm^2/pulse, respectively.

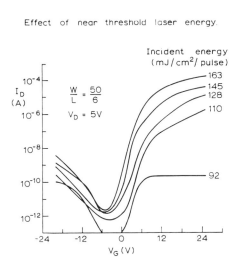

FIG.10 Influence of excimer laser energy on TFT transfer characteristics

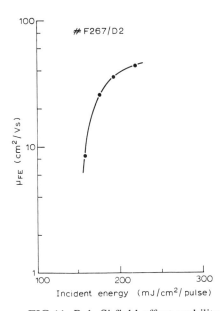

FIG.11 Poly-Si field effect mobility versus incident excimer laser power

6. OUTPUT CHARACTERISTICS

A common feature found with all poly-Si TFTs is degradation in the output characteristic at sufficiently large drain bias as shown in figure 12. Also illustrated in this figure is the strong dependence of the effect on device channel length for devices with L=6, 10, 20µm. This effect is not simply drain breakdown, which for avalanching should be channel length dependent. Indeed, what is assumed to be avalanche breakdown can be measured in the off-state and the values were between 40V and 50V, with only a weak channel length dependence. For smaller channel lengths of 2 and 3µm much lower breakdown voltages of 15-20V were found which could be due to punch-through. A possible explanation for the effect seen in figure 12 is the parasitic bipolar effect as reported for single crystal SOI devices[13]. This interpretation requires weak avalanching of the channel carriers and the holes thus injected into the floating TFT body result in further electron injection by an amount which corresponds to the bipolar gain of the device. For gain limited by recombination in the base, the gain and hence amount of excess current would be base width (channel length) dependent. This model has also been invoked for poly-Si by other workers[14]. One problem with this model is that the gain values inferred from the results in figure 12 would require a minority carrier recombination lifetime of ~1µs. This is considerably larger than the generation lifetimes of <1ns estimated from leakage currents[1]. Further evidence that the current degradation involves avalanching, albeit at a weak level, is that hot carrier degradation of the device occurs at the same time[16]. The TFT degradation and its correlation with the enhanced current are further discussed in an accompanying paper[17]. In view of the uncertainties of the parasitic transistor model it needs to be supported by detailed numerical modelling.

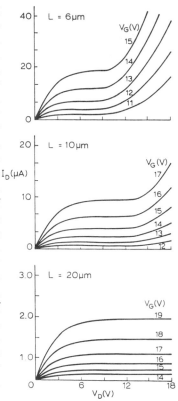

FIG.12 Variation of output characteristic degradation with TFT channel length

6 CONCLUSION

Poly-Si TFTs afford the opportunity to extend the range of large area electronics on glass applications due to their higher mobility than the currently dominant α-Si:H TFT technology. In particular, the integration of scanning and driving circuits in active matrix addressed displays will simplify the mounting and assembly of the display module. The various technological approaches to the fabrication of poly-Si TFTs are reviewed in this paper and some of the common issues such as field enhanced leakage currents and output characteristic degradation are discussed. In addition, because the different technologies result in different grain structure there are features of the device behaviour which are particular to a given grain structure. Examples given of this are the higher leakage currents which can occur in large grain material and sub-threshold slope degradation in laser annealed devices if the depth of recrystallisation is too shallow.

ACKNOWLEDGEMENTS

I would like to acknowledge the contribution of Mrs A Gill for sample fabrication, Dr D J McCulloch for the laser annealed samples and Dr J P Gowers and Mr J B Clegg for the TEM and SIMS analyses, respectively. In addition, there have been numerous valuable discussions with my close colleagues in this work, Dr N D Young and Mr J R Ayres.

REFERENCES

1. S D Brotherton, J R Ayres and N D Young. To be published in Solid State Electronics, 1991.
2. A Mimura, N Konishi, K Uno, J-I Ohawda, Y Hosokawa, Y A Ono, T Suzuki, K Miyata and H Kawakami. IEEE Trans. ED-32, 258, (1985).
3. T Noguchi, H Hayashi and T Ohshima. Jap. Jnl. Appl. Phys., 25, L121, (1986).
4. R B Iverson and R Reif. Jnl. Appl. Phys., 62, 1675, (1987).
5. K Sera, F Okumura, H Uchida, S Itoh, S Kameko and K Hotta. IEEE Trans ED-36, 2868, (1989).
6. T Sameshima, M Hara and S Usui. Jap. Jnl. Appl. Phys., 28, 1789, (1989).
7. M Yuki, K Masumo and M Kunigita. IEEE Trans ED-36, 1934, (1989).
8. J G Fossum, A Ortiz-Conde, H Shichijo and S K Banerjee. IEEE Trans. ED-32, 1878, (1985).
9. S K Madan and D A Antoniadis. IEEE Trans ED-33, 1518, (1986).
10. D W Greve, P A Potyraj and A M Guzman. Solid-St. Electron., 28, 1255, (1985).
11. G Vincent, A Chantre and D Bois. Jnl. Appl. Phys., 50, 5484, (1979).
12. K Winer, G B Anderson, S E Ready, R Z Bachrach, R I Johnson, F A Ponce and J B Boyce. Appl. Phys. Letts., 57, 2222, (1990).
13. M-H Gao, J-P Colinge, S-H Wu and C Claeys. ESSDERC '90, 445, (Ed. W. Eccleston, P Rosser, Published by Adam Hilger, 1990).
14. M Hack and J G Shaw. Extended abstracts for 22nd Conference on Solid State Devices and Materials, Sendai, 1990, p.999.
15. N D Young and A Gill. Semiconductor Sci. and Technol., 5, 728, (1990).
16. N D Young, S D Brotherton and A Gill. Accepted for publication in the proceedings of INFOS 91.

Paper presented at INFOS '91, Liverpool, April 1991
Contributed Papers, Section 1

Modelling of individual interface states in MOSFETs

M Schulz

Institut für Angewandte Physik, Universität Erlangen-Nürnberg,
Staudtstr. 7, D-8520 Erlangen

ABSTRACT: The potential perturbation of a single, individual interface point defect a few nanometers deep in SiO_2 near the Si interface is analyzed by a computer simulation. A repulsively charged trap punches a hole into the carrier density of the channel. The potential perturbation is only weakly (60%) screened up to carrier densities of $n = 2 \times 10^{12} cm^{-2}$ in the channel

1. INTRODUCTION

Random telegraph switching (RTS) of the source-drain current in μm-sized MOSFETs provides a unique opportunity to study individual defect centers in the SiO_2-Si interface (Ralls 1984, Kirton 1986). The present understanding of the random telegraph switching is reviewed in several recent papers (Kirton 1989, Schulz 1990, Schulz 1991). The interface states causing the random switching reside in the oxide a few nanometers away from the interface to silicon. Most frequently only a single electron or hole is captured and re-emitted by these states to and from the MOSFET channel. It has been shown that the 1/f noise is generated through a superposition of random telegraph signals of many individual defects (Kirton 1989).

A rather complex dependence on the gate bias voltage is observed for the trapping and emission rate constants (Ralls 1984, Schulz 1990). The transfer rates are strongly ($\Delta E > 700 meV$) temperature activated (Karmann 1989). The energy depth of the level occasionally strongly increases with increasing gate bias voltage and the trapping rate is reduced with increasing carrier density in the channel. Such a behavior is in strong contrast to the behavior of traps in bulk semiconductors.

Computer simulations of the potential and the charge distributions in MOSFETs around an individual defect center are presented in this paper. These simulations help to visualize the complex effects occurring in the interface region. The computations are performed for a Coulomb-repulsive defect center which is in a neutral state when unoccupied, or in a repulsive charge state, when a mobile carrier is trapped from the channel. Coulomb-attractive centers, for which a bistability is observed, are discussed in a separate paper of this volume (Karmann 1991).

2. COMPUTATIONAL DETAILS

The potential distribution of a point charge in a MOSFET is determined by solving Poisson's equation numerically, using the finite element method. A single, isolated point charge q is assumed a distance $t = 1..4 nm$ into the oxide in front of the plane SiO_2-Si interface. Because of the cylindrical symmetry of the problem, the potential $V(r,z)$ and the carrier concentration $n(r,z)$ are only a function of the radius r in plane of the interface

© 1991 IOP Publishing Ltd

and of the depth z normal to the interface. The minimum size of the finite elements in the computation is 1x1nm^2 in the regions where the potential strongly varies, i. e. around the point charge and in the channel area. A classical Boltzmann type distribution of mobile carrier density is assumed in the channel. Calculations are performed for room temperature T=293K where effects of two-dimensional subbands and quantum effects only induce corrections but do not change the general trend of the results. Further details of the program and the computational method will be published elsewhere (Kroneder 1991).

3. POTENTIAL AND CHARGE DISTRIBUTIONS

A typical potential profile normal to the interface in a cross section through a positive point charge is shown in Fig. 1. The electron and hole potentials are plotted as a function of depth z from the gate by following the band edges E_C, E_V of the conduction and valence band, respectively. The band offsets at the interface are not depicted for simplicity. The Fermi energy E_F shown is calculated for an n-type substrate doping of $N_D = 3.3 \times 10^{16} cm^{-3}$. A p-channel is induced at the interface for sufficiently high gate bias voltage. The unperturbed voltage far away from the point charge is shown for comparison (dashed line). The reference potential for the Coulomb well of the point charge (dotted line) and the trap level (full dot) are also shown. It is noticed that the interface potential and the trap potential are lowered by the perturbation. Holes are thus repelled from the channel area facing the trap.

A complex feedback mechanism occurs for the trap level occupation and the potential variations at the trap and at the interface. The trap is rapidly filled when the level is raised through the Fermi energy by an increasing gate voltage. The energy separation $E_T - E_{V_C}$ of the trap level from the band edge at the interface is then increased with the gate bias voltage. For a low voltage drop between the trap level and the interface, this increase of the energy separation does not occur. A low voltage drop is expected when the channel is fully depleted in the vicinity of the trap, thus leading to a low electric field normal to the interface.

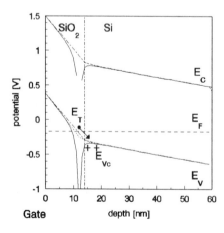
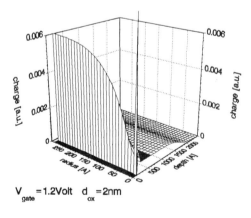

Fig. 1 Energy diagram normal to the interface through an elementary positive point charge t=2nm off the interface.

Fig. 2 Space charge distribution in the MOSFET with an elementary positive point charge t=2nm off the interface. $V_{Gate} = 1.2V$

A two-dimensional plot of the space charge distribution is shown in Fig. 2. The typical constant doping charge is visible in the space charge region. A sheet of mobile carriers is induced in the channel region. The point charge is depicted by the sharp spike in the oxide layer. The magnitude of the spike is reduced to 10% of its real value to approximately fit into the scale. The point charge in the oxide punches a hole into the carrier density of the channel. The radius of the hole is approx. r=200A. The drop in carrier density may be several orders of magnitude at the interface, depending on the distance of the trap in the oxide. It is obvious that the point charge strongly perturbs the charge carrier density in the channel. The Fermi energy generally is pinned to the trap rather than to the channel in the region around the point charge.

4. SCREENING

The potential induced at the interface by the point charge is partially screened by the mobile charge carriers, which rearrange in the channel to counterbalance the local fields. The screening of the two-dimensional channel, however, is much weaker than the well-known screening of a three-dimensional electron gas in a bulk semiconductor. The two-dimensional electron gas of the channel cannot surround the point charge. Electric fields therefore can spread unscreened in the oxide.

The Fig. 3 shows a computed contour plot of the potential perturbation induced in the MOSFET by the point charge. The uniform field gradient of the gate bias voltage $V_G = 1.6$ Volts is subtracted to depict the perturbation voltage ΔV only. In spite of a carrier density of $n = 6.5 \times 10^{11} \text{cm}^{-2}$, the potential still penetrates into the semiconductor even at a radial distance $r = 100..150 \text{A}$ away from the point location at $r = 0$ and depth $z = 130 \text{A}$. The kink in the potential contour indicates partial screening.

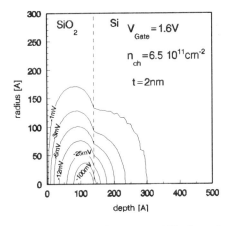
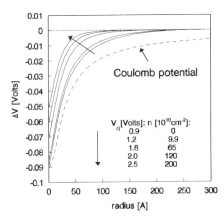

Fig. 3 Potential contours ΔV of a point charge perturbation in the MOSFET. The calculation data are indicated in the inset. The interface is marked by the dashed line. The vertical line indicates the position of the interface.

Fig. 4 Potential perturbation ΔV in the interface. Parameter is the carrier density in the channel, which causes screening. The dashed curve shows a Coulomb potential.

The screened potential distribution may be described by fractional image point charges behind the gate and behind the interface. Due to the parallel plate configuration of the gate and the interface, an infinite number of image charges occurs at even multiples of the oxide thickness. It can be shown that the trap energy is perturbed by its own image charge potential. The radial decay of the potential perturbation is more rapid than a Coulomb potential, because the multiple image charges create dipoles and multiple linear quadrupoles, for which the potential decays at a high power of the inverse radius. The screening is ineffective on these steep potential tails in contrast to earlier estimates which did not take into account the multiple image charges (Ando 1982).

The quantitative effect of screening is depicted in Fig. 4. This figure shows the radial decay of the potential perturbation in the interface. For a point charge $t=2$nm in the oxide, the potential maximum at $r=0$ is $V=90$mV. The perturbation decays below $V=1$mV at a radial distance $r=300$A. The Coulomb potential decays much slower (dashed curve). A carrier density in the channel of $n=10^{11}$cm^{-2} barely affects the potential. For a high carrier density of $n=2\times10^{12}$cm^{-3}, the interface potential still reaches 60% of its unscreened value and reaches out to $r=100$A. This weak screening effect is due to the strong localization of the potential perturbation and due to the restricted mobility of the carriers in the two-dimensional plane.

5. CONCLUSIONS

A computer simulation is performed to visualize the properties of an individual trap in the oxide a few nanometers deep from the interface in the MOSFET. The computations show, that the point charge punches a hole a few tens of nanometers wide into the mobile charge carrier density of the channel. The surface potential reaches a magnitude of $V\simeq 100$mV$\simeq 4$kT at room temperature $T=293$K for a point charge location at $t=2$nm from the interface. The potential perturbation is barely screened by mobile carrier densities in excess of $n=10^{12}$cm^{-2}. Due to multiple image charges, the potential decays rapidly in the interface plane.

REFERENCES

Ando T, Fowler A B, and Stern F 1982 *Rev. Mod. Phys.* **54** 437
Karmann A and Schulz M 1989 *Proc. Of the Int. Conf. on Insulating Films on Semicond.* eds F Koch and A Spitzer (Amsterdam: North-Holland) 500
Karmann A and Schulz M 1991 *this volume*
Kirton M J and Uren M J 1986 *Appl. Phys. Lett.* **48** 1270
Kirton M J and Uren M J 1989 *Adv. Phys.* **38** 367
Kroneder C and Schulz M to be published
Ralls K S, Skocpol W J, Jackel L D, Howard R D, Fetter L A, Epworth R W, and Tennant D M 1984 *Phys. Rev. Lett.* **52** 228
Schulz M and Karmann A 1990 *Proc. of the 5th Cong. of the Brasilian Society of Microelectronics SPIE* **140** 2
Schulz M and Karmann A 1991 *Physica Scripta* accepted for publication
Schulz M and Karmann A 1991 *Appl. Phys.* **A-52** 104

Tunneling resonance by barrier symmetrisation tuning at the Si–SiO$_2$–interface

T. Poppe[1], M. Bollu[2], F. Koch
Physikdepartment, Techn. Universität München, D-8046 Garching, FRG

ABSTRACT: Electrically stressed n-channel MOSFETs are used to investigate resonant tunneling via single localized states at the Si-SiO$_2$-interface. A given state can be observed as a resonance in two distinct modes: As a conductance peak in a gate voltage sweep, and as a peak in the current-drain voltage-characteristic. Here we will focus on the second mode.

1. INTRODUCTION

The well developed silicon technology provides a good tool for studying transport in very small systems where only several conduction paths are active. Resonant tunneling via single localized states has been observed in the channel conductance of short silicon MOSFETs (Fowler et al. (1986), Kopley et al. (1988)). These experiments are done by a gate voltage sweep. Resonance is achieved if the energy level of a proper localized state at the interface matches the Fermi-level. The drain voltage is thereby kept smaller than the linewidth of the resonant state devided by the elementary charge.

In Koch et al. (1988) we identified conductance peaks appearing in a gate voltage sweep of electrically stressed MOSFETs as tunneling resonances. In analogy to double barrier quantum wells, one expects at a fixed gate voltage V_G, close to a resonance, a resonance condition also by choosing a proper drain voltage. Then, in addition to the energetical tuning, a redistribution of the potential drop along the two confining barriers of the resonant state can modify the current-drain voltage-characteristic. This effect has been discussed theoretically by Ricco and Azbel (1984) for the one dimensional case, and by Zou et al. (1988) for a double barrier 2-dimensonal quantum well.

2. EXPERIMENTS

We used conventional silicon n-channel MOSFETs with a 42nm thick dry oxide and a poly-silicon gate. The effective channel length and width are 1.8um and 10um, respectively.

Electrical stressing of these devices at high drain voltages and moderate gate potential leads predominantly to the creation of interface states close to the drain contact (Asenov et al (1987)). At temperatures in the liquid helium regime all interface states are negatively charged. This results in the formation of a barrier, about 100nm in length, for the channel electrons. We work with a stressing intensity to produce an interface state density of the order of $10^{12}/cm^2$. By this means we fabricate devices with low temperature effective channel lengths of about 100nm.

3. RESULTS AND DISCUSSION

A typical conductance vs. gate voltage characteristic at 4.2K of a stressed device is plotted in Fig.1a. The conduction onset of the unstressed devices is at $V_G=0.9V$. After stress, we observed spikes like that one at $V_G=3.18V$ up to 4 orders in magnitude. Looking at the temperature dependence of the spike height and shape, we explained in Koch et al. (1988) each spike by resonant tunneling via a zero-dimensional electronic state within the barrier. The peak height decreases, and the flanks become less steep in a manner as predicted by Stone and Lee (1985) when the temperature is increased. The peak structures in our samples are well pronounced up to temperatures of about 40K, which is much larger than usually reported (Fowler et al.(1986), Kopley et al. (1988)). The reason for this is the small effective channel length of our stressed devices, which favours resonant tunneling compared to variable range hopping.

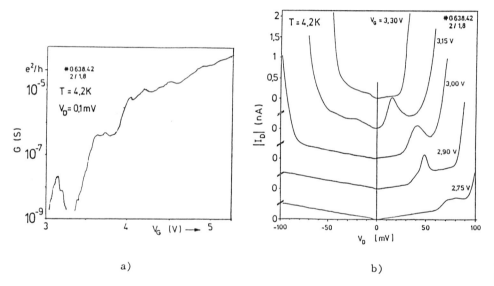

Fig.1: a) Channel conductance vs. gate voltage of a stressed n-channel MOSFET b) corresponding current-drain voltage characteristics. Curves for different gate biases are offset.

Choosing a fixed gate voltage just below an isolated peak the current-drain voltage-curve also reflects the resonance. For the peak in Fig.1a at 3.18V this characteristic is shown in Fig.1b. The parameter herein is the gate voltage. If it is chosen smaller than the actual peak position, a resonant tunneling structure appears at a positive drain voltage, whereas at $V_G=3.3V$ it does not. This observation is what one expects in analogy to 2-dimensional double barrier structures. The asymmetry, that no peak appears at negative drain voltages, is not surprising. A localized state need not lie in the centre of the barrier (Fig.2). The strong increase of the current for higher drain voltages in Fig.1b is due to overall transmission over the barrier.

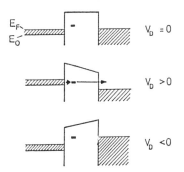

Fig.2: Scematic diagram of the tunneling paths in a stressed MOSFET. The gate voltage is adjusted just below the a resonance peak in the V_G-scan. Arrows indicate the paths which can lead to a peak in current-drain voltage-sweep. The position of the localized state is qualitatively chosen as to explain the data in Fig.1.

The picture depicted above is not complete. The peak height in the drain voltage sweep is about one order of magnitude larger than expected from the integral over the line broadening of the localized state whereby the linewidth is known from the temperature behavior of the peak in the V_G scan. In a drain voltage sweep not only the energy of the localized state is shifted, but also the effective barrier thicknesses on both sides of the state may be modified differently. The total transmission is proportional to the ratio T_{min}/T_{max} of the transmission coefficients of these barriers. Therefore, a localized state closer to the channel than to the drain will enlarge the total transmission if a positive drain bias is applied by the so called barrier symmetrisation (Ricco and Azbel (1984), Zou et al. (1988)). The specific form of a resonant tunneling structure in a drain voltage sweep depends on the interplay of the energy resonance condition and the barrier symmetrisation. The current at a drain voltage which enables energetical resonance is then determined by the degree of barrier symmetrisation at this particular drain bias. Of course both mechanisms are coupled in a way that by increasing the drain bias, the barrier partition is changed, and hence the speed at which the localized state follows the drain bias, too.

The unexpected large peak current in Fig.1b gives a strong evidence for barrier symmetrisation and has been observed in most of our devices which exhibited pronounced peaks in the V_G-sweep. For the example of Fig.1, a resonant state feeling a lower barrier to the channel side must be assumed (Fig.2). The applied biases for observing the resonance in both modes shown in Fig.1 are consistent with the known physical device parameters like the 2-dimensional density of states, which is a lower limit for the energy shift of the resonant state with the gate voltage, and the Fermi energy in the channel of 8meV at V_G=3V. Stressed devices with resonances at negative drain voltages have also been observed. The much wider energy window of the degenerate drain region usually broadens the peaks appearing at negative drain voltages.

Due to the zero-dimensional nature of our resonant states, no bistability as described by Goldman et al (1987) for a two-dimensional quantum well is expected. This is in accordance with our findings. Nevertheless, in about 10% of the stressed devices we found a quite regular peak sequence which we interpret as multiple occupation of one potential minimum in the barrier. In Fig.3 an example is given where at the two longest stress times a quasi periodicity is obvious.

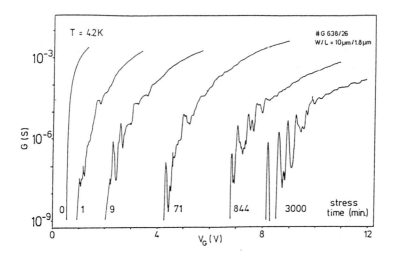

Fig. 3: Conductance vs. gate voltage for a MOSFET after successive stress cycles

4. CONCLUSIONS

At liquid helium temperatures, electrically stressed MOSFETs act as devices with an effective channel length of about 100nm. They have been used to study resonant tunneling via single localized states at the Si-SiO_2-interface. We found a strong evidence for enhanced resonance transmission by barrier symmetrisation.

We thank M. Raikh for valuable discussions, and the Siemens AG for financial support.

Asenov A, Bollu M, Koch F, Scholz J 1987 Appl. Surf. Sci. 30 319
Fowler A B, Timp G L, Wainer J J, Webb R A 1986 Phys. Rev. Lett. 57 138
Goldman V J, Tsui D C and Cunningham J E 1987 Phys. Rev. Lett. 58 1256
Koch F, Bollu M, Asenov A 1988 MOSFETs under electrical
   stress, in Springer Series in Solid State Sciences No.83
   ed. H.Heinrich et. al., Springer Berlin 1988 p.253
Kopley T E, McEuen P L, Wheeler R G 1988 Phys. Rev. Lett. 61 1654
Ricco B, Azbel M Ya 1984 Phys. Rev. B29 1970
Stone A D, Lee P A 1985 Phys. Rev. Lett. 54 1196
Zou J, Xu J and Sweeny M 1988 Semicond. Sci. Technol. 3 819

1) now with Bosch, Reutlingen, FRG
2) now with Siemens AG, Research and Development, D-8000 München 83, FRG

Hole trap analysis in SiO_2/Si structures by electron tunneling

M Schmidt (a) and H Köster jr. (b)

(a) Zentralinstitut für Elektronenphysik, Rudower Chaussee 5, O-1199 Berlin
(b) Gerling-Institut, Am Friesenwall 89, W-5000 Köln 1

> Abstract: A concept based on tunneling between extended Si states and localized SiO_2 states was developed to explain the logarithmic time dependence, gate voltage dependence and temperature dependence of hole trap depopulation. A hole trap level at E = 6.3 ±0.2 eV above the SiO_2 valence band edge was found. The hole trap concentration amounts to N = 2 ... $5*10^{19}$ cm^{-3} inside a small (50 Å) near interface region. Tunneling takes place during hole trap population, too, and leads to a dynamic trapping behaviour with severe consequences for the determination of the capture cross section and trap concentration.

1.   INTRODUCTION

The misfit at the Si/SiO_2 interface implicates broken bonds, hydrogen saturated bonds, bond angle decrease etc. in a thin interlayer. This causes electron and/or hole trap states and interface states or their precursors. Especially the hole interaction with such foregoing states leads to interface state generation and positive oxide charge generation (Powell R J et al 1971, Lai S K 1983). Between these insulator trap states and the adjacent contact (Si) a tunnel transfer takes place (Manzini S et al 1983, Lakshmanna V et al 1988); this effect will be considered taking into account consequences for trapping kinetics.

2.   EXPERIMENT

MOS capacitors were fabricated on <100> n-type Si substrate (20 Ωcm) growing 330 nm thick oxide and evaporating semitransparent metal gates (Al). The experimental equipment consistet of a VUV source, a HF-CV (1 MHz) set up and a photo-IV arrangement. We used a VUV light source (10.2 eV) to create electron-hole pairs at the (semitransparent) metal gate side of the oxide (absorption depth about 140 Å). Applying a positive gate bias during the irradiation the holes are driven to the Si/SiO_2 interface while electrons are swept out of the gate. The change of the near-interface positive oxide charge and the generation of interface states can be monitored by the measurement of flatband voltage U_{fb} and midgap voltage U_{mg}. The latter one represents the oxide charge because charge neutrality of interface states is realized at this point of band bending. Applying a positive gate bias during a 30 minute VUV irradiation a hole current $j_h = 2 * 10^{-8}$ Acm^{-2} was flowing. Thus, the positive oxide charge and the generated interface states reached their saturation level. The decay $\Delta U_{mg}(t)$ after switching off the VUV light has been investigated at different gate voltages (Figure 1).

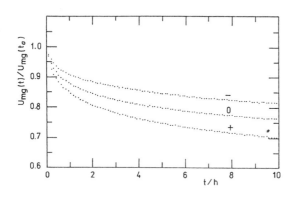

Fig. 1.
Relative change of midgap voltage vs. time for different external field strengths F_{ex}.

(+) $F_{ex}=1*10^6$ Vcm^{-1}
(o) $F_{ex}=0$
(−) $F_{ex}=-9.9*10^5$ Vcm^{-1}

3. BASIC EQUATIONS

In our case the midgap voltage is essentially determined by the distribution of positive charged traps $N(x,t)$ in the oxide layer. The neutralization of the hole traps can be described by a rate equation taking into account direct tunneling of holes from localized insulator gap states to extended semiconductor band states (Schmidt M 1990).
Using the golden rule of the transition probability for the tunneling transfer (Harrison W 1961), the two band $k(E)$ relation (Flietner H 1972) and the WKB-approximation the total tunneling rate constant at temperature T, energy E, and distance x from the interface can be written as

$$P_h(E,T,x) = (f(E,T)/\tau_o(E)) \exp \left(2i \int_0^x K(E) dx'\right) \qquad (1)$$

where denotes $E = E(x)$ the transfer level, $K(E(x))$ the x-component of the imaginary wave vector, $f(E)$ the occupation probability of the final states and $\tau_o(E)$ the tunneling time constant depending on the k-vectors and effective masses in Si and SiO$_2$, respectively. $P_h(E,T,x)$ is graphically presented in figure 2. Our experiments (cf. Figure 1) point out that with increasing gate voltages, from negative to positive, i. e. decreasing values of the effective trap energy $E(x)$, the tunneling rate $P_h(E)$ increases by more than 3 orders of magnitude. This is qualitatively fulfilled at $E_t = 6.3 \pm 0.2$ eV only and was supported by model calculations of ΔU_{mg} vs. U_g for different trap energies. The corresponding wave vector at $E = E_t$ amounts to $|K| = 0.61$ Å^{-1} and $\tau_o = 3 * 10^{-16}$ s.
For $E > E_F$ from eq. 1 follows

$$P_h(E,T,x) = (\exp - 2Kx/\tau_o(E)) \exp - ((E(x) - E_F)/kT). \qquad (2)$$

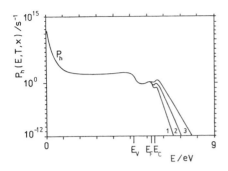

Fig. 2.
Tunneling rate constant $P_h(E,T)$ vs. E at x = 20 Å;
(1) T = 300 K,
(2) T = 400 K,
(3) T = 500 K
for <100> n-Si
(SiO_2: m_c=0.5, m_v=5;
Si: m_c=0.916, m_v=0.546)

Taking into account only tunneling on the basis of eq. 2 we can explain the "thermal activation energy" ($E(x)-E_F \approx 0.7 \pm 0.3$ eV) for thermally stimulated hole trap depopulation and their gate voltage dependence, as found experimentally by Lakshmanna in 1988. Furthermore, the solution of the rate equation using eq. 1 yields a logarithmic time dependence of $\Delta U_{mg}(t)$. This was confirmed by our experiments.

4. CONSEQUENCES OF TUNNELING FOR TRAPPING

For near interface localized hole (or electron) trap states - depending on their energetic position relative to the Fermi level of the adjacent contact and their actual filling state - the influence of tunneling on trap population and trapping kinetics can be important and must not be neglected. Regarding the two processes of both hole capturing and subsequent tunneling of electrons to hole trap states, the temporal change of the concentration of occupied traps in the insulator is determined by the capture rate constant $P_c = j\ \sigma_h / q$; (j - current density of injected holes, σ_h - capture cross section, q - elementary charge) and the tunneling rate constant $P_h = (1/\tau_{oh}(E))\ \exp(-2|K|x)$. Here $\tau_{oh} = \tau_o(E)/f_e(E)$ is the effective hole tunneling time constant. The complementary processes of filling (hole capturing) and emptying (hole tunneling) lead to a stationary equilibrium position x_g (trap occupation factor $f(x_g) = 0.5$) at the Si/SiO$_2$ interface side of the oxide charge distribution, which is given by

$$x_{gh} = -(1/2|K|) \ln (\sigma_h\ j\ \tau_{oh} / q) \qquad (3)$$

x_{gh} sharply separates the region bleached by tunneling from the region where hole capturing dominates. The same considerations are valid for electron traps too and yield the corresponding value x_{ge}. This is shown in figure 3 using the two band K(E) relation ($m_c = 0.5$, $m_v = 5$). Generally one can say, the smaller the current density of holes (or electrons) through the insulator, the deeper will be the decharged (not stationarily chargeable!) region within the insulator. For a given energetic and spatial distribution of traps the value of x_g and consequently the quasi-saturation level of occupied traps is determined by j only and can be reversibly shifted depending on j. This is clearly demonstrated for hole traps in figure 4. The dynamic trapping model (Schmidt M et al 1991) gives an

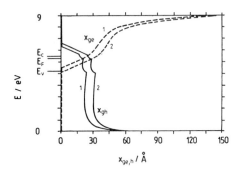

Fig. 3.
Separation between filled and unfilled hole trap (dotted line) and electron trap states (dashed line)
(1) $j_{e,h} = 10^{-6}$ Acm^{-2}
(2) $j_{e,h} = 10^{-10}$ Acm^{-2}
(T = 300 K, $\sigma_{e,h} = 1*10^{-14}$ cm^2)

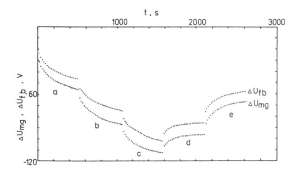

Fig. 4.
Transients of $\Delta U_{mg}(t)$ and $\Delta U_{fb}(t)$ for different hole current densities j_h.
(a) $1.97*10^{-8}$ Acm^{-2}
(b) $3.61*10^{-8}$ Acm^{-2}
(c) $5.45*10^{-8}$ Acm^{-2}
(d) $4.1\ *10^{-8}$ Acm^{-2}
(e) $2.58*10^{-8}$ Acm^{-2}
(d_{ox} = 350 nm, 10 Ωcm n-Si <100>)

unambiguous algorithm to determine real values for capture cross sections and effective trap concentrations N_{eff} which can deviate by more than one order of magnitude from values determined by using simple first order kinetics which neglect tunneling phenomena. Thus, for the hole trap we deduced corrected values of $\sigma_h = 1.49 * 10^{-14}$ cm^2 and $N_{eff} = 1.47 * 10^{13}$ cm^{-2} from the experimental data in contrast to incorrect values of $\sigma_h = 1.62 * 10^{-14}$ cm^2 and $N_{eff} = 5.29 * 10^{12}$ cm^{-2}.
The tunneling process and its effect on trap occupation allowed to characterize the dominating hole trap in the Si/SiO$_2$ system in a comprehensive manner. These results agree with EPR experiments (Witham et al 1987) and confirm that the E' centre (oxygen vacancy) constitutes the hole trap as studied here.

REFERENCES
Flietner H 1972 phys. stat. sol. (b) 54 201
Harrison H 1961 Phys. Rev. 123 85
Köster H jr. and Schmidt M 1989 Mat. Sc. Forum 38/41 1331
Lai S K 1983 J Appl. Phys. 54 2540
Lakshmanna V and Verengurlekar A S 1988 J. Appl. Phys. 63 4548
Manzini S and Modelli A 1983 Proc. INFOS 83 ed J F Verweij D R Wolters (Amsterdam:North Holland)pp 112-5
Powell R J and Derbenwick G F 1971 IEEE Trans.Nucl.Sci. NS-18 99
Schmidt M 1990 Dissertation B (Berlin present adress)
Schmidt M and Köster H jr. 1991 phys. stat. sol. submitted
Witham H S and Lenahan P M 1987 Appl. Phys. Lett. 51 1007

Interface trap measurements using 3-level charge pumping

N.S. Saks[a], M.G. Ancona[a], and Wenliang Chen[b]

a) Code 6813, Naval Research Laboratory, Washington, D.C. 20375
b) Yale University, New Haven, Ct. 06520

ABSTRACT: We describe an improved charge pumping technique with three voltage levels for measuring interface trap D_{it} parameters in MOS devices. This 3-level technique enables accurate determination of electron and hole trap cross sections for capture and emission as a function of trap energy. Results are compared to standard measurement techniques, including capacitance-voltage and ac conductance.

1. INTRODUCTION

Charge pumping (CP) is now a widely accepted technique for measuring interface trap parameters in MOS devices. The standard 2-voltage-level charge pumping technique [1] permits determination of trap densities D_{it} over a wide range of trap energies E_t with high sensitivity, and is easy to implement. Recently, charge pumping has been further improved using a 3-voltage level technique [2,3]. This improved technique enables both $D_{it}(E_t)$ and the electron and hole capture-cross sections (σ_e and σ_h, respectively) to be determined as a function of energy E_t [3]. Here we extend our previous work [3], describing how 3-level CP can be used to determine $\sigma_e(E_t)$ and $\sigma_h(E_t)$ for emission *and* capture of carriers. In addition, we compare these results with those obtained from other (standard) measurement techniques, and present some additional numerical simulations.

In the charge pumping technique, an ac pulse which oscillates between MOS accumulation and inversion is applied to the MOSFET gate (Fig. 1). In standard 2-level CP [1], the rise and fall times of the ac pulse are varied to obtain $D_{it}(E_t)$. In 3-level CP, this waveform is modified to include a third voltage level (V_3) which places the surface potential (ϕ_s) within the silicon bandgap. $D_{it}(E_t)$ may then be determined by varying V_3 [2,3]. Furthermore, by varying the duration time t_3 of V_3, the energy-dependent cross sections may be determined as shown in [3].

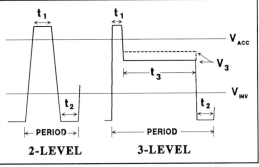

Fig.1 Comparison of ac waveforms used in standard 2-level and 3-level charge pumping.

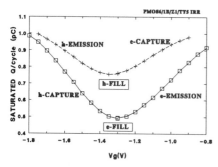
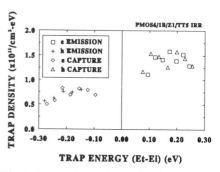

Fig.2 Recombined charge per ac cycle vs. gate voltage after filling traps with holes ("h-fill") or electrons ("e-fill").

Fig.3 Comparison of trap density vs. energy obtained from capture and emission data.

2. EXTENDED ANALYSIS--"CAPTURE" REGIME

In [3], we showed that if the interface traps are first filled with electrons, then the saturated (i.e., at long t_3) value of the charge loss per cycle (Q/cycle) will decrease as V_3 falls towards mid-gap. This is shown in Fig. 2 for the lower "e-fill" curve for V_3 between -0.8V (near accumulation) and -1.3V (near mid-gap). Q/cycle decreases because, as V_3 pushes ϕ_s towards mid-gap, more traps emit their electrons back to the conduction band, thereby reducing the number of remaining trapped electrons available for recombination. This behavior is determined by electron emission rates from the traps ("e-emission" region in Fig. 2). The density of traps $D_{it}(E_t)$ above mid-gap can be obtained from these data [3]. Similarly, $D_{it}(E_t)$ below mid-gap can be determined from the hole emission data ("h-emission" in Fig. 2). $D_{it}(E_t)$ calculated from emission data is shown in Fig. 3 for an irradiated p-channel MOSFET test device. The density of traps above mid-gap is larger than below mid-gap, as is typically observed in irradiated oxides.

Under certain conditions, the interface trap occupancy can be limited not by carrier emission, but rather by capture of the opposite carrier type. This occurs for electron traps, for example, when ϕ_s moves below mid-gap towards inversion ($V_3 < -1.3V$). In this case, trap occupancy is determined not by emission of the trapped electrons but rather by capture of free holes. As shown in Fig. 2 for trapped electrons ("e-fill" data), Q/cycle does not continue to fall as ϕ_s decreases below mid-gap but rather increases again. This increase occurs because hole capture by a trapped electron contributes to the charge pumping current, while electron emission does not. This behavior clearly shows the non-equilibrium character of charge pumping; obviously, this regime is *not* accessible to conventional equilibrium techniques, e.g., C-V and G-V. Interface trap densities determined in the "capture" regime are shown also in Fig. 3. Agreement with "emission" data is within experimental error.

In addition to new $D_{it}(E_t)$ data, temporal data from the capture regime can be used to obtain new trap cross section information. Previously, ref [3] demonstrated that the electron and hole *emission* cross sections, σ_e and σ_h, can be determined above and below mid-gap, respectively, using time dependent data from 3-level charge pumping. The procedure is to determine the equilibrium electron and hole *emission* time constants as a function of ϕ_s (by varying V_3). Analogously, the *capture* time constants can be obtained when the time constant for removal of a trapped carrier is determined by capture rather than emission. σ_e and σ_h *for capture* can then be determined from a knowledge of the density of the free carriers at the surface (determined by ϕ_s, whose value is set by V_3) and the associated measured time constant. It should be noted that the energy of the traps being measured is no longer at ϕ_s, which somewhat complicates the analysis in this regime. Electron and hole cross sections determined from both capture and emission data are shown in Fig. 4. Note that cross section values for capture and emission are nearly the same above mid-gap, while the values below mid-gap are

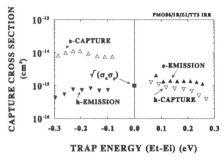

Fig.4 Comparison of capture and emission cross sections.

not. It is not clear at present if this results from measurement errors or a fundamental difference in the nature of the cross sections. $\sqrt{(\sigma_e \sigma_h)}$ obtained from surface generation currents is also shown in Fig. 4 [4]. Values obtained for $\sqrt{(\sigma_e \sigma_h)}$ are generally found to be in good agreement with the *emission* values from charge pumping as they should.

To aid our understanding of the e-capture and h-capture processes, and to evaluate our simplified analysis method, we have performed additional numerical simulations using the approach of Ghibaudo and Saks [5]. These simulations generally confirm our understanding of the dynamics of 3-level CP as presented above [6]. One illustration of this is shown in Fig. 5 which shows calculated electron and hole currents as a function of time. The labels "capture" and "emission" indicate net cur-

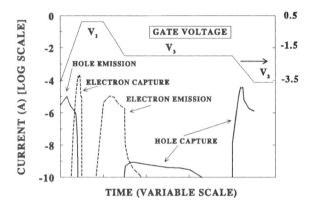

Fig.5 Numerical simulation of transient electron and hole current (at bottom) vs. the 3-level gate pulse (top of Fig.).

rent to or away from the surface, respectively. The critical part of this simulation occurs after the traps are filled with electrons and the gate voltage falls from V_1 to V_3, *pushing ϕ_s below mid-gap.* The simulation verifies that electron emission dominates during the gate voltage fall time, with hole capture dominant thereafter.

3. COMPARISON TO OTHER TECHNIQUES

To assess the accuracy of the 3-level CP technique, we have undertaken comparisons with other techniques, including capacitance-voltage (C-V) and ac conductance (G-V) [7]. All measurements were made on the same identical large area MOSFET. In the C-V measurements, $D_{it}(E_t)$ is obtained from a comparison of high vs. low frequency capacitance [8]. Because of the lower sensitivity of the C-V and G-V techniques, test MOSFETs with large areas and relatively high trap densities ($\geq 10^{11}$) are required. Results are presented here for one irradiated p-channel MOSFET with 9 μm length x 1000 μm width. The same device was used in Figures 2-4 above. (Results presented for this one sample are representative of the three different devices measured in detail to date.) In Fig. 6, $D_{it}(E_t)$ is shown for C-V, G-V, and 3-level CP techniques. Reasonably good agreement, especially above mid-gap, is obtained. The 3-level CP data is consistently smaller than the G-V results, which are less accurate then would normally be obtained (using a large capacitor) because of edge effects and uncertainties in the MOSFET parasitic capacitances. In Fig. 7, both σ_e and σ_h are consistently a factor of 3-5 smaller than the 3-level CP results, which is within the uncertainty of the two measurements. There is the possibility that the different techniques may not be measuring all the same traps. C-V and G-V are sensitive to all interface traps. Charge pumping is sensitive only to those traps through which electrons and holes can recombine. Therefore, CP will be insensitive to tunneling-type interface traps which trap only one type of carrier.

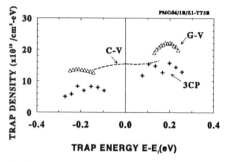

Fig.6 Comparison of interface trap densities *all measured on a single MOSFET* using three different measurement techniques.

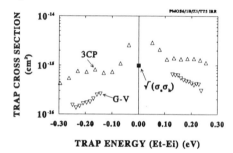

Fig.7 Comparison of trap cross sections obtained by three different techniques.

4. SUMMARY/CONCLUSIONS

We have performed several new experiments extending 3-level charge pumping into a "capture" regime where trap occupancy is determined not by emission of the trapped carriers, as analyzed previously [3], but rather by capture. Numerical simulations have been performed which validate our overall interpretation of the data. Analysis in the capture regime is somewhat more difficult to perform and thus the emission regime will typically be preferred for routine 3-level CP measurements. We have also compared 3-level CP results with data from standard D_{it} measurement techniques. In general, reasonably good agreement is obtained.

5. REFERENCES

[1] G. Groeseneken et al., IEEE Trans. Elec. Dev. ED-31, 42, Jan. 1984.
[2] W.L. Tseng, J. Appl. Phys. 62, 591, July 1987.
[3] N.S. Saks and M.G. Ancona, Electron Dev. Letts. 11, 339, Aug. 1990.
[4] A.S. Grove and D.J. Fitzgerald, Solid-St. Electron. 9, 783 (1966).
[5] G. Ghibaudo and N.S. Saks, J. Appl. Phys. 64, 4751 (1988).
[6] Simulations of emission were described at the 1990 Device Research Conference, Santa Barbara, CA.
[7] H.S. Haddara and M. El-Sayed, Solid-State Electron. 31, 1289 (1988).
[8] G. Gildenblat, J.M. Pimbley, and M.F. Cote, Appl. Phys. Letts 45, 558 (1984).

Individual attractive defect centers in the Si–SiO$_2$ interface

A Karmann, M Schulz

Institut für Angewandte Physik, Universität Erlangen-Nürnberg,
Staudtstr. 7, D-8520 Erlangen

ABSTRACT: The alternate capture and emission of holes into and from an individual attractive interface trap is measured by the switching of the source-drain current in a micron-sized p-channel MOSFET. Anomalous switching is observed with a high channel conductance, when a hole is trapped at the defect center and a low channel conductance after re-emission of the hole. The measured results are interpreted by a tunneling transfer of a hole from a bound state in the channel to an attractive center 2.4nm deep in the oxide and vice versa.

1. INTRODUCTION

Individual interface traps are studied in micron-sized MOS structures having a state density of $D_{it} = 10^9 - 10^{10} cm^{-2} eV^{-1}$ at the Si-SiO$_2$ interface. The low interface state density leads to a total number of 1 to 10 defect states per device. The statistical occupation of these few states induces a Random Telegraph Switching (RTS) in the channel conductance (Ralls 1984, Kirton 1986). The emission transient after a perturbation of the trap occupation is quantized into steps (Karwath 1988).

RTS was so far mainly discussed for defect centers, which are repulsively charged for the mobile carriers in the MOSFET channel. The switching observed is between a high resistance, when the charge carrier is trapped and a low resistance, when the carrier is re-emitted. The change in resistance is predominantly caused by the change of the number of mobile carriers in the channel. The charge transfer into the defect is strongly activated. The activation is explained by the Thermionic Emission Model presented at INFOS 89 conference in Munich (Karmann 1989, Schulz 1990).

In this paper, we present the study of a defect center which shows an anomalous switching. An observation of anomalous switching was first reported in 1984 (Ralls 1984). For this switching, the channel conductance is high, when the charge carrier is trapped, and the channel conductance is low, when the charge carrier is re-emitted. The temperature and gate bias dependence is different compared to that of repulsive centers. The experimental results are interpreted in this paper by a tunneling transfer of a charge carrier into an individual attractive defect center and by a mobility change in the channel conductivity.

2. EXPERIMENTAL DETAILS

The experiments are performed on a p-channel MOSFET fabricated by VLSI technology. The device was kindly provided to us by the Siemens Research Lab, Munich. The technical data of the device are: oxide thickness $t_{ox} = 18nm$, channel length $l = 0.7\mu m$, channel width $w = 0.8\mu m$, threshold voltage $V_{Th} = -2.4V$ at temperature $T = 107K$. The gate voltage V_G applied is in the range -2.1V to -2.6V. The effective channel conductance is measured for a source-drain voltage about $V_D \approx 10mV$ with a constant current I_D in the range 5nA to

© 1991 IOP Publishing Ltd

150nA. For this low source-drain current, the carrier density in the channel is negligibly affected by the lateral voltage drop. The random telegraph signal is recorded in the voltage drop for constant current by a transient recorder. The signal then represents the channel resistance. The switching time constants in the RTS are averaged for up to 300 switching events.

3. EMISSION AND CAPTURE RATES

The averaged emission and capture rate constants of the center showing the anomalous switching are shown in Fig. 1 as a function of the gate bias voltage. The arrows indicate the switching direction in the random telegraph signal of the channel resistance. We attribute the rate constant increasing with gate voltage to a capture process and the decreasing rate to the emission process. In general, the emission and capture cannot be distinguished in the symmetrical RTS. Our attribution is made by reasoning that an increasing probability of the trap occupation occurs when the gate voltage is increased into strong inversion; however, the resistance change is anomalous in this case, because the capture and thus an immobilization of a hole by a transfer into the defect induces an unexpected anomalous decrease of the resistance.

The anomalous trapping behavior is observed in a wide temperature range from 90K up to 250K. The activation energies observed are of the order of a few meV. This is a small value compared to those for repulsive centers, where activation energies of hundreds of meV are found.

4. TRAPPING MODEL

The observed properties of the anomalous switching can be described by an attractive defect center located in the oxide near the interface. A complete description of the experimental evidences and a full discussion of the model are presented in the original paper (Schulz 1991). In this conference presentation, we only explain the main features.

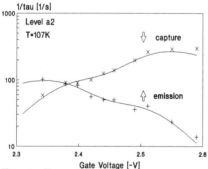

Fig. 1: Transfer rates of anomalous switching versus gate voltage.

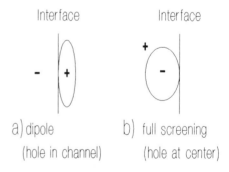

Fig. 2: Charge distribution: a) hole located in the channel bound to the defect. The level energy is E_2 b) hole located at the defect center in the oxide. The level energy is E_1.

The Fig. 2 illustrates the charge distribution around the attractive defect center. The diagrams are drawn for an acceptor-type defect center in accordance with the p-channel MOSFET studied. In Fig. 2a, the hole is located in the channel but is still bound to the negatively charged acceptor center in a potential well. The charge distribution forms a dipole consisting of the negative acceptor charge and of the bound positive hole in the channel. In Fig. 2b, the hole is assumed to be trapped at the defect in the oxide. The charge of the acceptor is fully screened by the hole.

The switching process is interpreted by a transfer of the hole from the **bound** state at the acceptor in the oxide to the **bound** state in the well of the channel. The hole itself is never free to participate in the conductivity of the channel. The switching in the channel conductance is induced by the mobility change of the holes in the channel. The change is between dipole scattering when the hole is in the channel, and the weak scattering of the neutral acceptor center. Dipole scattering was already invoked earlier for complex normal switching to explain a residual resistance change (Uren 1989); in the present case, we explain the anomalous switching sign by a change in mobility. Dipole and neutral scattering are the most likely scattering mechanisms deduced from the model considerations.

The trap energy may be described by a two-level system of bound states. The level energy E_1 represents the ground state of the hole located at the defect center in the oxide (Fig. 2b). E_2 is given by the level energy of the hole in the potential well of the channel (Fig. 2a). The system is occupied by a single hole which alternatively occupies level E_1 or level E_2. The level offset $U=(E_2-E_1)/2$ is induced by the spatial separation of the hole from the defect center. It is linearly dependent on gate voltage and is given by the fraction of the gate voltage drop over the distance t_{trap} between the interface and the defect center:

$$dU/qdV_G = t_{trap}/t_{ox} \qquad (1)$$

The emission and capture rates are given by

$$1/\tau_e = P(V_G,T) \cdot g^{1/4} \cdot \exp(-U/2kT) \qquad (2)$$

$$1/\tau_c = P(V_G,T) \cdot g^{-1/4} \cdot \exp(U/2kT) \qquad (3)$$

The energy dU is the change of level offset U, q is the elementary charge and dV_G is the change of the gate voltage. $P(V_G,T)$ is the transfer probability and g is the degeneracy factor of the two levels. A possible entropy contribution to the rate constants is included in the transfer probability $P(V_G,T)$. The transfer probability $P(V_G,T)$ and the level offset U are determined by forming the product and the ratio of the measured emission and capture rates, respectively.

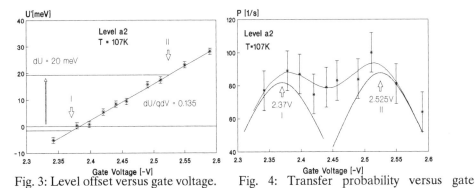

Fig. 3: Level offset versus gate voltage.   Fig. 4: Transfer probability versus gate voltage.

The Fig. 3 shows the level offset U as a function of the gate voltage at a constant temperature T=107K. In accordance with eq. (1), we find a linear relation. A depth in the oxide of t_{trap}=2.4nm is deduced from the slope.

The Fig. 4 shows the measured transfer probability versus gate voltage. The continuous line represents two maxima of the transfer probability given by tunneling into a ground state of the defect in the oxide and into an excited state 40meV above the ground state. The width of the maxima is approx. $\sigma=kT=9$meV. The superposition of the two maxima fits the experimental data. The tunneling probability of $P\simeq 80 s^{-1}$ for each energy level is in good agreement with a tunnel distance of $t_{trap}=2.4$nm into the oxide (Manzini 1983). The left maximum corresponds to the ground state when the two energy levels line up at $U=0$eV (Fig. 3).

5. CONCLUSION

Random telegraph signals and emission transients are studied for the anomalous switching observed at a micron-sized p-channel MOSFET. For anomalous switching, the channel conductance is high, when the defect is occupied and the channel conductance is low after re-emission of the trapped hole.

The anomalous switching is interpreted by tunneling of a hole bound in the channel to an attractive defect center located in the oxide. The switching originates in a change of scattering mechanism between dipole scattering and weak scattering of the neutral defect center. In contrast to repulsive centers, the switching originates in the change of the mobility of the charge carriers in the channel rather than by the change of the carrier density.

The observed tunneling transfer for the attractive defect center is negligibly activated in contrast to the transfer observed for repulsive centers where strongly activated thermionic emission is reported.

ACKNOWLEDGEMENT

The authors are grateful to the Siemens Research Lab which provided the micron-sized MOSFETs for the research. The work is partially funded by Siemens Co.

REFERENCES

Karmann A and Schulz M 1989 *Proc. of the Int. Conf. on Insulating Films on Semicond.* eds F Koch and A Spitzer (Amsterdam: North-Holland) 500
Karwath A and Schulz M 1988 *Appl. Phys. Lett.* **52** 634
Kirton M J and Uren M J 1986 *Appl. Phys. Lett.* **48** 1270
Manzini S and Modelli A 1983 *Proc. of the Int. Conf. on Insulating Films on Semicond.* eds J Verwey and D Wolters (Amsterdam: North-Holland) 112
Ralls K S, Skocpol W J, Jackel L D, Howard R D, Fetter L A, Epworth R W, and Tennant D M 1984 *Phys. Rev. Lett.* **52** 228
Schulz M and Karmann A 1990 *Proc. of the 5th Cong. of the Brasilian Society of Microelectronics SPIE* **140** 2
Schulz M and Karmann A 1991 *Appl. Phys. A* **52** 104
Uren M and Kirton M *Proc. of the Int. Conf. on Insulating Films on Semicond.* eds F Koch and A Spitzer (Amsterdam: North-Holland) 479

Paper presented at INFOS '91, Liverpool, April 1991
Contributed Papers, Section 1

Electron trapping-induced conductance- and noise-dynamics in ultra-thin metal−oxide−silicon tunnel diodes

M O Andersson, K R Farmer and O Engström
Dept. of Solid State Electronics, Chalmers Univ. of Technology, S-412 96 Göteborg, Sweden

ABSTRACT: A negative charging effect, seen as a gradual decrease in the magnitude of the tunnel current over time, is studied in small, ultra-thin metal-oxide-silicon diodes biased at constant voltage. Along with the charging, the current noise power decreases. Results are described from measurements of the temperature and voltage dependence of the current transients. A model is presented in which filling and emptying of electron traps in the as-grown oxide cause the noise, and the conductance decrease is caused by the transformation of these traps into fixed negative charge storage centers. Results from low-temperature annealing experiments suggest that the traps are water-related defects.

1. INTRODUCTION

Using sub-micron sized metal-oxide-silicon (MOS) tunnel diodes, we have previously investigated large (~0.01% to ~10%) discrete conductance fluctuations associated with individual oxide defects (Farmer *et al* 1987, Andersson *et al* 1990, Farmer and Buhrman 1989). We have also reported a study of macroscopic charging, both positive and negative, in similar devices (Farmer *et al* 1990). One question that remains is whether or not the macroscopic charging and trapping effects are related to the defects which are causing the large fluctuations. To address this question, we have made an extensive study of the effects of *negative* charging on the conductance and the noise in the tunnel current through these diodes. We find that the decreasing current is due to trapping to what is suggested from low-temperature annealing-experiments to be deep, water-related electron traps. The band-to-trap transport occurs in these thin oxides through direct tunneling from the gate. We attribute the observed $1/f$ noise to the superposition of many small-amplitude discrete conductance fluctuations, each one caused by the capture and emission of charge at a single trap site. On the other hand, the *large* individual fluctuators previously found in these devices may not be caused by the same water-related traps; the negative charging is inhibited at low temperatures (< 180 K) whereas large fluctuators can be active at least down to 50 K. This suggests that there are at least two different noise sources in these oxides, and that while the origin of the defect causing the large-amplitude fluctuations remains unknown, the small-amplitude fluctuations may be correlated to the presence of the water-related traps.

2. EXPERIMENTS

The devices used in this study were made on a 5 Ω-cm, p-type, <100> oriented silicon wafer. The active areas, ranging from 0.008 to 20 μm^2, were formed by etching circular holes in a 1000 Å field oxide, and the thin oxide was subsequently grown at ~700 °C in dry O_2 to a thickness of ~20 Å. The metal gate was formed by evaporating 2500 Å of aluminum which was also deposited as a back side contact. The wafer was initially not subjected to post-metallization annealing treatment. Previously unbiased devices showed current-voltage (I-V) characteristics which reproducibly scaled well with the device area.

The measurements were made using either a probe-station at room-temperature or a cryo-unit at lower temperatures. The device was biased at a negative gate voltage, V_g, (ranging from

© 1991 IOP Publishing Ltd

−0.42 to −3.1 V). The resulting slow current transient was automatically recorded over a period of 3-5 days using a current amplifier, a digital voltmeter and a computer. At 14-19 predetermined instances during the acquisition, the current-sampling was interrupted (without changing V_g), and instead the noise power in the current was measured for frequencies, f, between 0.5 and 12.5 Hz using an HP 35660A spectrum analyzer.

A typical charging transient is shown in Fig. 1, where the fractional change in current is plotted as a function of time for a device at T = 295 K and $V_g = -2.84$ V. As shown in Fig. 2, where the fractional change in current at three times is plotted as a function of temperature at $V_g = -2.84$ V, the strength of the charging effect decreases strongly with decreasing temperature. The voltage dependence of the charging is rather weak, as shown at T = 295 K in Fig. 3; we have noticed that charging occurs at gate voltages as low as $V_g = -0.42$ V, corresponding to an oxide field of ~0.5 MV/cm. This is in contrast to the results reported by Nagai and Hayashi (1984), who found a negative charging threshold at $V_g = -2.0$ V, or 3.7 MV/cm, for similar devices. In no case do we observe the charging to saturate, even after a 5-day experiment.

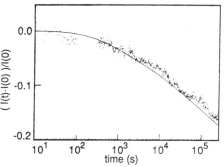

Fig. 1. Measured (dots) and calculated (solid curve) fractional decrease of device current vs. time at T = 295 K and $V_g = -2.84$ V.

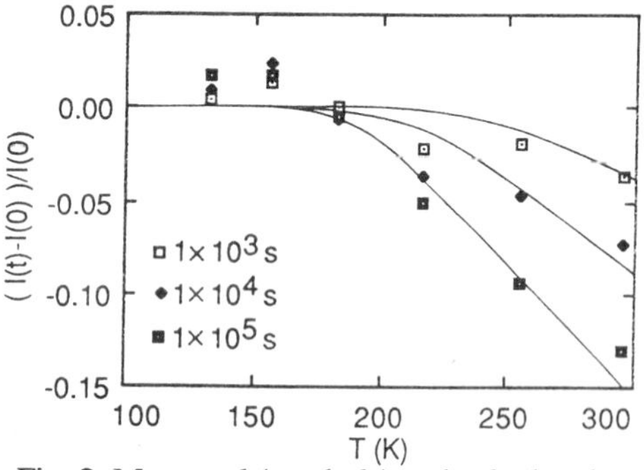

Fig. 2. Measured (symbols) and calculated (solid curves) fractional current decrease vs. temperature at $V_g = -2.84$ V.

In a separate set of experiments, we found that subjecting *previously unbiased* devices to a 350 °C anneal in N_2 for 15 minutes completely disabled the charging process; in these devices, little or no decrease in device current at room temperature could be seen even after 60000 seconds at $V_g = -3.1$ V. In addition, the current noise power S_I was considerably smaller after this treatment, as shown in Fig. 4. In contrast, subjecting a *previously negatively charged* device to a 200 °C anneal in room air for 16 minutes restored the conductance to its initial value (before charging). This device could then be charged negatively again, the new current vs. time trace tracking the original curve very well. A second 200 °C anneal again restored the conductance. We could thus cycle the charging process back and forth at least twice.

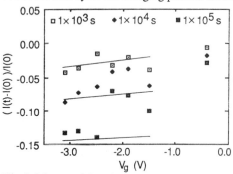
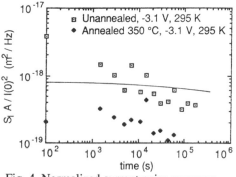

Fig. 3. Measured (symbols) and calculated (solid curves) current decrease vs. gate voltage at T = 295 K.

Fig. 4. Normalized current noise power vs. time. Solid curve: calculation for an unannealed device.

3. RESULTS AND DISCUSSION

I-V characteristics taken before and after the charging show that the conductance decrease is due to oxide charge and not to the removal of a trap-assisted tunnel path (as has been suggested to explain the large discrete current fluctuations (Andersson et al 1990)), because the most significant change occurs in the depletion plateau region of the I-V curve (Green et al 1974). The oxide charge, slowly built up on a time scale of thousands of seconds, remains fixed for as long as the device is not subjected to an elevated temperature as described in the previous section. In contrast, the observed current noise power, which tracks the charging curve, involves much shorter time constants—of the order of seconds or less. We propose that the measured features of the noise power result from a sum of independent Lorentzian spectra, each of which is caused by a small increase in the local barrier height when an electron is captured to a trap, and a corresponding decrease when the electron is emitted again. If both the noise and the charging are related to the same defects, then the defects must undergo some change that turns off or disactivates the rapid fluctuations, eventually leaving a number of the traps in a state of permanent negative charge. The strong temperature dependence of the charging, shown in Fig. 2, suggests that the slow charging process is thermally activated.

We have attempted to verify this physical model quantitatively. The traps are considered to be independent and uniformly distributed throughout the oxide, but at one energy level relative to the oxide conduction band. The effect of a single point charge on the direct tunnel current in a metal-insulator-metal structure was originally calculated by Schmidlin (1966). We have adapted his method to our MOS devices. Essentially, we regain the same order of magnitude as Schmidlin for the relative change in the current $\Delta I_1 / I(0)$ due to a single charge.

For the calculations, the oxide traps are placed in L=17 equidistant planes, in each of which the initial concentration of traps is denoted N_T [cm^{-2}]. The total concentration of traps, D_{ox}, is then $L \times N_T$. The concentration n_T of disactivated, negatively charged traps at a certain time t in a given layer at a distance x_i from the metal-oxide interface is described by first-order kinetics:

$$n_T(x_i, t) = N_T \left[1 - \exp\left(-t / \tau_{dis}(x_i)\right) \right], \tag{1}$$

where $\tau_{dis}(x_i)$ is the time constant for disactivation of the traps at $x = x_i$. The disactivation process probably involves energy being dissipated when the tunneling electron is captured to the trap, resulting in some permanent damage. Thus, the process cannot be completely described by an elastic tunneling-calculation. However, the time constant for disactivation should increase with the tunnel distance from the electron source, i.e., the metal. We use a WKB-barrier penetration factor to account for this dependence. Since the traps are located at one discrete energy level, we must also incorporate the probability that an electron occupies a corresponding energy level in the metal. Also, we include a factor which accounts for the fact that the charging is thermally activated. Altogether, the time constant for charging is given by

$$\tau_{dis}(x_i) = \tau_{dis,0} \frac{1}{f(E_T)} \exp\left[2 \int_0^{x_i} \kappa(x') \, dx' \right] \exp\left[\frac{E_{dis}}{kT} \right], \tag{2}$$

where $\tau_{dis,0}$ is a constant, $f(E_T)$ is the metal Fermi function at the energy corresponding to the trap level, $\kappa(x)$ is the imaginary part of the wave vector in the oxide and E_{dis} is the activation energy. The relative change in the current at a given time can be expressed as

$$(I(t) - I(0)) / I(0) = \sum_{i=1}^{L} \left[\gamma \left(\Delta I_1(x_i) / I(0) \right) n_T(x_i, t) \, A \right], \tag{3}$$

where A is the active device area and γ is a fitting parameter which we describe below. The current noise power is calculated as a sum of Lorentzians (Kirton and Uren 1989):

$$S_I(f, t) = 4 \, I(0)^2 \sum_{i=1}^{L} \frac{\left(\gamma \Delta I_1(x_i) / I(0) \right)^2 (N_T - n_T(x_i, t)) \, A}{\left[\tau_c + \tau_e \right] \left[\left(1/\tau_c + 1/\tau_e \right)^2 + (2\pi f)^2 \right]}, \tag{4}$$

where τ_c and τ_e are the active-trap time constants for capture and emission, respectively.

In fitting the model to the measured data, we use the variables D_{ox}, $\tau_{dis,0}$, E_T, E_{dis}, γ, τ_c and τ_e. We find that $E_{dis} = 0.3$ eV reproduces the measured temperature dependence of the current transients well, as shown by the solid curves in Fig. 2. The trap level E_T is hard to extract with precision and in reality the levels are probably not so well-defined. Nevertheless, the traps must be deeper than 3.0 eV, or they could not be accessed from the metal at the experimental gate voltages, but not much deeper, or they would already be filled at t=0 at zero bias. We use $E_T = 3.1$ eV. The effect of the time constant $\tau_{dis,0}$ is to displace the theoretical current transient in time. Using the value $\tau_{dis,0} = 5 \times 10^{-4}$ s, the total time constant $\tau_{dis}(x)$ varies between ~10^2 s close to the aluminum and ~10^{10} s close to the silicon at T = 295 K. It should be stressed that a distribution of time constants is needed to reproduce the curvature of the slow charging transients; a single time constant gives a much too abrupt charging curve. The product γD_{ox} governs the magnitude of the calculated charging effect. For S_I measured at $f=1$ Hz, the expression for S_I in Eq. (4) is dominated by contributions from fluctuators with time constants of the order 0.1 to 10 s. To obtain an estimate, we set $\tau_c = \tau_e = 1$ s. The best fits are then obtained using $\gamma = 40$ and $D_{ox} = 6.2 \times 10^{11}$ cm^{-2}. In this approximation, we cannot predict the bias or temperature dependences of S_I. Using this set of parameter values, a good overall fit is obtained for the the charging process, as shown by the solid lines in Figs. 1, 2 and 3. We also obtain an estimate for the time dependence of S_I, as shown in Fig. 4.

A striking result from the calculations is that $\Delta I_1 / I(0) \approx 10^{-5}$ in a 0.3 µm^2 device as obtained from the Schmidlin model is not large enough to explain the magnitude of S_I. $|\Delta I_1 / I(0)|$ must be increased by a factor $\gamma \approx 40$ in order to produce a reasonable fit to the measured S_I. We conclude that the model of noise and charging due to the Coulombic effects of a point charge in an otherwise perfect oxide cannot explain our data. A reasonable suggestion is that local oxide inhomogeneities, where the current density is large, can coincide with the traps.

We propose that the active traps consist of Si-OH groups in the oxide network. We further suggest that the disactivation of some traps, resulting in the "permanent" negative charge, occurs (with an activation energy $E_{dis} = 0.3$ eV) as the H atom is released from the complex in the electron capture process, resulting in an Si-O$^-$ defect. The de-charging in the 200°C annealing experiments would then occur as the H atom re-binds to the Si-O$^-$ defect and thus re-activates the trap. According to calculations by Pantelides (1982), Si-OH defects in SiO$_2$ form acceptor levels at 4.5 ± 2 eV from the oxide conduction band, and Si-O$^-$ is a deep acceptor close to or even in the oxide valence band. The de-charging at 200°C is strikingly similar to what Nicollian et al (1971) found for water-related traps in their investigation of much thicker oxides. It has been shown that OH-groups are present in as-grown, nominally dry oxides, and that they anneal out at 350 °C forming water (Hartstein and Young 1983). These findings are consistent with our 350 °C annealing-data on the charging and the noise.

ACKNOWLEDGEMENTS

This work was sponsored by a grant from the Swedish National Board for Technical Development. We wish to acknowledge the important contributions of S. Norrman and Z. Xiao who fabricated our devices.

REFERENCES

Andersson M O, Xiao Z, Norrman S and Engström O 1990 *Phys. Rev. B* **41** 9836
Farmer K R, Andersson M O and Engström O 1990 *Proc. 20th Int. Conf. Phys. Semicond.* eds E M Anastassakis and J D Joannopoulos (Singapore: World Scientific) pp 391-4
Farmer K R and Buhrman R A 1989 *Semicond. Sci. Technol.* **4** 1084
Farmer K R, Rogers C T and Buhrman R A 1987 *Phys. Rev. Lett.* **58** 2255
Green M A, King F D and Shewchun 1974 *Solid State Electron.* **17** 551
Hartstein A and Young D R 1981 *Appl. Phys. Lett.* **38** 631
Kirton M J and Uren M J 1989 *Adv. Phys.* **38** 367
Nagai K and Hayashi Y 1984 *Appl. Phys. Lett.* **44** 910
Nicollian E H, Berglund C N, Schmidt P F and Andrews J M 1971 *J. Appl. Phys.* **42** 5654
Pantelides S T 1982 *Thin Solid Films* **89** 103
Schmidlin F W 1966 *J. Appl. Phys.* **37** 2823

Paper presented at INFOS '91, Liverpool, April 1991
Contributed Papers, Section 1

Interface state distributions and the center of reconstruction model

H Flietner

Zentralinstitut für Elektronenphysik, O-1199 Berlin, Germany

ABSTRACT: The center of reconstruction(CR) model together with the complex band structure gives a **complete and unified** description of the defect structure at interfaces of homopolar bonded semiconductors and a relation between electronic states and atomistic configuration.

1. INTRODUCTION

The complex band structure model together with a disorder assumption and some total energy considerations brought some insight into the nature of defects at interfaces of semiconductors with predominant homopolar bonding (Flietner 1972, 74, 85, 88). U-shaped disdributions and Fermi-level-pinning could be explained at Si- as well as at compound semiconductor-interfaces. Recently this line was also followed by Hasegawa et al (1983, 86, 89) but without total energy considerations. EPR-experiments clearly demonstrated the dangling bond(DB)-character of interface states at Si/SiO_2 and their amphoteric character (Lenahan et al 1984, Pointdexter et al 1984, Articles 1987) and gave a new impulse to model considerations concerning the nature of the defects. The world wide huge amount of experimental material (reported and analyzed by Sah 1976, Nicollian, Brews 1982, Lang, Chen 1985, Fahrner, Klausmann 1986) gave a good background for a critical reexamination of the models. Many so far and even today cited models may be ruled out because of making no allowance for that important amphoteric character. Especially those models are obviously concerned which are using two different kinds of defects for the lower and upper part of the gap irrespectively of claiming to explain interface distributions or Fermi-level pinning. The essential point of the amphoteric character as a chemical species is the total energy consideration. Interesting that at the very beginning of semiconductor surface physics (Kingston 1957) and in the ball and stick time (Kooi 1967) scientists had a good flair for it.

Up to now no theoretical model describes the experimental interface state distributions correctly. The Si/SiO_2-interface gives a good chance for a general understanding as will be demonstrated below and suchwise stimulating new ideas. Total energy considerations and a new application of the CR model leads for the first time to a direct relation between energetic position in the spectrum and atomic configuration.

2. THE EXPERIMENTAL BACKGROUND AND SOME REMARKS ON MODELS

The continuous distribution of interface states at Si/SiO_2 is not of uniform chemical character and certain groups of states with different behaviour in technological processes and stress are detected (Flietner 1985, 88):
- U_T: Exponential tails adjacent to the band edges leaving a gap of $(3/4)\Delta E$ free of states, insensitive to technological processes;
- U_M: Camel hump like distribution with its minimum E_0 somewhat below midgap and small gaps of about $(1/8)\Delta E$ at each side of the distribution up to the band edges, $E < E_0$ donorlike, $E > E_0$ acceptorlike states;
- P_L, P_H: Gaußian peak like distributions low and high in the gap.

© 1991 IOP Publishing Ltd

- E'-centers with an energetic position above the conduction band edge found by EPR and acting as positive fixed charge in the oxide near the interface.
The mean value of energy for the donor states of these distributions are ordered in the way $E_M < E_L < E_H < E'$ with decreasing width of these distributions $\sigma_M > \sigma_L > \sigma_H > \sigma'$.

During **all** processes so far exerted each group changes uniformly, no part of the spectrum has any preference. This statement expresses the chemical uniform character of the group. Strictly speaking: All defects have the same chemical potential and therefore all neutral defects the same total energy. Pure incomplete splitting of bonding/antibonding states (Hasegawa 1986, 89) can neither explain this property nor the long known difference between U_T and U_M nor the existence of the other groups nor the difference in chemical activity of the different groups.

It is the lucky advantage of the Si/SiO$_2$-System that one may observe **all** kinds of such defects (all groups cited above) and that EPR clearly identfied the DB character of U_M, P_L, and E' as Si-DB with no, one and two Si-O back bonds. 3-5 and other compounds only exhibit U-shaped distributions (Hasegawa 1983, 89, Shinoda 1981, Hirota 1982). Relying on the results for Si the unified defect model (Spicer et al 1982, Newman et al 1986, Lindau et al 1986) contradicts the observed amphoteric character of U-shaped distributions and the correlated Fermi-level pinning.

3. THE CR-MODEL AND ITS APPLICATION TO INTERFACES

DB were the first canditates to explain gap interface states at homopolar semiconductors since bond breaking releases the gap creating bonding/anti-bonding splitting. This bond breaking process may be illustrated by a simple total energy consideration: Displacing one Si-atom from its equilibrium position in the ideal lattice gives a localized distortion the energy change of which must not exceed the bonding energy E_{Si-Si}, otherwise bond breaking will appear. This is the reason why the strained bond group U_T leaves a gap of $(3/4)\Delta E$ free of states (Flietner 1985). The generated DB-defects devide their energy into an electronic and a lattice part, $E_e + E_L$, the total energy being the same for all defects of the group with same occupancy. The electronic part of the energy changes as seen in the spectrum. So the width of the spectrum σ is a measure of the lattice deformation or of the ability for reconstruction.

The total energy of the DB-defect may be given in a simple way by the CR-model (Harrison 1976) based on the BOM(**b**ond **o**rbital **m**ethod) approximation. This concept has found a widespread application for clean surfaces to explain superstructures. For isolated defects it was applied to give an estimate of the amphoteric character of DB (Ngai, White 1981).The total DB-energy is composed of the equilibrium hybrid energy E_h, the change of hybrid energy due to deformation described by Λ and x, the elastic energy and the electron repulsive energy:

(1) $E(x;n) = E_h n - \Lambda x(n-1) + (1/2)[Cx^2 + Un(n-1)]$.

n = 0,1,2 are the numbers of electrons attached to the DB- defect. The essential point is that the dehybridization energy may become negative to reduce the repulsive energy and in consequence the distance between the donor and acceptor level. Calculating the minimum of (1) for n = 1 and n = 2 gives

(2) $E(1) = E_h + \Lambda^2/(2C)$ $E(2) = 2E_h + U$ with $E(0) = 0$.

Apparently the repulsive energy of the electrons $E(2) - 2E(1)$ is reduced to

(3) $U_{eff} = U - \Lambda^2/C = U(1 - \lambda^2)$ with $\lambda^2 = \Lambda^2/(UC)$.

This corresponds to a postive or negative Anderson-U depending on $U_{eff} >$ or < 0. A negative Anderson-U was observed for vacancies in Si (Watkins 1984), therefore both possibilities may be expected for Si-interfaces. (3) gives the splitting of the two 1e-levels(donor/acceptor) E_1 and E_2. For a first approximation U and C may be kept constant for all defects of the group. λ is then characteristic for a special configuration of the defect. It has its maximum value for the ideal tetrahedral configuration and becomes smaller the larger the deviation. λ will be zero for the rectangular and the planar configuration where the DB wave functions are pure s- or p-type. According to the above cited strain situation at interfaces a distribution of DB-hybrids is to be expected. This gives a distribution in λ. The mean value $E_0 = (E_1 + E_2)/2$ of the symmetrically split 1e-levels E_1 and E_2 is independent of λ and characteristic for the group. It is the 2e-level of the group or the level for homopolar bonding. The analogue situation in chemical binding is: E_1 corresponds to the ionization energy and E_2 to the e-affinity. E_0 corresponds to the chemical potential and has a close relation to the electronegativity and therefore also to the binding energy (Pauling 1964). So far equations and statements hold for all kinds of DB-defects. This model of CR in combination with the complex band structure fully describes the observed distributions, Fermi-level pinning and the chemical behaviour.

While the donor/acceptor splitting for the CR-model is given by (3), the corresponding splitting for the U_M-group in the complex band structure model is given for the strict symmetrical case ($m_v = m_c$, density of states effective masses; E_0 symmetry point) by (Flietner 1985):

(4) $E_2 - E_1 = \Delta E(1 - \kappa^2)^{1/2}$ $\kappa = K/K_0$, $K_0 = \Delta E m/(2\hbar^2)$.

Equating (3) and (4) gives the desired relation between λ and κ:

(5) $U(1 - \lambda^2) = \Delta E(1 - \kappa^2)^{1/2}$

This is the first time that a direct relation is established between a parameter characterizing the state in the spectrum (κ) and a parameter characterizing the atomic configuration of the defect (λ).

The complex band structure allows to implement the non symmetrical case $c^2 = m_v/m_c \neq 1$ into the CR by the simple linear transformation:

(6) $(E_c - E_1)/(E_1 - E_v) = c^2(E_2 - E_v)/(E_c - E_2)$.

This relation describes the deviation from symmetry in the spectra. E_0 becomes

(7) $E_0 = (E_v + E_c)/2 + \tfrac{1}{2}\Delta E(1-c)/(1+c)$.

Within the framework of BOM this would have been difficult to implement.

If U exceeds $(3/4)\Delta E$ then reconstruction cannot afford to compensate totally the e-e repulsive energy and a Hubbard gap will appear between the donor and acceptor states. The homopolar tetrahedral bond is the most sensitive to reconstruction as stated above. Any deviation from this bonding character lowers the energy which may be supported via dehybridization. This situation applies to the groups P_L, P_H and E' with increasing extent. While U_M being a continuous donor/acceptor spectrum, P_L, P_H and E' are the donor parts of a spectrum with a Hubbard gap. This different behaviour may be described in the CR model by the parameters Λ_n, C_n, U_n corresponding to Si-DB defects with n Si-O back bonds. The relation between the parameters is: $\Lambda_0 > .. > \Lambda_3$, $C_0 < .. < C_3$, $U_0 < .. < U_3$.

In conclusion the character of the groups is given by:
- U_T: strained Si-Si-bonds
- U_M: DB of character $DB_0 = Si_3Si-$ with E_0-level E_{00}
- P_L: DB of character $DB_1 = Si_2OSi-$ with E_0-level E_{01}
- P_H: DB of character $DB_2 = SiO_2Si-$ with E_0-level E_{02}
- E': DB of character $DB_3 = O_3Si-$ with E_0-level E_{03}.

Clearly the corresponding values of E_0 are ordered in the row $E_{00} < E_{01} < E_{02} < E_{03}$ and describe the differences in chemical activity while the decreasing width σ of the spectra is explained by the increasing rigidity due to increasing numbers of Si-O back bonds.

The E_0-value of a surface does not depent on defect creation but on the chemical character and as a consequence on crystal orientation and compound of the surface. Bringing together two surfaces like in a hetero-structure the bands are aligned without any shift if the two values of E_0 are equal. Otherwise an additional dipole component arises and a shift of bands which corresponds to the ionic character of the chemical bond. This may help for a better understanding of the band offset problem (Tersoff 1985, Flores 1987).

CONCLUSION: Even this rough one-parametric model of defect distributions gives a clear picture of the involved nature of the defects and a description of the involved phenomena. This may be a step in for a more detailed description of defect behaviour and elementary processes (recombination etc).

REFERENCES

Articles 1987 in Zs Phys Chem (NF) **151** 165...235
Fahrner W R, Klausmann E 1986 in Barbottin G, Vapaille A, Eds: Instabilities in Silicon Devices, North Holland
Flietner H 1972 phys stat sol (b) **54** 201; - 1974 Surface Science **46** 251; - 1985 phys stat sol (a) **91** 153; - 1988 Surf Sci **200** 463
Flores F, Indurian J C, Munoz A 1987 Physica Scripta T**19** 102
Harrison W A 1976 Surface Science **55** 1
Hasegawa H et al 1983 Thin Solid Films **103** 119; - 1986 J Vacuum Science and Technol B**4** 1130; - 1989 SPIE Vol 1144: InP and related Mat for advanced Electronic and Optical Devices p 150
Hirota Y, Kobayashi T 1982 J Appl Phys **53** 5037
Kingston R H Edit 1957 Semiconductor Surface Physics, Univ Pennsylvania Press
Kooi E 1967 Surface Properties of Oxidized Silicon, Eindhoven
Lang D V, Chen M C 1985 Proc 17th Int Conf Phys Semicond, Springer
Lenahan P M, Dressendörfer P V 1984 Appl Phys Lett **44** 96; J Appl Phys **55** 3495
Lindau I, Kendelewicz T 1986 CRC Crit Rev Solid State and Mat Science **13** 27
Newman N, Spicer W E, Kendelewicz T, Lindau I 1986 J Vac Sci Technol B**4** 931
Ngai K L, White C T 1981 J Appl Phys **52** 320
Nicollian E H, Brews J R 1982 MOS Physics and Technology, Wiley, N.Y.
Pauling L 1964 Natur der chemischen Bindung, Weinheim, p 91
Pointdexter E H et al 1984 J Appl Phys **56** 2844
Sah C T 1976 IEEE Trans Nucl Sci **23** 1563
Shinoda Y, Kobayashi T 1981 J Appl Phys **52** 6386
Spicer W E, Eglash S, Lindau I, Su C J, Skeath P R 1982 Thin Sol Films **89** 47
Tersoff J 1985 Phys Rev B**32** 6968
Watkins G D 1984 Festkörperprobleme **24** 163

**Dynamics of silicon oxide growth: molecular simulation and test**

V.V.PHAM, R.RAZAFINDRATSITA, G.SARRABAYROUSE, J.J.SIMONNE
Laboratoire d'Automatique et d'Analyse des Systemes
7, avenue du Colonel Roche 31077 Toulouse Cedex France

ABSTRACT: A molecular model is used to simulate the early stages of silicon oxidation at atomic scale (<10Å). Interactions of oxygen atoms with the two first silicon atomic layers are involved. A fast rate followed by a slow rate gives a linear-logarithmic relationship, consistent with Morita's experiments indicating a layer by layer growth of native oxide in the air.

1. INTRODUCTION

It is obvious that questions on proposed mechanisms for native oxide formation in silicon oxidation will continue to feed research programs, according to the survey paper presented by Deal B.E.. This statement has been emphasized by the shrinking of silicon devices sizes which enlarges the problem to the growth of very thin oxides, problem on which a microscopic simulation approach is presented.

2. POSITION OF THE PROBLEM

If we goes back to the well-known linear-parabolic model proposed by Deal, (Deal 1963 1965 1988) we observe that it has accurately predicted the effect of process variables on the oxidation and was extensively used as long as oxide thickness was centered around thousand angströms and more. When microelectronics became concerned with smaller oxide thicknesses, new models on dry O_2 oxidation were presented and are listed in (Deal-1988). Among them, one based on parallel mechanisms proposed by Han (Han.1987) used the sum of the linear-parabolic oxide growth rates; combining the single diffusion/parallel interface model (Ghez 1972)(Blanc 1978) and the parallel diffusion/parallel interface model (Rayleigh 1966), the model, where two oxidizing species are involved, proposed a diffusion of molecular oxygen through former grown SiO_2, a dissociation into atomic oxygen to react with silicon, leading to an oxygen deficient defect which diffused up to the surface and reacted with adsorbed O_2 (Helms 1988). This very attractive model is consistent with experiment from 20 Å to 300 Å and up. The interest of our paper is focused on the very beginning of the dry oxidation process on ideal silicon (thickness < 10 Å).

3. ATOMIC SCALE MODEL

The basic method of this model is a computer simulation whose result depends only on the potential energy of the system. Two ways are proposed:

- The molecular mechanics method(Ghaisas 1986), which relies on the chemical bond concept more adapted to heteroepitaxy where a crystal growth upon a crystal is considered.
- For amorphous silicon oxide growth on silicon, we have preferred to use the three-body model (Tiller 1984), which can account for large deviations from the natural crystal organization.In the many-body interaction model, the total potential energy is, for a system of n particles, expressed through n-body potentials. The two-body term is a LEONNARD-JONES type; the three body term is a AXILROD-TELLER type.

4.ANALYSIS OF THE PROCEDURE,RESULTS AND INTERPRETATIONS

The procedure would consist in the deposition of one atom on the free silicon surface (which presents dangling bonds). These bonds rearrange themselves to minimize the energy of the surface. The atoms move from their initial position until they get a new steady state. This is valid when the outer atom is larger than the silicon one, and we get effectively one solution to the problem. However,when the diameter of the outer atom is much smaller than the silicon one, and this is the case with oxygen, you may get a variety of solutions which depend on the exact location where the oxygen atom has been initially deposited on the silicon surface.

We have therefore deposited on an ideal silicon surface a package of oxygen atoms and let them interact with the first and the second silicon layers. After relaxation through the METROPOLIS test (Metropolis 1943) we get the final configuration similar to that presented in Fig. 1

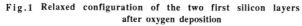
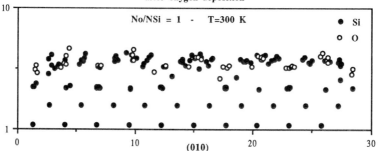

Fig.1 Relaxed configuration of the two first silicon layers after oxygen deposition

Then, we measure the thickness in which the oxygen is present and get the curves of the Fig. 2 as a function of the number of oxygen atoms deposited on a unit area of the silicon surface, the procedure being conducted with various silicon crystal orientations.

Concerning the silicon oxide formation, we perform a simulation which exhibits atoms movements towards the three space dimensions. However, for the oxide growth analysis, we only take care of the vertical displacement of the atoms. We observe that most of the silicon atoms move upwards. As the number of oxygen atoms goes along increasing, the distance between the two silicon layers is extended

If we examine in fig.3 the extreme positions of the oxygen atoms in the structure, it is shown that, with respect to the initial position of the surface, some of them go upwards to oxidize silicon atoms which are attracted in that way, while others go deeper into the substrate.

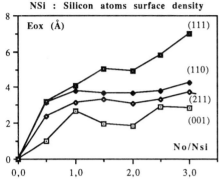

Fig.2 Oxide thickness versus No/NSi
No : oxygen atoms deposited per silicon surface unit area
NSi : Silicon atoms surface density

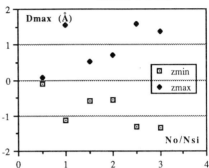

Fig.3 Vertical maximum location changes (up and down) of oxygen atoms [Si(001) T=300K]

5. OXYGEN PENETRATION DEPENDENCE ON THE CRYSTAL ORIENTATION

The oxygen penetration D_p into the substrate from the initial position of the surface before oxidation is presented in fig.4 for different silicon crystal orientations.

We observe that, when the number of oxygen atoms deposited N_O does not exceed the silicon atom density per unit surface area N_{Si}, the inequalities

$$D_p(001) < D_p(\bar{2}11) < D_p(111) < D_p(110)$$

follow the same comparative values of N_{Si}:

$$N_{Si}(001) < N_{Si}(\bar{2}11) < N_{Si}(111) < N_{Si}(110)$$

For $1<(N_O/N_{Si})<2$, the new inequalities become

$$D_p(111) < D_p(\bar{2}11) < D_p(001) < D_p(110)$$

similar to the progression of the distance d between two adjacent silicon layers:

$$d(111) < d(\bar{2}11) < d(001) < d(110)$$

Beyond $(N_O/N_{Si})>2$, Dp appears insensitive to the crystal orientation.
This behaviour is consistent with the simulation conditions, allowing the two first silicon layers to move upwards while they interact with oxygen, the silicon layers underneath being considered fixed.

6. APPLICATION TO EXPERIMENTAL DATA

Few experimental data concerning the early stages of silicon oxidation at atomic scale are available in the litterature and are generally focused on the growth of native oxide on Si surfaces exposed to air (Taft-1988). The reason is that oxygen molecules cannot start alone a thermally grown oxide (Morita 1990) and the oxidation mechanism must be initiated by a reaction of oxygen molecules with ionic species (in presence of water for example) (Raider 1975). Identification of these chemical species involved in this reaction is out of our scope;

however considering that oxygen atoms are the result of the reaction and act in the first phase of oxidation, comparison with our data of native oxidation experiments such that recently published by Morita et al(Morita 1990)becomes significant.
The growth generally follows a fast-slow law which can be reasonably simulated by a linear-logarithmic time dependance, each time a silicon oxide layer is grown (Fig 5). Parameters used to fit the linear part of the growth simulation still verify the logarithmic part of the curve. In our example,the curve is limited to one step due to the number of mobile silicon atomic layers, restricted to two in this program. The dynamic of oxidation is, in that way, described by an analytical function.

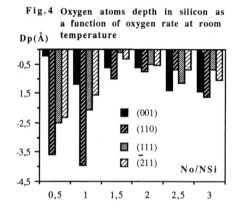

Fig.4 Oxygen atoms depth in silicon as a function of oxygen rate at room temperature

Fig.5 Oxide thickness as a function of oxygen exposure time

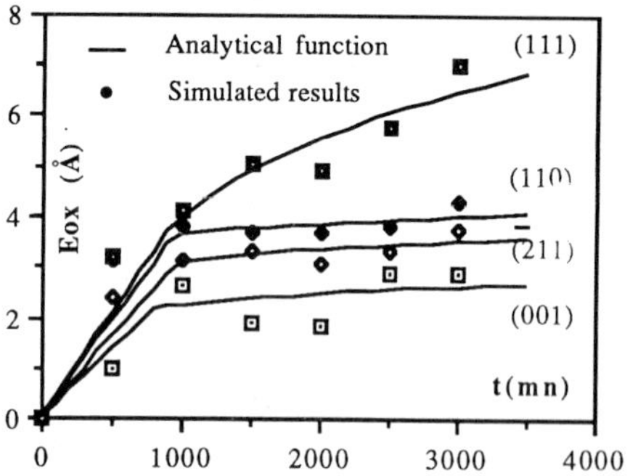

REFERENCES

Blanc J.1978 Appl.Phys.Lett.,33,424
Deal Bruce E. 1963 J.Electrochem.Soc.,110,527
Deal Bruce E. and Grove A.S. 1965 J.Appl.Phys., 36,3770
Deal Bruce E. 1988 The physics and chemistry of SiO_2 and the
  Si-SiO_2 interface Ed.C.R.Helms and B.E.Deal p 5-16
Ghaisas S.V.and Madhukar A. 1986, Phys.Rev.Lett,56,1056
Ghez R.and Van der Meulen Y.J. 1972 J.Electrochem.Soc.,119,1100
Han C.J. and Helms C.R. 1987 J.Electrochem Soc.,134,1297
Helms C.R. and de Larios J.1988 The physics and chemistry of
  SiO_2 and the Si-SiO_2 interface
    Ed.C.R.Helms and B.E.Deal p 25-34
Metropolis N. et al. 1943 J.Chem.Phys.11,299
Morita M. Ohmi T. Hasegawa E. Kawakani M. and Ohwada M. 1990
  J.Appl.Phys.68,1272
Raider S.I. Fitsch R. and Palmer M.J. 1975
  J.Electrochem.Soc. 122,413
Rayleigh D.O. 1966 J.Electrochem.Soc. 113,782
Taft E.A. 1988 J.Electrochemical Soc., 135,1022
Tiller W.A. 1984 J.of Crystal Growth 70,13

Effect of preliminary atomic hydrogen treatment on Si oxidation

I.A.Aizenberg, A.V.Andrianov, S.V.Nosenko, V.A.Khvostov

Institute of Microelectronics Technology Academy of Science, 142432 Chernogolovka, Moscow District, USSR

Abstract: It has been found that preliminary atomic hydrogen treatment of Si above 723K considerably affects the process of further Si oxidation. The changes of the structural properties of $Si-SiO_2$ interface and the mechanism of the effect observed are discussed. The correlation between the oxidation rate and interface roughness has been obtained.

Great interest has been recently aroused in studies of the use of atomic hydrogen in microelectronics technology, mainly for low temperature passivation (T<400°C) of electrically active defects in semiconductors (Pearton et al 1987). Besides, we have shown (Aizenberg et al, 1989) that atomic hydrogen can be used for cleaning of Si surface by etching of Si and SiO_2 at elevated temperatures.

The present paper is concerned with the study of the effect of preliminary atomic hydrogen treatment on Si oxidation and some properties of $Si-SiO_2$.

Two Si samples located sequentially along the gas flow were treated by hydrogen under the pressure of 10Pa at 723-1343K. Then within one minute the atmosphere in the reactor was substituted by dry oxygen and thermal oxidation of silicon was carried out at 10^5Pa. The kinetic data of Si oxidation as a function of the conditions of preliminary atomic hydrogen treatment is displayed in Fig.1. Firstly, let us consider the case when the temperature of hydrogen and oxygen treatments was the same. It can be seen that the oxidation rate of hydrogen-pretreated samples is significantly less, essentially at the initial stage, than that of control samples (without hydrogen treatment), and the shape of the kinetic curves is considerably different. The rate of oxidation slightly increases but only at the initial stage and remains unchanged at longer oxidation time if the time of hydrogen treatment decreases from 30 to 15 min. Besides, in this case the whole shape of the curve is more smooth. It should be noted that the shapes corresponding to the first and second samples are similar while the points of the second curve are located somewhat higher. The concentration of atomic hydrogen near the surface of the second sample is

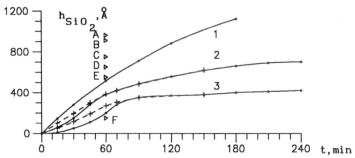

Fig.1 Kinetics of oxidation: • -is the temperature of oxidation and hydrogen treatment $T_{ox}=T_h=1243K$, time of hydrogenation $t_h=30min$; ×-$t_h=15min$; 1-control sample, 2 and 3- second and first samples in the case of preliminary hydrogen treatment; points A-F correspond to oxidation at $T_{ox}=1323K$ and hydrogenation at T_h:B-723K, C-873K, D-1023K, E-1173K, F-1323K, A-control sample (without hydrogen)

lower owing to homogeneous and heterogeneous recombination and, therefore, it is natural to suppose that the value of oxidation delay is in inverse proportion to the dose of hydrogen introduced into silicon. Such suggestion is confirmed by the dependence of grown oxide thickness on atomic hydrogen treatment temperature at fixed temperature and time of oxidation (points A-F Fig.1). Preliminary hydrogen treatment influences the oxidation kinetics at temperatures up to 723K. It is interesting that the effect of oxidation rate retardation does not depend on: i) additional high temperature vacuum annealing of hydrogenated samples; ii) the presence of oxide film on the Si surface during hydrogenation and further cooling (down to room temperature) prior to oxidation. Thus, we can expect that at these conditions atomic hydrogen penetrates effectively into the Si bulk where it is in a very stable state.

The investigation of these samples by AES spectroscopy confirms this suggestion. The most characteristic AES spectra for the investigated samples are shown in Fig.2.

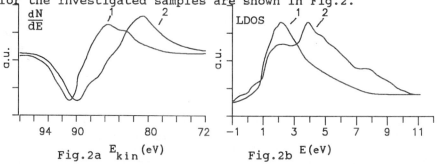

Fig.2 Si(LVV) Auger spectra (a) and LDOS vs binding energy for Si (b) before (1) and after (2) hydrogenation

According to the data obtained by Zajac and Bader (1982), such spectrum points to the presence of Si-H bonds rather than $Si-H_2$ or $Si-H_3$. Their concentration decreases when getting away from the interface in both Si and SIO_2 bulks, and increases with increasing temperature of hydrogen treatment and decreases with longer oxidation time. Besides, AES spectroscopy data reveal the changes in the SiO_2 film stoichiometry of the SiO_2 film on Si treated by atomic hydrogen. It is natural to associate the observed oxidation retardation effect with the saturation of Si by hydrogen during this treatment. Upon further oxidation hydrogen starts to diffuse from the Si bulk and is accumulated as Si-H or SiO-H groups at the $Si-SiO_2$ interface which is the oxidation front. From the RFS data it is known that accumulation of hydrogen as Si-H bonds at the $Si-SiO_2$ interface the increased concentration of which is also revealed by IR spectroscopy takes place even during oxidation in dry oxygen (Revesz 1979). Distinct accumulation of deuterium at the interface after annealing of $Si-SiO_2$ structures in atomic deuterium at 503K was revealed by SIMS (Revesz 1979). The thermal stability of Si-D bonds was also revealed, namely, plenty of deuterium disappeared at the $Si-SiO_2$ interface upon annealing above 873K. Groups of Si-H in Si have the same thermal stability (Wu and Qin 1987). Groups of SiO-D and SiO-H have a higher thermal stability (1073-1273K) (Mikkelsen 1981). It is quite natural that the presence at the $Si-SiO_2$ interface of a high concentration of such hydrogen-containing groups as Si-H and SiO-H the bond energy of which is higher than in the Si crystal ($E_{Si-H} \approx$ 70Cal/M and $E_{SiO-H} \approx$ 110Cal/M compared to $E_{Si-Si} \approx$ 55Cal/M) will considerably retard oxidation of Si, particularly at the initial stage when the oxidation rate is limited by the rate of the chemical reaction at the interface.

Nonmonotonic changes of $Si-SiO_2$ interface roughness during oxidation of usual Si and Si pretreated by atomic hydrogen were revealed by the ellipsometric method that we have developed (Aizenberg *et al* 1989). Fig.3 shows the experimental values of the angular change of $\delta\Delta$ (the difference $\delta\Delta=\Delta_e-\Delta_m$ between the experimental value of polarization angles Δ_e and the computed value Δ_m) as a function of oxidation time (temperature 1123K). As it is evident from fig.3, after a definite oxidation time, the monotonic curve $\delta\Delta$ becomes a function with several extrema, and the angular position of the extrema of various curves is intact but the absolute value $|\delta\Delta|$ is changed. Thus, the behavior of the curves points to the changes of the surface roughness of the $Si-SiO_2$ interface during oxidation of Si.

Fig.4 displays the curves of the model calculations for a

homogeneous transparent film on the rough surface. Calculations were carried out for the following values of roughness parameters : the rms deviation of the values of $Z(X,Y)$ from the mid-plane is $\sigma=2\div50\text{\AA}$, the medium distance between the micro-peaks is $\gamma=300\text{\AA}$, the value of residual oxide is $d=20\text{\AA}$. As seen from comparison between fig.3 and fig.4, the positions of the extrema of the calculated and experimental values of $\delta\Delta$ are closely located and by the absolute value. This enabled us to use the high sensitivity of ellipsometric parameter Δ in the vicinity of the pseudo-Brewster angle for characterization of Si surface roughness.

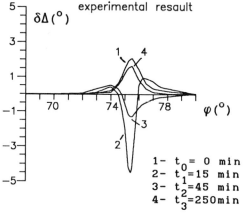

Fig.3   Fig.4

The characteristic changes of σ as $Si-SiO_2$ interface roughness were observed upon oxidation or annealing. Fig.5 shows the value σ as a function of time, temperature and the treatment media. The correlation between the oxidation rate (curves 2,3 Fig.1) and interface roughness (curve 4 Fig.5) has been found.

Fig.5 Oxidation of Si at temperature $T_1=1323K$ (curve 1) and $T_2=1243K$ (curve 2) ; 3 - vacuum annealing at T_2; 4 - oxidation after hydrogen treatment (curve3 Fig.1)

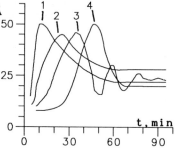

References.
Aizenberg I A, Moskvina I R, Nosenko S V, 1989, Appl. Surf. Sci., v39, pp 473-6
Mikkelsen J C, 1981, Appl.Phis.Lett., v39, pp 601-3
Pearton S J, Corbett J M, Shi T S, 1987, Appl. Phys., v.A 43, pp 153-42
Wu J, Qin G G, 1987, Appl. Phys. Lett., v.50, pp 1355-3
Zajac G, Bader S D, 1982, Phys. Rev. B, v.26, pp 5688-6
Revesz A G, 1979, J.Electrochem.Soc., v.126, pp 122-9

An effective oxidation technique for the formation of thin SiO_2 at $\leqslant 500°C$

Vishal Nayar* and Ian W. Boyd

Department of Electronic and Electrical Engineering, University College London, Torrington Place, London, WC1E 7JE. (* V. Nayar is presently at the Royal Signals and Radar Establishment, St. Andrews Rd., Malvern, Worcs. WR14 3PS)

Abstract A new photochemical oxidation technique for the growth of SiO_2 on c-Si is described. Oxide formation was found to dependent upon substrate temperature and the ozone concentration in the reaction chamber. A Cabrera-Mott type growth process is implied from the analysis of the oxidation data. Oxides formed in this manner were incorporated into MOS capacitors and studied by the high frequency capacitance voltage and current voltage methods. The average breakdown field for the oxide was found to be $\approx 10 MV/cm$ excluding low field breakdowns and nearly ideal Fowler-Nordheim tunnelling was observed.

1 Introduction

The oxidation of silicon until recently has required high temperatures (>800°C and more typically $\approx 1050°C$). However as integrated circuits have become smaller and more densely packed the necessity for low temperature processes have become self-evident. Thin oxide formation is a critical step not only from the point of device performance but also because its thermal budget is large, and this can lead to undesired side effects, such as wafer warpage, unwanted dopant diffusion and defect generation. During the recent years some progress has been made to try to reduce oxidation temperatures by using novel techniques. The most effective of these have been the plasma afterglow (to generate atomic oxygen) method of Vinckier et al. (1987), ultraviolet (UV) irradiation by Nayar et al. (1989), growth in a liquid phase oxidant mixture of sulphuric and nitric acid by Uchida et al. (1986) and UV photo-dissociation of N_2O, Uno et al (1988) and Ishikawa et al. (1989). Our recent experiments with UV irradiation and the work of Vinckier et al. on atomic oxygen oxidation have suggested that a combined process should be more effective. An alternative to oxide growth is deposition however it has been shown recently by Nguyen et al. (1990) that the performance is very dependent upon the thin "grown" oxide at the interface. This implies that only a grown oxide is capable of producing the good interface properties necessary for microelectronics. In the following a simple technology which is capable of controllably growing large areas of very thin oxide on silicon at temperatures lower than 550°C is described.

2 Experimental

The experimental system is shown schematically in figure 1. It consists of a vacuum chamber containing a (20x20cm^2) low pressure mercury lamp grid which emits 184.9nm and 253.7nm radiation. Silicon samples were placed on a substrate heater under the lamp where temperatures up to 550°C could be induced. Electro-grade II (from BOC Ltd.) dry oxygen was passed through the chamber at a low flow rate. The reaction scheme is as follows:

$O_2 + h\nu(184.9nm) \rightarrow 2O$
$O + O_2 + M \rightarrow O_3 + M$ (M is a third body)

$$O_3 + h\nu(253.7nm) \rightarrow O_2 + O$$

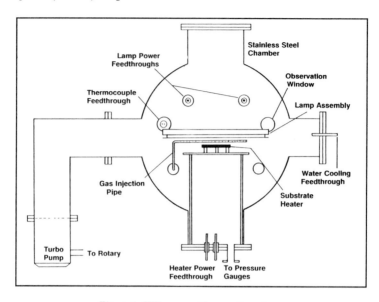

Figure 1: UV/ozone oxide growth system.

The 184.9nm radiation photo-dissociates oxygen molecules to ground state radical atoms. These almost immediately react with other oxygen molecules to form ozone (O_3). However the second and more intense wavelength 253.7nm is very strongly absorbed by O_3 and leads to its destruction. Thus a dynamic equilibrium forms with a reservoir of reactive oxygen species in the form of O_3 and O radicals. It is important to note here the simplicity of our technique and the many inherent advantages it offers. The oxidation is carried out at atmospheric pressure and is based solely on oxygen chemistry. Very high purity oxygen is easily obtainable whereas other gases are commonly not so well refined. Only the sample is heated and therefore contamination from chamber walls is negligible. Additionally Ruzyllo et al. (1987) have reported that UV/ozone treatments reduce surface contamination. The system can be scaled up to oxidise Si wafers of any dimension simply by increasing the size of its components. In particular the grid lamp area can be easily enlarged and its output intensity improved.

The processing chamber was evacuated with a turbo pump to $\approx 10^{-6}$mbar and then filled to atmospheric pressure with oxygen and the flow rate was set. The samples were irradiated initially at near room temperature for five minutes. After this period the substrate heater was switched on. The growth time noted below is the time at temperature excluding the heat up period. P-type (boron doped), 0.2-1Ωcm, Si substrates with <100> surface orientation were oxidised at a variety of growth temperatures. Prior to oxidation the Si samples were given two different surface treatments. Following a dip in dilute HF to remove the native oxide the Si was given either a rinse in DI water (13MΩcm) for 5 minutes or a H_2SO_4/H_2O_2 immersion for 5 minutes and a DI water rinse (samples with the first process are suffixed HF and those with the second BOMB). Oxide thickness was measured using a single wavelength ellipsometer (Rudolph Auto ELII) assuming a fixed refractive index of 1.462 and is the average of at least 12 measurements taken over each sample.

The UV/ozone oxides were incorporated into metal oxide semiconductor structures by thermal evaporation of Al through a shadow mask to form contact dots. The nominal Al gate capacitor area was 1.96×10^{-3}cm^2. All electrical data was collected without any post oxidation or forming gas anneals. Current-voltage (IV) characteristics were measured using a voltage staircase with negative gate bias, to determine oxide breakdown fields, defined as those which caused a catastropic failure of the oxide. IV data was also used to study the Fowler-Nordheim (F-N) tunnelling behaviour.

3 Results

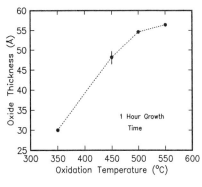
Figure 2a

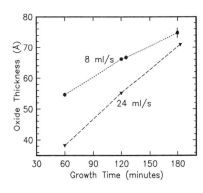
Figure 2b

Figure 2a shows the oxide grown during one hour of exposure time at a gas flow rate of 8ml/s for different substrate temperatures. It is important to note that thermal oxide growth in pure O_2 at these temperatures is negligible. Control experiments to determine the thickness of oxide grown with the lamp switched off confirmed this to be the case. The growth rates here are much improved in comparison to our previous study, primarily because of increased lamp intensity, but were also strongly influenced by the flow rate of gas through the reaction chamber. Figure 2b shows the growth of oxide at 500°C for two gas flow rates (8 and 24ml/s) where it is clear that the growth rate is lower for higher gas flow condition. The ozone concentration generated inside the chamber was measured by absorption spectroscopy in separate experiments and was also found to decrease when the oxygen flow rate was increased. These results strongly link the oxide growth rate with the ozone concentration in the reaction chamber.

Spectroscopic ellipsometry, infrared spectroscopy, x-ray photoelectron spectroscopy all confirmed that stoichiometric SiO_2 was grown by the UV/ozone process. Averaged results from the IV characteristics collected are shown in the table for oxides grown at 500°C. The majority of capacitors break down at high fields. At least 25 capacitors on each sample were analysed. The number of breakdowns at low fields (i.e. those showing no F-N behaviour) over all the samples analysed was around 20%. The breakdown field averages in table I exclude these low field failures and are calculated to be around 10MV/cm. Good thermal oxides of similar thicknesses have average breakdown fields of around 12-13MV/cm.

Sample	612HF	612BOMB	512HF	512BOMB
Mean E_{BD} (MV/cm)	10.9 ± 0.7	9.5 ± 1.1	10.9 ± 1.5	9.5 ± 1.7
Oxide Thickness (Å)	70.0 ± 1.6	79.4 ± 1.7	82.2 ± 4.9	82.0 ± 2.7

Close inspection of the table reveals that the different chemical processes prior to oxidation apparently did not have an impact on these particular oxide properties. It was however generally found that that samples which were processed in H_2SO_4/H_2O_2 had slightly thicker oxides. The origin of this is the very thin chemical oxide that is grown during this chemical treatment. It should be emphasised, however that the UV/ozone oxidation apparatus was situated in a non-clean room environment and that chemical processing was carried out with only electronic grade reagents rather than the higher purity chemicals commonly used in VLSI surface preparation.

Characteristic F-N tunnelling was observed as is shown in Figure 3, where $\log_{10}(I/V^2)$ is plotted against $-1/V$ and a straight line is appropriately fitted. A weak oscillation of the form reported by Maserjain (1988) for thinner thermal oxides in the F-N curve is evident. The oxide thickness in the present case is just at the limit of the value ($\approx$7nm) for which oscillations are predicted to appear and therefore is likely to be affected by scattering in the oxide. However its presence indicates that the interface between the Si/SiO$_2$ is abrupt and the transition region is thin.

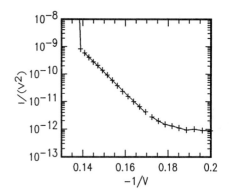

Figure 3 Typical F-N tunnelling characteristic.

4 Conclusion

To conclude, we have demonstrated enhanced oxide growth at 500°C by UV/ozone stimulation which would require a thermal oxidation temperature of at least 775°C. Oxide growth rates reported here are significantly greater than in our previous study, Nayar et al. (1990). An improvement of the growth rates beyond the present results should certainly be possible hence reducing the oxidation period further. Given that there were no post-oxidation anneal steps it is both surprising and encouraging that oxides displaying reasonable electronic performance were grown. The good electrical and materials properties which have been demonstrated indicate that with the other advantages of the UV/ozone technique it may become a candidate for ultra-thin oxide formation for microelectronic uses.

5 Acknowledgements

This work was partially funded under SERC contract GR/F 02229 and the EEC/SCIENCE initiative. V. Nayar would like to acknowledge the Royal Signals and Radar Establishment and SERC for a joint CASE studentship. We would like to thank Prof. M. Green (Imperial College London) for use of the single wavelength ellipsometer, Parthiv Patel for assistance with some of experiments.

6 References

Ishikawa Y., Takagi Y., and Nakamichi I.: Low temperature thermal oxidation of silicon in N$_2$O by UV-irradiation, Japanese. J. Appl. Phys., 1989, 28(8), pp.L 1453-1455.

Maserjain J., "Historical perspective in tunnelling in SiO$_2$," in The physics and chemistry of SiO2 and the Si-SiO$_2$ interface, Plenum Press, p497, 1988.

Nayar V., Boyd I. W., Arthur G. and Goodall F.: Low temperature oxidation of crystalline silicon using excimer laser irradiation, Appl. Surf. Sci., 1989, 36, pp.134-140.

Nayar V., Patel P. and Boyd P., "Atmospheric pressure, low temperature (< 500°C) UV/ozone oxidation of silicon" in Electronics Letters vol.26 no.3, p205, 1990.

Nguyen S. V., Dobuzinsky D., Dopp D., Gleason R., Gibson M. and Fridmann S., "Plasma-assisted chemical vapour deposition and characterisation of high quality silicon oxide films" in Thin Solid Films, vol.193/194, p595, 1990.

Ruzyllo J., Duranko G. T. and Hoff A. M.: Preoxidation UV treatment of silicon wafers, J. Electochem. Soc., 1987, 134(8), pp.2052-2055.

Uchida Y., Jin-hai Y., Kamase F., Suzuki T., Hattori T. and Matsumura M.: Low temperature oxidation of silicon, Jap. J. of Appl. Phys., 1986, 25(11), pp.1633-1639.

Uno K., Namiki S., Zaima T. and Nakamura T.: XPS study of the oxidation process of Si < 111 > via photochemical decomposition of N$_2$O by an UV excimer laser, 1988, Surf. Sci., 33, pp.321-335.

Vinckier C., Coeckelberghs P., Stevens G., Heyns M., and De Jaegere S.: Kinetics of the silicon dioxide growth process in afterglows of microwave induced plasmas, J. Appl. Phys., 1987, 62(4), pp.1450-1458.

© Crown Copyright 1991/MOD, with the permission of the controller HMSO.

Paper presented at INFOS '91, Liverpool, April 1991
Contributed Papers, Section 2

Effects of cooling rate on MOS interface properties

K. Heyers, A. Esser, H. Kurz and P. Balk*

Institute of Semiconductor Electronics
Technical University, 5100 Aachen, Fed. Rep. Germany

*DIMES, TU Delft, 2600 GB Delft, NL

ABSTRACT:

Rapid cooling after the final high temperature treatment of MOS systems leads to an increased interface state density and a reduced minority lifetime in the Si substrate. Measurements on MOS-transistors showed an improved electron mobility at high gate fields for high cooling rates. The observed effects are explained by mechanical stress caused by the rapid cooling.

1. INTRODUCTION

Rapid Thermal Processing (RTP) is an attractive approach to high temperature treatments in the ULSI technology, like activation of dopants, silicidation and formation of insulating films. However, recent studies showed that RTP of layered systems, like Si-SiO_2, also has some disadvantages (Vandenabeele et al 1989). During RTP large lateral temperature gradients exist across wafers with patterned oxide layers which give rise to stress related defects. To achieve the short process times at high temperatures, which are typical for RTP, very high heating and cooling rates are necessary. Earlier studies have shown, that especially the cooling steps are critical (Heyers et. al 1990). In the present investigation we studied the effects of the cooling rate after the final high temperature treatment on the interfacial properties of the Si-SiO_2 system.

2. EXPERIMENTAL

Interface state density and minority carrier lifetime after different cooling steps were determined on MOS-capacitors with Al or poly-Si gate. The oxides (thickness 15 nm) were prepared by Rapid Thermal Oxidation (RTO) or by conventional furnace oxidation of 0.1 Ω cm p-type (100) Si wafers in dry O_2. Both methods produce identical results regarding the properties discussed in this paper. In the case of Al gate samples Post Oxidation Annealing (POA) was done by Rapid Thermal Annealing (RTA) in N_2 with 5% O_2 at 1100 °C or 1200 °C and different cooling rates. In the case of the poly-Si gate capacitors the POA was performed with the maximum cooling rate obtainable in our apparatus (300 °C/s); the cooling rate after Post Implantation Annealing (PIA) was varied. All samples had e-beam evaporated Al-metallization on front- and backside.
Transconductance measurements were performed on sub-µm LDD n-channel MOSFETs with gate length 0.9 µm. In this case the oxides were produced by RTO/RTA in the same manner as for the poly-Si capacitors, mentioned above. All lithography steps were done optically, except for the gate definition, which was carried out by e-beam lithography.
The interface state density of the MOS capacitors was determined by quasi-static CV measurements. The electrical determination of the minority carrier lifetime was performed by the steady-state ramp method after Pierret and Small (1975). The carrier lifetime was also determined from optical reflectance measurements (Opsal et al.1987). Although at the present time we are only able to extract qualitative results on the carrier lifetime from this technique, it has the advantage of being a contactless method which is very sensitive to the properties of the interface.

© 1991 IOP Publishing Ltd

3. RESULTS AND DISCUSSION

First the correlation between cooling ramp rates and the interface state density of MOS capacitors was investigated. To exclude the effect of oxide defects caused by oxygen deficiency during annealing at high temperatures POA was performed in N_2 with 5% O_2 using Al-gate capacitors. A distinct increase in interface state density in the upper half of the band gap, especialy around 0.7 eV above the valence band edge, was observed with increasing cooling ramp rate. The rest of the density distribution was unaffected (figure 1).

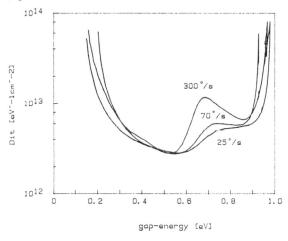

Figure 1: Interface state density of Al gate capacitors after cool down from RTA (1200 °C) at different rates

A similar experiment was carried out with poly-Si gate capacitors, where we varied the cooling ramp during PIA, while the cooling during POA was done at the maximum rate. Here a very similar influence on the interface state distribution was observed. However, in this case an additional increase of the density of states around midgap was found for faster cooling.

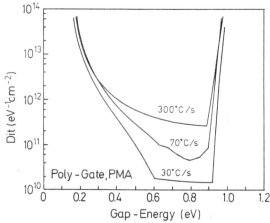

Figure 2: Interface state density of poly gate capacitors after PMA for different cooling rates at PIA (1100 °C)

After annealing in forming gas at 400 °C this increased interface state density could not be completely removed. Only the slowly cooled samples showed a low interface state density (in the order of 10^{10} cm^{-2} eV^{-1}) (figure 2). Our experiments further showed that the interface degradation is not affected by the number of cooling steps. However, the peak in the interface state distribution vanishes when rapidly cooled samples are annealed once more but using a slow cooling ramp. This suggests that the observed results are related to mechanical stress at the interface. If this is the case, the stress should also show up in the results of charge injection experiments. Avalanche injection of electrons indeed removes a large part of the structure (including the peak at 0.7 eV) from the interface state distributions for the rapidly (300°C/s) cooled samples. The distribution for the more slowly cooled samples is much less affected and only a small overall increase in density was obtained.

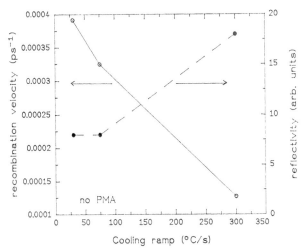

Figure 3: Optical and electrical measurements of minority carrier lifetime after cooling from RTA (1200 °C) at different rates

To see whether the silicon substrate is also affected by the cooling procedure we determined the minority carrier lifetime as a measure of the defect density in the silicon. A comparison of the data from electrical measurements, which give information on the behavior of defects in the space charge region deeper in the substrate, to optical data, which are dominated by defects very close to the interface, revealed that only the region near the Si-SiO$_2$ interface is degraded by rapid cooling (figure 3).

However, it is exactly in this damaged region that the current transport of surface channel MOS transistors takes place. To check the influence of the cooling step on the transistor characteristics the transconductance of devices cooled down at different rates during PIA was compared. Figure 4 shows that the peak value of the transconductance is reduced for the transistors cooled down at a high rate. This result can be explained by the increase in carrier scattering caused by the larger interface state density, as observed on MOS capacitors. Surprisingly, the transconductance at high gate voltages shows the opposite behavior. In this case the rapidly cooled samples exhibited a distinctly higher transconductance. Momose et al. (1990) found a very similar behavior on MOSFETs with nitrided gate oxides. They were able to correlate the effect with stress at the interface. Therefore, it appears likely that also in the case of our rapidly cooled samples mechanical stress near the Si-SiO$_2$ interface is responsible for the observed transconductance behavior. It was suggested (Momose 1990) that this stress possibly changes the Si band structure which could lead to an increased mobility.

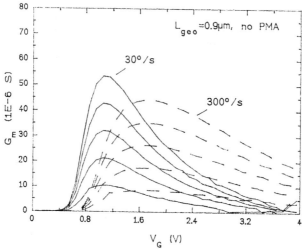

Figure 4: Transconductance of LDD NMOS transistors cooled down from RTA (1100 °C) at different rates

Our findings suggest that the cooling rate particularly affects the redistribution of the stress in the system by the viccous flow of the SiO_2. This effect is presently under investigation.

4. CONCLUSIONS

Rapid cooling in the final RTP high temperature treatment causes serious deterioration of the interfacial properties of the MOS system. With increasing cooling rate the interface state density and the defect density in the Si substrate close to the interface increase. For rapidly cooled transistors the maximum transconductance is reduced whereas the transconductance values at high gate voltages are improved. The observed effects are most likely due to mechanical stress induced by rapid cooling. Our findings also show that the cooling rate after high temperature treatment is a limiting factor in the application of RTP in the MOS technology.

LITERATURE

Heyers et al. 1990, 21th IEEE SISC, San Diego, Ca.
Momose et al. 1990 in IEDM Tech. Dig. p. 65,
Opsal et al.1987, J. Appl. Phys. 61, pp 240-8
Pierret et al. 1975, IEEE Trans. Elec. Dev. ED-22, 1051
Vandenabeele et al.1989, Mat. Res. Soc. Proc. 146, p 149

Paper presented at INFOS '91, Liverpool, April 1991
Contributed Papers, Section 2

Origin of reliability differences in oxides grown by RTO in HCl/O_2 and pure O_2 ambients

K. Barlow[*] and V. Nayar[#]

[*]GEC Plessey Semiconductors Ltd., Roborough, Plymouth, Devon PL6 7BQ
[#]Royal Signals and Radar Establishment, St. Andrews Rd., Malvern, Worcs. WR14 3PS.

Abstract Rapid thermal oxidation (RTO) is currently under investigation as an alternative to furnace oxidation (FO) in the production of gate quality oxides. Here we report the electrical and optical properties of RTO layers formed in pure O_2 and O_2/HCl ambients in comparison to standard furnace oxides. Both interface trap density and the charge to breakdown were found to degrade up on the addition of HCl to the growth ambient. The origin of this may be the increased roughness of RTO surfaces as indicated by spectroscopic ellipsometry on similarly oxidised wafers.

1 Introduction

Rapid Thermal Processing (RTP) of silicon is an attractive technique for the growth of thin layers of silicon dioxide. The technique offers the advantage of growing thin layers of oxide at very high temperatures (>1000°C) whilst maintaining good uniformity and reproducibility (Nulman et al., 1985). RTP is carried out in a quartz chamber in a tightly controlled atmosphere. The rapid temperature ramp up characteristic of the technique is provided by an array of tungsten-halogen lamps, emitting mainly in the infra-red and capable of rise times of almost 300°C/second. As the heating is essentially isothermal with heaters above and below the wafer temperature stabilisation is not required to prevent distortion. This gives very accurately controlled process times and reduces dopant diffusion to a minimum. The main thrust of the work has been in growing oxides in the 16nm range, to produce a layer that has comparable properties to a furnace grown oxide, which is the conventional processing method used for growing the gate oxide. Data will be presented on growth characteristics and reproducibility and on the electrical properties of the oxides grown. It has been suggested that HCl can lead to roughening of the oxide interfaces. In order to study the physical properties of these oxides spectroscopic ellipsometry (SE) was used. Nayar et al. (1991) have identified density changes in thin thermal oxide layers implying that oxide density is high near the Si/oxide interface and falls towards the surface. A low oxide surface density would suggest that a high level of voids or roughness exists. We attempt to correlate SE with electrical results.

2 Growth Characteristics and Device Fabrication

The RTP work was performed in an AG4100 heatpulse system using a susceptor. The susceptor removed the need to strip the wafer backs of any layers that otherwise cause severe problems with temperature measurement and control. In order to produce good quality dielectrics by this method a range of growth conditions were investigated. Fig 1 shows the RTO growth-time characteristics for a range of temperatures and ambients. Oxide thickness variation across a wafer was found to be <3%, with a similar figure for reproducibility. The characteristics show, as expected that incorporation of HCl into the ambient leads to an increase in the growth rate, as does increasing the growth temperature. It was found that prior to ramping the temperature a stabilisation period of at least 25seconds was required for ambient gas mixing, to ensure good uniformity. In order to assess the bulk and interface properties of the RTO oxides polysilicon gate capacitors were fabricated using the procedure outlined in Table I.

© Crown copyright 1991/MOD

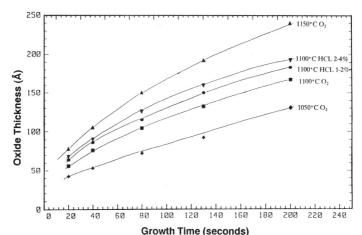

Figure 1: RTO growth-time characteristics

3 Electrical Characterisation of the Oxides

A range of electrical tests were performed to assess the quality of the layers produced. The breakdown voltage of the various oxides was obtained on large area capacitors (2.5mm*2.5mm) by voltage ramping to a predetermined current (1mA). This current did not necessarily give destructive breakdown for each device, but enabled a comparison of the various growth conditions being investigated. High/low frequency Capacitance-Voltage (CV) measurements were also used to obtain the interface trap density at midgap. The high frequency CV plot was performed at 100KHz, whilst quasistatic CV was used for the "low frequency" measurement. The measurements were taken simultaneously to remove possible errors. An assessment of the oxides resistance to stress was performed using charge injection, Q_{bd}. This was done using a current density of 100mA.cm^{-2}, with electron injection from the substrate. The time to breakdown, during the measurement was defined as the time for the voltage across the oxide to fall by 20% of its initial value.

Start material p-Si
Oxidation to grow 20nm
Implantaion set bulk doping level to $2 \times 10^{16} cm^{-3}$
Drive-in anneal
Nitride deposition, 150nm
Active area mask definition
Field oxidation
Active area strip
Gate oxidation
Poly Si deposition and doping
Poly Si definition and interlayer deposition
Reflow anneal, contact via etch
Metallisation and sinter anneal

Table I

Table II Q_{bd} and Interface Trap Density Values as a function of Growth Conditions

Temp. (°C)	%HCl	T_{ox} (nm)	Q_{bd} (Ccm^{-2})	D_{it} ($*10^{10}$.cm^{-2}eV^{-1})
1100	0	17.2	42.8 ± 3.4	1.5
1100	2	16.5	33.1 ± 9.4	2
1150	2	16.5	39.3 ± 8.3	8
1100	4	16.1	32.3 ± 9	8
900	3	16.1	43.1 ± 5	3 Furnace

Table II shows the range of RTO conditions investigated, along with the Q_{bd} and interface trap density (D_{it} midgap) values. Also shown are the results obtained for a furnace grown oxide that was used as a reference. It can be seen that the use of HCl in the RTO oxidising ambient leads to a reduction in Q_{bd}

and an increase in the midgap trap density. It was also found that the spread of values was larger for RTO oxides grown in a HCl ambient, as well as them having a lower mean Q_{bd} than the non-HCl case. Additionally it was noted that whilst the furnace oxides had the tightest breakdown distributions the RTO oxides also had reasonably good distributions with, in fact a slightly higher mean breakdown field. Table III shows the mean breakdown field (E_{av}) and the densities for the furnace and RTO oxides. All the ambients show low defect densities for type A and type B defects, in fact the RTO oxides have slightly lower levels than the furnace oxides indicating the cleanliness of the heatpulse system and the effectiveness of the procedure used for RTO growth.

Table III Breakdown Fields and Defect Densities for RTO and Furnace Oxides

Temp. (°C)	%HCL	E_{av} (MV/cm)	A (cm^{-2})	B (cm^{-2})
1100	0	12	0.8	0.1
1100	2	12.8	0.6	0.2
1150	2	12.4	0.6	0.3
1100	4	12.5	0.6	0.3
900	3	11.8	1	0.7

Type A defect: $0 < E < 1 MV/cm$, type B defect: $1 MV/cm < E < 5 MV/cm$

4 Spectroscopic Ellipsometric Analysis

Ellipsometric data over the wavelength range 250 to 800nm was collected using a commercial ellipsometer (from SOPRA). Typically 110 wavelengths were measured. Details of the instrument and data analysis have been reported previously by Nayar et al. (1991). To simulate the density grading a three layer model for the oxide was used for comparison with the single layer model. It should be noted that the three layer structure cannot truely represent the oxide, however it can be indicative of high levels of density grading. A number of wafers were studied these included two furnace oxides grown at 950°C (labelled A and B) in 3%HCl/O_2. One of these (B) was etched back with HF leaving approximately half of the original thickness (which was 18nm). For comparison two RTO layers (C and D), one grown in pure oxygen and the other in 2%HCl/O_2 at 1100°C, were measured. The results from single layer fits (with their 90% confidence limits) and the level of improvement with the graded model are compiled in table IV. Previously, in our experience, the relative density (calculated using the single layer model) of thermal oxides was not found to be lower than that of fully relaxed vitreous silica. The multilayer result for the sample C shows a high level of grading. A very dense near interface region is compensated by a lower density surface region so that the average density calculated in the single layer model approached that of fully relaxed SiO_2.

Table IV Ellipsometric results

Sample	Process Ambient	T_{ox}(nm)	Rel. Density	Graded fit Improvement
A	FO 950°C 3%HCl	18.3 ± 2	0.997 ± 0.021	None
B	FO 950°C 3%HCl	8.9 ± 1	1.093 ± 0.479	None
C	RTO 1100°C 2%HCl	20.0 ± 2	0.978 ± 0.018	Significant
D	RTO 1100°C 0%HCl	19.6 ± 2	0.986 ± 0.018	Marginal

Result for sample C:

	T_{ox}(nm)	Relative Density
Surface	6.5 ± 1.8	0.72 ± 0.17
Middle	9.0 ± 2.3	1.0 fixed
Lower	5.0 ± 1.9	1.20 ± 0.13

The fit quality as determined by the unbiased estimator was improved from 0.0056 (single layer) to 0.0052 (graded layer). This change while small is significant and these results tentatively suggest that the oxide surface is either rough or has a large density of voids. In the calculation the relative density of the middle region was fixed to 1. Varying this parameter did not improve the fit. It is important to recognise that small density changes in adjacent oxide regions are difficult to deduce simply because their optical constants are very similar.

5 Discussion and Summary

An assessment of rapid thermal oxidation, to grow oxides in the 16nm range has shown that highly uniform and reproducible oxides can be grown. Measurements have shown that breakdown fields and defect densities are comparable to those of a good quality furnace oxides. Interface trap density levels and Q_{bd} values were found to be dependent on the growth temperature and ambient, with HCl incorporation reducing Q_{bd} and increasing the spread of values. This effect may be associated with HCl increasing surface roughness and/or introduce voids in the near surface region, as the results from SE have indicated on similarly oxidised wafers. The cause of the lower surface density requires further investigation. It may be due to the RTO process itself, the presence of HCl or even the nature of the native oxide present. For example during furnace oxidation there is normally a period when a thin thermal oxide can form at a temperature lower than the defined growth temperature. This oxide can then withstand or reduce the etching effects of HCl by reducing its arrival rate at the Si, however in RTO this period is only a few seconds. Thus the highly activated Cl radicals can reach and attack the Si or even defects in the oxide causing inhomogeneous oxide growth.

To conclude the fabrication of RTO oxides with broadly comparable interface and bulk properties to thermal oxides has demonstrated the cleanliness of the RTP chamber and that good procedures are in place. Further work is required to look at long term reliability aspects of the RTO oxides and to assess their performance in a full device process, however these results do indicate that RTO is a feasible alternative to furnace technology for the growth of ultra thin SiO_2 layers (<10nm).

6 Acknowledgements

The work contained within this paper was performed under the IED 2/1/1769 project. We are grateful to C. Pickering, A. M. Hodge and M. J. Uren for helpful discussions.

7 References

Z. Lilienthal, O. L. Krivanek, S. M. Goodnick and C. W. Wilmsen, Mat. Res. Sym. Proc., 37, p193, 1985.
V. Nayar, C. Pickering and A. M. Hodge, Thin Solid Films 195, p185, 1991.
J. Nulman, J. Kruisus and A. Gat, IEEE Electron Device Letts. EDL-6, No 5, May 1985, p205.

© Crown copyright 1991/MOD, published with the permission of the controller HMSO.

Paper presented at INFOS '91, Liverpool, April 1991
Contributed Papers, Section 2

Low pressure thermal oxide applicable as bottom oxide in oxide–nitride–oxide triple layers

E.P. Burte and A. Bauer

Fraunhofer Arbeitsgruppe für Integrierte Schaltungen
Artilleriestraße 12, D-8520 Erlangen, FRG

Abstract: Silicon oxide layers up to 6nm thickness were thermally grown at low pressure in a LPCVD silicon nitride hotwall system. Growth rates of these thin oxide films were studied with pressure and with oxygen flow at a fixed oxidation temperature of 950°C. Densities of fast interface traps and of fixed interface charges were determined by capacitance voltage measurements. In order to study dielectric breakdown, current voltage charakteristics were monitored. Furthermore first investigations on oxide-nitride-oxide triple layer structures processed in a single run were performed.

1. Introduction

Oxide-nitride-oxide (ONO) triple layer dielectrics have become important for storage cells of dynamic memories. Conventionally, the bottom oxide is thermally grown in diluted dry oxygen; the silicon nitride layer is deposited by a low pressure chemical vapor deposition (LPCVD) process, while thermal steam reoxidation of the surface of the nitride layer is applied in order to form a top oxide. This ONO-process suffers from the fact that wafers have to be transferred from an oxidation tube via a LPCVD reactor back to an oxidation tube.
Using fully CVD and establishing the entire ONO process in one single LPCVD reactor, this disadvantage of the conventional ONO process can be overcome [1]. Thermal oxide as bottom layer, however, yields a superior Si-SiO_2 interface, and results in a lower density of low-field breakdown defects of ONO structures than CVD high temperature bottom oxide [1].
By growing the thermal bottom oxide in a LPCVD silicon nitride reactor at low pressure, the complete ONO process can be performed in one single tube. In this work the characteristic properties of low pressure thermal oxide (LPO) films are reported. Furthermore this LPO films have been applied to an intra chamber ONO-process.

2. Experimental Procedure

100mm (100) 1-2Ωcm p-type FZ silicon wafers and 1-5Ωcm n-type CZ silicon wafers were used for the experiments. After cleaning the wafers, the low pressure oxidation process was performed in a standard LPCVD silicon nitride hotwall system slightly modified by adding an dry oxygen gas line. The oxidation process was performed at oxygen flow rates of 1,3,5 and 10 slm. The growth pressure was ranging from p=133Pa up to p=670Pa. The growth temperature was kept constant at T=950°C. The thickness of the LPO films was deter-

mined by ellipsometry using a fixed refractive index of n=1.46.
The level of metallic contaminants existing in the LPO films was investigated by vapor phase decomposition (VPD) atomic absorption spectroscopy (AAS). Combined high and low frequency capacitance voltage measurements were applied to aluminum-LPO-silicon capacitors in order to determine densities of fixed interface charges and fast interface traps. By using a Keithley SMU 237 current-voltage characteristics of the above metal-LPO silicon structures were monitored.

3. Results and Discussion

3.1. Metal Contamination

Metal traces are wellknown to influence interfacial properties and dielectric breakdown of thermally grown silicon oxide layers [2,3]. To study the level of metallic contaminants transferred onto the wafer surface or incorporated into the LPO film during processing in the LPCVD reactor, the surface densities of the most important elements, like Na, K, Ca, Zn, Fe, Cr, Cu, and Al were determined by use of VPD atomic absorption spectroscopy. They turned out to be as low as those of native oxide films after a cleaning procedure. The concentrations of Al and Cu were too low to be detected by AAS. The negligible differences in surface densities of the above elements between the investigated samples reveal, that the LPCVD system does not influence the level of metal traces of processed wafers.

3.2 Growth Kinetics

To get insight into the growth kinetics of thermal silicon oxide at low pressure the growth rate with respect to pressure and oxygen flow rate was investigated. Fig. 1 shows film thickness d versus processing time t at a fixed pressure of p=650Pa and an oxygen flowrate of F_{ox}=1slm. The linear relationship between d and t reveals a reaction controlled oxide growth at the above pressure. An increase in pressure at constant oxygen flowrate results in an increase in the growth rate, as can be seen in Fig. 2. The growth rate, thereby, depends linearly on the total pressure which is equal to the oxygen partial pressure.

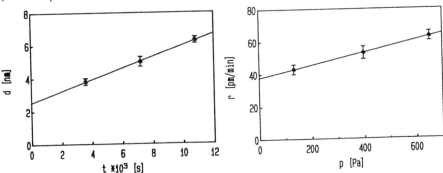

Fig 1: Film thickness d versus processing time t

Fig 2: Growth rate r versus pressure p

3.3 Electrical Characterization of Low Pressure Oxide Films

In order to obtain a certain specific capacitance of storage capacitors in advanced DRAM concepts thermal oxide layers with thicknesses ranging from 5nm

up to 10nm are of high technical importance.
Fig. 3 shows the current density-electric field characteristics of a MOS capacitor with a 5.5nm low pressure oxide layer. Both negative and positive bias correspond to electron injection from the silicon substrate and the aluminum gate, respectively.
At lower values of the electric field the displacement current dominates until at about 8 MV/cm the current density increases drastically due to the onset of Fowler Nordheim injection (see below). Under both bias conditions, non-destructive breakdown is evident at values of electric field strength ranging from 8 MV/cm up to 10.5 MV/cm. Repeated ramping of the applied electric field up to strengths of 12.5 MV/cm results in very similar curves exhibiting no destructive breakdown, as Fig. 4 reveals. Destructive breakdown occurs after electric fields of strength higher than 13 MV/cm have been applied to the samples.

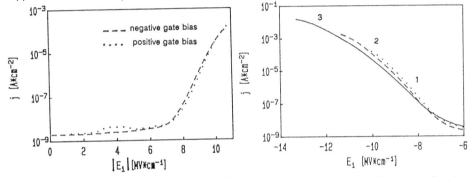

Fig 3: Current density-electric field charakteristics of a MOS capacitor with a 5.5nm LPO layer

Fig 4: Repeated ramping of the applied electric field up to strengths higher than 13 MV/cm

The current transport mechanism at fields higher than 8 MV/cm can be consistently described by Fowler-Nordheim tunneling of electrons. Fig. 5 shows a Fowler Nordheim plot of a typical ramped current density-electric field relationship. Evaluating the slope of the linear curve of Fig. 5, the tunneling barrier height of the silicon-LPO interface was determined to be $\phi_B = (2.9 \pm 0.2)$ eV in accordance with published data for silicon-silicon oxide interfaces [4]. Typical high (1MHz) frequency and quasistatic capacitance voltage curves of metal-LPO-silicon samples annealed at 450°C in forming gas atmosphere are shown in Fig.6.

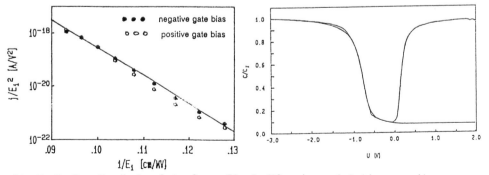

Fig 5: Fowler Nordheim plot of a LPO layer

Fig 6: Hf and quasistatic capacitance voltage curves of a LPO layer

Using the measured value, C_I, of the capacitance in accumulation and the known doping level of the substrate, the ideal values of midgap capacitance C_{MG} and midgap voltage V_{MG} were calculated according to a standard procedure [5]. Fixed charge density was calculated using the difference in voltage between the actual and the ideal midgap voltage [6]. By means of the formula $\Delta V_{MG} = \phi_{MS} - Q_F/C_I$ (ϕ_{MS} = work function difference between the metal gate and the silicon substrate), the density of fixed interface charges has been calculated to be $Q_F/q = (4\pm2)\times10^{11}\text{cm}^{-2}$. By applying the combined high-low frequency capacitance method [5], the midgap interface trap density of the above sample was determined to be $D_{IT}=(5\pm3)\times10^9\text{eV}^{-1}\text{cm}^{-2}$. This interface state density is comparable to those measured for thermally grown conventional oxide films.

Assuming a dielectric constant of 3.9, the thickness of the LPO films was evaluated from the measured accumulation capacitance C_I. This value agrees to that determined by ellipsometry using a fixed refractive index of 1.46, revealing a fairly correct measurement of the accumulation capacitance of the above sample.

3.4 ONO-Structures Using LPO as Bottom Oxide

In order to form ONO structures, a LPCVD silicon nitride layer of thickness 9.3 nm at 820°C and a LPCVD high temperature oxide layer of thickness 4.4nm at 910°C were deposited onto a LPO layer of thickness 4.8nm in one single process tube without breaking the vacuum. Dielectric breakdown of these structures occurs at an electric field of 12 MV/cm. These first results reveal that LPO films are appropriate for the bottom oxide of ONO structures. Furthermore an intra-chamber ONO process is possible.

4. Conclusions

In this work, silicon dioxide grown at low pressure has been studied. Within the investigated thickness range (below 10nm) a reaction controlled linear growth has been observed.
Trap densities at the low pressure oxide-silicon interface are below $8\times10^9\text{eV}^{-1}\text{cm}^{-2}$. Destructive dielectric breakdown occurs at electric field strengths higher than 12.5 MV/cm. These results reveal the high quality of low pressure thermal oxide films.
It has been successfully demonstrated that low pressure oxide is applicable as bottom oxide in oxide-nitride-oxide (ONO) triple layer structures, whereby an intra-chamber ONO process has been realized.

5. References

[1] W. Hönlein and H. Reisinger, Appl. Surf. Sci. **39** (1989), 178
[2] R. Falster, J. Appl. Phys. **66** (1989), 3355
[3] K. Honda, T. Nakanishi, A. Ohsawa and N. Toyokura
     Inst. Phys. Conf. Ser. No.87:Section 6
     Paper presented at Microsc. Semicond. Mater. Conf., Oxford, (1987), 463
[4] R. Baunach and A. Spitzer, Appl. Surf. Sci. **30** (1987), 180
[5] E.H. Nicollian and J.R. Brews MOS Physics and Technology, Wiley, New York (1982)
[6] E.P. Burte, Solid State Electronics. Vol. 31, No. 12, (1988), 1663

Electrical characteristics of Cl-implanted thin gate oxides

S. Verhaverbeke*, M.M. Heyns and R.F. De Keersmaecker

IMEC v.z.w., Kapeldreef 75, B-3001 Leuven, Belgium

ABSTRACT : It is the purpose of this study to clarify the role of Cl in thin thermal oxide layers by investigating the effect of the introduction of a low concentration of Cl by means of ion implantation. Chlorine was implanted at an energy of 20 KeV with three different doses: $10^{11}/cm^2$, $10^{12}/cm^2$ and $10^{13}/cm^2$. The mid-field breakdowns can be reduced by the implantation and subsequent annealing of Cl in a Si/SiO$_2$/poly-Si structure. The density of the hole trap with a capture cross section of $4\text{-}6 \cdot 10^{-13}$ cm^2 is decreased when Cl is implanted, while no effect on the electron trapping was observed.

1. INTRODUCTION

The importance of chlorine in the fabrication of thin oxide layers has been considerably changed over the last few years. While initially the passivation of mobile ions (Janssens et al 1978a) and the annihilation of stacking faults were targetted (Claeys et al 1977), recent work has focussed on the improved degradation resistance and radiation hardness which can be reached when only minute amounts of TCA are used in the oxidation ambient (Wang et al 1988). It is the purpose of our study to clarify the role of Cl in thin thermal oxide layers by investigating the effect of the introduction of a low concentration of Cl by means of ion implantation.

2. EXPERIMENTAL CONDITIONS

After a modified RCA-clean <100>Si wafers were oxidized in dry O$_2$ to a thickness of 12 nm. To protect the oxide a 10 nm poly-Si layer was deposited immediately after the oxidation. Chlorine was implanted at an energy of 20 KeV with three different doses: $10^{11}/cm^2$, $10^{12}/cm^2$ and $10^{13}/cm^2$. As source CoCl$_2$ (Cl^{35} ions) was used, so the injected ions were Cl$^+$. The projected range based on the LSS-theory was 2.2 nm (10 nm poly + 12 nm SiO$_2$) so that the maximum of the implanted profile is located at the Si/SiO$_2$ interface. The implantation was followed by a two-step annealing of 750°C for 120 min and 950°C for 60 min in N$_2$. The first step allows the reaction of Cl with defect sites in the oxide before being outdiffused to either the anneal ambient or the Si-substrate during the second step (Ling Z.M. et al 1988) which is needed to anneal the implantation damage. After deposition of the remaining part of the poly-Si gate electrode and solid source doping at 900°C, a drive-in was performed at 975°C during 20 min. Capacitor structures were defined by conventional wet lithography.

3. RESULTS

The breakdown histograms, shown in fig.1, were constructed from the breakdown points determined on the current-vs-voltage curves after correction for the series resistance. The series resistance was obtained from comparison of the low-frequency CV-curve with the high-frequency CV-curve in accumulation. The implantation of $10^{11}/cm^2$ Cl ions increases the number of mid-field breakdowns. When the implantation dose is increased the mid-field

---

* Research Aspirant of the Belgian National Fund for Scientific Research
© 1991 IOP Publishing Ltd

breakdowns are suppressed and for a dose of $10^{13}/cm^2$ a clear improvement in the breakdown characteristics over the control sample is observed. The maximum breakdown field is not affected by the Cl implantation. These results were reproduced on a second batch of wafers. The reduction of mid-field breakdowns by the removal of metal ions from the oxidation ambient through the addition of Cl is a well-known phenomena. It is, however, a rather suprising result that the number of mid-field breakdowns can be reduced by the implantation and subsequent annealing of Cl in a Si/SiO_2/poly-Si structure where the volatilization of metal compounds is almost impossible.

The interface state densities, as determined from quasistatic and high-frequency CV-curves, ranged from 5 to $9 \cdot 10^9$ $cm^{-2}.eV^{-1}$ and did not systematically vary with implanted Cl dose. The same was observed for the fixed oxide charge density.

The electron and hole trapping in the oxides was investigated using avalanche injection. The effective concentrations and capture cross sections of the traps were obtained using a first-order kinetic model to fit the flatband voltage shift versus time curves recorded with a fixed current density. In our case a best fit of the data was always obtained when two different hole trapping sites were assumed, each characterized by its own capture cross section and effective density. For the avalanche hole injection highly doped n-type wafers were prepared using a two-step Phosphorous implant to get a doping level of $8 \cdot 10^{17} cm^{-3}$ near the Si/SiO_2 interface which extends deep into the Si-bulk. Von Schwerin et al (1990) have shown that the hole trapping is not dependent on the oxide electric field. Only a very small dependence of the capture cross section on the oxide field was observed. No evidence was found for the generation of traps during hole injection. Therefore, avalanche hole injection, where the oxide field during injection is not controlled, is still a valid technique to investigate the hole trapping properties of SiO_2 layers.

The results from the hole injection experiments are given in table 1. In all samples two types of traps were observed. The density of the trap with a capture cross section of $4\text{-}6 \cdot 10^{-13}$ cm^2 is decreased when Cl is implanted. This is also illustrated in fig.2. The decrease is only a fraction of the density of implanted ions. Therefore, either only a limited number of defects can be annealed by Cl or the diffusion of Cl towards the defects and the reaction at these defects is the limiting step. The density of the other hole trap is not affected by the Cl implantation.

No electron trapping could be observed on the samples (detection limit $\approx 10^{10}/cm^2$) both during injection at room temperature and at 140°C which was used to lower the positive charge build-up by slow trapping instabilities (Heyns et al 1988). This indicates that the implanted Cl does not generate electron traps.

4. DISCUSSION

It must be remarked that the Cl content used in these experiments is extremely small. A dose of $10^{13}/cm^2$ Cl ions, which is the highest implantation dose used, corresponds roughly to a HCl content of 0.01% in the oxidation ambient. In the early days, when typically 2% HCl was added to the oxidation ambient, peak concentrations of Chlorine as high as $2 \cdot 10^{21}$ cm^{-3} were measured (Deal et al 1978) and concentrations per unit area, averaged over the total oxide thickness were in the range of $10^{15}\text{-}10^{16}$ cm^{-2} (Butler et al 1977).

From the breakdown histograms it is clear that Cl implantation can give an improved dielectric strength in the mid-field regime. The improved breakdown characteristics which are observed when Cl is added to the oxidation ambient is normally attributed to the formation and volatilization of metallic chlorides during the oxidation. However, on the basis of the vapor pressure of the metallic chlorides, some authors (e.g. Janssens 1978b) have already argued, that the volatilization is certainly not the only mechanism which can improve the dielectric breakdown of SiO_2 layers upon addition of Cl. An alternative explanation is that Cl reduces the oxidation induced stacking faults (Claeys et al 1977) and, thereby, reduces the precipitation centra for metals. This effect can not be excluded in our experiments due to the long annealing times which were used.

Greeuw et al (1981) and Vengurlekar et al (1985) have studied the diffusivity of Cl when implanted in SiO_2. Although they found different activation energies and a different pre-exponential factor D_0, when evaluating the diffusivity at the annealing temperatures of 750°C and 950°C the values they obtained are of the same order of magnitude. Both authors conclude that the implanted Cl has a diffusivity different from that of Cl in HCl-grown oxides and that the diffusion mechanism is interstitial. During the first annealing cycle (750°C for 120 min) The Cl diffuses over a distance of approximately 20 nm. During the second cycle the diffusion distance is of the order of 70 nm and all of the non-bonded Cl is diffused out.

When a simple random walk model is assumed for the motion of Cl atoms through the oxide the total travel distance of a Cl atom in the oxide during the anneal treatment at 750°C can be calculated. After one step the probability distribution is a gaussion distribution with a spreading given by $\sigma = l/\sqrt{3}$, with l the typical step distance which is assumed to be of the order of 0.2 nm (interatomic distance). The factor $\sqrt{3}$ accounts for the effect that random motion is possible in three directions. After n (independent) steps we can write $\sigma = \sqrt{n}.l/\sqrt{3}$. This spreading is equal to the diffusion distance $\sqrt{2D.t}$. From this we can calculate the random velocity of the Cl atoms (v_{Cl}) during the annealing :

$$v_{Cl} = n.l/t = 6.D/l = 1.5 \text{ nm/s when D(Temp=750°C) is taken as } 5*10^{-16} \text{ cm}^2/\text{s}$$

From this the probability of capture of the Cl atom by a hole trapping center can be calculated by assuming first order kinetics :

$$\frac{dn_t}{dt} = \sigma_c(N-n_t).n_{Cl}.v_{Cl}$$

with n_t the number of Cl atoms trapped permanently on a hole trap defect, σ_c the capture cross section of the defect to trap a Cl atom, N the total number of trapping defects available and n_{Cl} the concentration of Cl atoms. The solution to this equation is :

$$n_t(t) = N(1 - e^{-t\sigma_c v_\sigma n_q})$$

The only unknown in this equation is the capture cross section and a value of approximately $2\cdot 10^{-17}$ cm^2 can be calculated for this. This capture cross section gives a measure for the probability of a hole trapping defect to capture a Cl atom which inactivates it as a hole trap.

5. CONCLUSION

These results indicate the importance of small amounts of Cl in thin thermal oxide layers on the electrical characteristics of these layers. They demonstrate that some benefits can be obtained by implanting Cl as an alternative to its introduction during the oxidation. In this respect the implantation of Cl has the advantage that no hydrogen is introduced in the oxide which can not be avoided when TCA is added to the oxidation ambient.

ACKNOWLEDGEMENT

This work was supported by the European Community in the framework of the ESPRIT Basic Research action PROMPT.

REFERENCES

Butler S.R., Feigl F.J., Rothatgi A., Kraner H.W. and Jones K.W. 1977, paper 77 presented at the Electrochem. Soc. Meeting, Philadelphia
Claeys C.L., Laes E.E., Declerck G.J. and Van Overstraeten R.J. 1977, paper 224 presented at the Electrochem. Soc. Meeting, Philadelphia
Deal B.E., Hurrle A. and Schulz M.J. 1978, J. Electrochem. Soc. 125, pp. 2024-2027
Greeuw G. and Hasper H. 1981 Proceedings of INFOS 81, Eds. M. Schulz and G. Pensl (Springer, Berlin), pp.203
Heyns M.M. and De Keersmaecker R.F. in "The Physics and Technology of Amorphous SiO$_2$", Ed. R.A. Devine (Plenum, New York, 1988), pp.411
Janssens E.J. and Declerck G.J. 1978a J. Electrochem. Soc. 125, pp. 1696-1703
Janssens E.J. 1978b ph. d. thesis, K.U. Leuven

Ling Z.M., Dupas L.H. and De Meyer K.M. 1988 Abs. No 146, Electrochem. Soc. Spring Meeting, Atlanta, may 1988
Vengurlekar A.S., Ramanathan K.V. and Karulkar V.T. 1985 J. Electrochem. Soc. 132, pp.1172-1177, May
Von Schwerin A. and M.M. Heyns 1990 J. Appl. Physics 67 (12), 15 June
Wang Y.,Nishioka Y.,Ma T.P. and Barker R.C. 1988 Appl. Phys. Lett. 52 pp. 573

impl. dose (cm^{-2})	Trap 1		Trap 2		Tot. density
	σ_c (cm^2)	N_{eff} (cm^{-2})	σ_c (cm^2)	N_{eff} (cm^{-2})	N_{eff} (cm^{-2})
no implant	6*10^{-13}	3.3*10^{12}	6*10^{-14}	0.7*10^{12}	4.0*10^{12}
1*10^{11}	4*10^{-13}	3.0*10^{12}	6*10^{-14}	0.9*10^{12}	3.9*10^{12}
1*10^{12}	5*10^{-13}	2.9*10^{12}	6*10^{-14}	0.7*10^{12}	3.6*10^{12}
1*10^{13}	6*10^{-13}	2.6*10^{12}	7*10^{-14}	0.7*10^{12}	3.3*10^{12}

Table 1 : Capture cross sections (σ_c) and effective concentrations (N_{eff}) of the hole traps as a function of the implanted Cl dose.

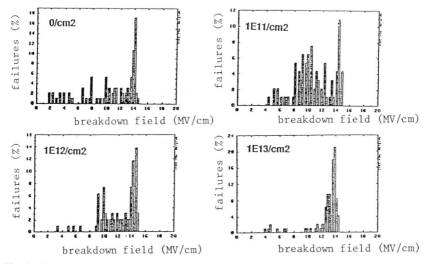

Fig. 1 : Breakdown histograms for the thin oxides as a function of the implanted Cl dose.

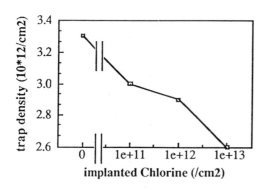

Fig. 2 : Density of the largest hole trap as a function of the implanted dose of Cl.

Paper presented at INFOS '91, Liverpool, April 1991
Contributed Papers, Section 2

Noise and other electrical characteristics of CMOS FETs fabricated with furnace, Anodic and rapid thermal oxides

D.C.Murray*, J.C.Carter, N.Afshar-Hanaii, A.G.R.Evans, S.Taylor[†], J.Zhang[†] and W.Eccleston[†].

Dept. of Electronics and Computer Science, The University, Southampton, SO9 5NH, U.K.
*Dept. 9800, PIDC, Philips Components Ltd., Millbrook, Southampton, SO9 7BH, U.K.
†Dept. of Electrical Engineering and Electronics, The University of Liverpool, Liverpool, L69 3BX, U.K.

ABSTRACT: A comparison between CMOS FETs with furnace, plasma anodic and rapid thermal gate oxides has been made. Plasma anodic oxides 40nm thick were grown at room temperature and 15nm and 25nm rapid thermal oxides were grown in an AG610 at 1150°C. These were compared to standard furnace oxides grown at 950°C to a thickness of 40nm. Threshold voltages, channel mobilities, average interface state densities and low frequency noise were extracted from measurements on n- and p-channel devices. Low frequency noise was measured as a function of drain current and a model used to extract a density of states as a function energy in the oxide. Although most extracted parameters were poorer for anodic and rapid thermal oxides the differences with the standard 40nm furnace oxide were small. However for all the oxides measured here, there was a sharp increase in the density of oxide states below the energy of the silicon valence band.

INTRODUCTION

Rapid thermal oxides (RTO) appear to be attractive for growth of thin oxides required by deep sub-micron CMOS devices since they offer a greater degree of control over processing parameters - in particular ramp up and cool down times (Moslehi 1988). Plasma anodic oxides have additional advantages in that the growth temperatures are very low. In this paper we compare CMOS FETs made using a standard furnace oxide with those made with RTO and plasma anodic gate oxides. Apart from measuring the standard parameters of threshold voltage, channel mobility and interface state density, low frequency noise is also measured. Low frequency noise is predominently due to the tunneling of channel carriers to traps within the oxide and from low frequency noise measurements we have extracted a density of oxide traps around the silicon valence and conduction band energies.

EXPERIMENTAL DETAILS

Our paper describes measurements made on two seperate batches. In the first batch the standard Southampton University self-aligned n-well CMOS process was used to compare 40nm furnace and plasma anodic oxides. In the second batch a modified process with n- and p-wells of higher doping was used to compare a 40nm furnace oxide with 15nm and 25nm rapid thermal oxides. In both batches the furnace oxides were grown at 950°C in oxygen+3% chlorine, at the end of the oxide growth the wafers were given an anneal at 950°C for 15 minutes in nitrogen. The anodic oxides were grown using the Liverpool University machine that has been described elsewhere (Taylor et al 1987). Oxide growth is achieved by D.C biasing the wafer to a positive potential in the presence of an oxygen plasma - negatively charged oxygen ions are drawn to the growing silicon-silicon dioxide interface. For the oxides grown here the substrate was unheated, but during the 17.5 minute growth time, the substrate temperature rose to 120°C. The D.C. bias used was 70V which resulted in a current density of $1.25 mAcm^{-2}$, the pressure of the chamber during growth was 150mT. Like the furnace wafers the anodic oxide wafers were given an anneal in nitrogen for 15 minutes at 950°C. In the second batch the rapid thermal oxides were grown in an AG610 in oxygen at 1150°C. The time taken for 15nm and 25nm growth was 30s and 100s respectively. Other differences between the two batches that may affect oxide quality were the source and drain anneal - in the first batch a furnace anneal at 1000°C in nitrogen was given after the arsenic implant and a 950°C anneal in wet oxygen after the boron difluoride implant; the second batch however received just one rapid thermal anneal at 1100°C for 1 second after the both source and drain implants had been capped with low temperature oxide. For both batches post metal alloys were carried out in forming gas at 450°C for 30 minutes.

Threshold voltage and channel mobilty were extracted from I_d-V_g curves corrected for field induced mobilty reduction and series resistance using the 1+θ method (Tsividis 1988). Average interface state density was determined using a charge pumping method of Groesenken (1984). However source and drain leakage currents were too high on the first batch for this measurement to be carried out. Low frequency noise was measured using the ACUMENS system developed at AERE Harwell (Cox 1982). Most measurements were made at a fixed drain current, however some devices were measured over a range of drain currents - this allowed a density of states in the oxide as a function of energy to be calculated.

RESULTS

Tables 1 and 2 show the comparison between furnace and anodic oxides for n and p-channel devices respectively. The threshold voltage measurements indicate the presence of more positive oxide charge for the anodic oxide causing shifts of ~0.25V relative to the furnace oxide, mobilities are also about 10% lower. Input referred noise levels at 30Hz measured on 100x4μm devices are slightly higher for the anodic oxides however the frequency indices are nearly identical. (The frequency index is γ where $E_n^2(f) \propto 1/f^\gamma$).

	V_t (V)	μ (cm^{2}V^{-1}s^{-1})	E_n (nVHz$^{-1/2}$)	γ
Furnace 40nm	0.58±0.08	590±30	365±50	0.97±0.04
Anodic 40nm	0.25±0.10	525±10	410±70	0.96±0.04

Table 1. Comparison between n-channel devices in batch 1. (Vt-threshold voltage, μ-channel mobility, En-gate referred noise at 30Hz and γ-frequency index).

	V_t (V)	μ (cm^{2}V^{-1}s^{-1})	E_n (nVHz$^{-1/2}$)	γ
Furnace 40nm	-0.68±0.10	305±10	170±50	1.11±0.04
Anodic 40nm	-0.96±0.07	290±30	210±40	1.16±0.04

Table 2. Comparison between p-channel devices in batch 1.

Tables 3 and 4 show a similar comparison between furnace and RTO oxides for n and p-channel respectively. Again there are no large differences between the different oxides. There is an expected decrease in threshold voltage as the oxide thickness is reduced. The 25nm RT Oxide does have slightly higher interface state density and a lower surface mobility for both n and p-channel devices. Because the oxides are not the same thickness it is not possible to directly compare the input referred noise - however in the following section a method is described that extracts oxide trap densities from low frequency noise measurements and the densities of these states can be compared.

	V_t (V)	μ (cm^{2}V^{-1}s^{-1})	E_n (nVHz$^{-1/2}$)	γ	D_{it} (cm^{-2}eV^{-1})
Furnace 40nm	0.46±0.01	700±50	370±40	0.98±0.03	2.5±0.2
RTO 25nm	0.44±0.01	520±50	300±30	0.96±0.03	3.3±0.2
RTO 15nm	0.24±0.03	640±50	195±20	0.98±0.03	2.0±0.2

Table 3. Comparison between n-channel devices in batch 2. (D_{it} x10^{10}).

	V_t (V)	μ (cm^{2}V^{-1}s^{-1})	E_n (nVHz$^{-1/2}$)	γ	D_{it} (cm^{-2}eV^{-1})
Furnace 40nm	-1.70±0.05	165±20	220±20	1.09±0.03	1.1±0.1
RTO 25nm	-1.37±0.03	150±10	270±25	1.09±0.03	1.7±0.1
RTO 15nm	-1.17±0.01	175±10	165±20	1.05±0.03	1.2±0.1

Table 4. Comparison between p-channel devices in batch 2. (D_{it} x10^{10}).

EXTRACTING OXIDE DEFECT DENSITIES

Flicker noise in MOSFETs is predominantly caused by gate oxide defects. These defects trap carriers and re-emit them at random causing variations in the channel current. The current variations arise due to fluctuations in the number of channel carriers and scattering induced fluctuations in the carrier mobility. For a uniform distribution of oxide defects a pure 1/f noise spectral shape is predicted. For the linear mode of operation ,the input referred noise spectral density S_v ($E_n^2 = S_v$) can be expressed as (Jayaraman and Sodini 1989, Hung et al 1990)

$$S_V(f) = \frac{kTN_{ox}\phi}{ZLf} \left\{ \frac{q^2}{C_{ox}^2} + \frac{2q\theta\mu V_{GT}}{C_{ox}} + \theta^2\mu^2 V_{GT}^2 \right\}. \qquad (1)$$

Where N_{ox} is the oxide defect density, ϕ is a characteristic tunneling length, θ is a scattering parameter and V_{GT} is the effective gate bias V_G-V_T. Equation (1) is also approximately valid for the saturation mode. It is defects with energies close to the surface fermi-level that contribute the most noise - where the capture and emission times (σ,τ) are equal. For oxide traps the capture and emission times are lengthened due to the tunnelling interaction

$$\sigma = \frac{1}{nv_{TH}a_T}\exp\{x/\phi\}. \qquad (2)$$

For a particular frequency, maximum noise occurs for

$$\sigma \approx \tau \approx 1/\pi f, \qquad (3)$$

implying that for any frequency of interest one can associate the noise with defects at a particular distance into the oxide, x_{opt}, usually <3nm. For n-channel transistors operated in strong inversion the surface fermi-level lies close to the conduction band. Due to oxide band-bending the 'noisey' oxide traps (at distance x_{opt}) will have energies close to and <u>often above</u> the silicon conduction band

$$E_T = E_{Fs} + \frac{V_{GT}}{t_{ox}}x_{opt}. \qquad (4)$$

Similarly for p-channel devices it is oxide defects with energies close to the silicon valence band that are important. Oxide defect densities can be extracted from noise data using equations (1) to (4). In this paper we use a modification of the Jayaraman and Sodini method (Jayaraman and Sodini 1990). Firstly noise vs V_{GT} data is collated. Then the data for a single frequency (30Hz) allows N_{OX} to be extracted for a sweep of energies close to the band edges. One has to assume a trap capture cross section and in this case we have chosen $10^{-16}cm^2$ - the effect of changing this value is to stretch or contract the energy scale - however the dependency is quite weak.

Figure 1 shows the extracted distribution for the furnace and anodic oxides from the first batch. The left hand graph was extracted from p-channel devices and the right hand curve from n-channel devices. The anodic oxide shows a large increase in defects that are responsible for low frequency noise in n-channel devices as the device is operated in weaker inversion - however in moderately strong inversion these defects do not lead to any increase in noise over the furnace oxide. For both furnace and anodic oxides there is a sharp increase in states below the energy of the silicon valence band.

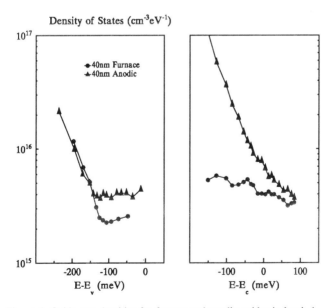

Figure 1. Oxide trap densities for furnace and anodic oxides in batch 1.

Figure 2 shows the similar curves for the furnace and rapid thermal oxides from the second batch. The lamp oxides have slightly more oxide states than the furnace samples at energies around the silicon conduction band. At valence band energies the 25nm lamp oxide and 40nm furnace oxides are similar but the 15nm lamp oxide has twice the number of defect states. However, as with the previous batch, all samples show an increase at energies below the silicon valence band edge.

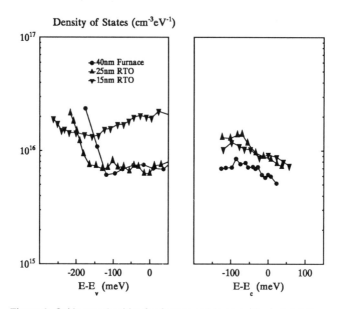

Figure 2. Oxide trap densities for furnace and lamp oxides in batch 2.

The defect densities calculated here are consistent with those obtained from random telegraph signals (RTS) (Uren et al 1985). The rise in the density of states at energies below the valence band is consistent with the frequency index γ of p-channel devices being greater than one. Recent work has shown that 1/f noise is linked to the radiation induced hole trap - the E′ centre (Fleetwood and Schofield 1990). A theoretical investigation of defects at the Si-SiO$_2$ interface suggests that the E′ centre may reside close to the silicon valence band edge (Chu and Beall Fowler 1990). It could be that the increase in oxide defects below the valence band edge seen in our work is related to the E′ centre.

CONCLUSIONS

Furnace, plasma anodic and lamp oxides have been used to fabricate CMOS FETs. Only small variations in threshold voltage, channel mobility and interface state density are seen between the different oxides. However all oxides have a characteristic sharp increase in the density of oxide states responsible for low frequency noise near the silicon valence band edge. It is tentatively suggested that this increase is related to the E′ centre.

REFERENCES

Chu A X and Beall Fowler W 1990 Phys.Rev.B **41** 5061
Cox C 1982 AERE Tech.Rep. R10298 HMSO
Fleetwood D M and Schofield J H 1990 Phys.Rev.Lett. **64** 579
Groeseneken G, Maes H, Beltran N and de Keersmaecker R 1984 IEEE Trans. Electron Dev. **ED-31** 42
Hung K K, Ko P K, Hu C and Cheng Y C 1990 IEEE Trans.Electron Devices **ED-37** 654
Jayaraman R and Sodini C G 1989 IEEE Trans.Electron Devices **ED-36** 1773
Jayaraman and Sodini 1990 IEEE Trans.Electron Devices **ED-37** 305
Moslehi M M 1988 Appl.Phys.A **46** 255
Taylor S, Barlow K, Eccleston W and Kiermasz A 1987 IEE Elec.Lett. **23** 309
Tsividis Y P 1988 *The Operation and Modelling of the MOS Transistor* (McGraw-Hill)
Uren M J, Day D J and Kirton M J 1985 Appl.Phys.Lett. **47** 1195

Paper presented at INFOS '91, Liverpool, April 1991
Contributed Papers, Section 3

Thin TiO$_2$ as a film with a high dielectric constant

A. Spitzer, H. Reisinger, J. Willer, W. Hönlein, H. Cerva, and G. Zorn

Siemens AG, Otto-Hahn-Ring 6, D-8000 München 83, Germany

Abstract. Thin TiO$_2$-films prepared by sputtering were investigated with respect to their chemical, structural, and electrical properties. Annealing the films above 700°C yields TiO$_2$-layers in the rutile phase with a dielectric constant above 100. Dielectric films with a SiO$_2$ equivalent thickness of 2.5nm were realized on silicon substrates. Compared to an equivalent SiO$_2$-layer the TiO$_2$-films show a reduction of the current densities to values suitable for an application in DRAMs.

1. Introduction

The development of future DRAM generations beyond 16Mbit requires alternative dielectric materials to reach charge storage densities that are larger than those of capacitors with a 5nm SiO$_2$-layer. Materials with a higher dielectric constant like tantalum pentoxide or yttrium oxide are considered as a possible solution but the net gain in charge storage density is limited to a factor of about 2 due to the lower breakdown fields [1]. However, dielectric constants of more than 100 are found for TiO$_2$ in the rutile phase with breakdown fields comparable to tantalum pentoxide. TiO$_2$ is used in commercial capacitors, it can be prepared in thin films, and it is physically and chemically stable [2]. Therefore TiO$_2$-films should offer the chance to increase considerably the charge storage density.

2. Experimental

The TiO$_2$-films investigated in this study were deposited on single crystal Si wafers using a dc magnetron sputtering system. The layers were reactively sputtered in Ar/O$_2$ atmosphere from a pure (99.9%) Ti target. Both, oxygen and argon flow rates, were controlled to maintain a pressure of 5mTorr and a flow ratio of 1.5. The target was oxidized under these gas conditions. A typical deposition rate of 2.3nm/min was obtained for a power density of 5W/cm^2. We have prepared two sets of TiO$_2$-films with dielectric thicknesses of about 25 and 50nm for the evaluation of the electrical properties. Some of the sputtered TiO$_2$-films were annealed under various conditions. Planar capacitors with evaporated Al gate electrodes and with areas between $2*10^{-4}$cm^2 and 0.1cm^2 were used for the electrical measurements.

3. Results

As-deposited films with a thickness between 20nm and 300nm were found to be shiny and of perfect optical quality. The composition of the thicker films was determined by electron microprobe analysis (EMA) yielding an oxygen to titanium ratio of 2:1. This indicates that the stoichiometry of the films was indeed TiO$_2$. It should be noted that no measurable amount of Ar could be detected by EMA. Structural information was obtained from X-ray diffraction analysis (XRD). The diffraction pattern of Fig.1 indicates

© 1991 IOP Publishing Ltd

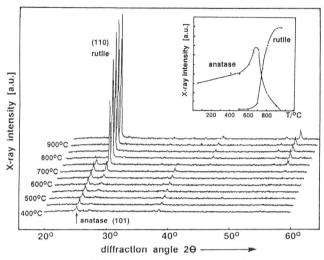

Fig.1: X-ray diffraction data of a TiO_2-film taken after annealing at the temperatures given in the figure. The inset shows the X-ray intensity of the (101)-peak of the anatase phase and the (110)-peak of the rutile phase as a function of the annealing temperature.

that as-deposited films have primarily anatase phase, showing major diffraction peaks at the (101)- and (112)-direction. However, as expected, this phase transforms into the rutile structure when the sample is annealed above 700°C. The recrystallization behaviour is indicated by a significant increase of the (110)-peak. This is also shown by the inset of Fig.1.

Information about the layer thicknesses was obtained by TEM analysis of various samples in the as-deposited as well as the annealed state. The micrographs show a native oxide layer at the interface to the Si substrate (thickness 1.8nm) which remains stabled during annealing in Ar-atmosphere. Annealing the 25nm TiO_2-films in oxygen diluted by argon increases the native SiO_2-layer to more than 5nm whereas the SiO_2-layer is stable for the 50nm TiO_2-films under the same annealing conditions. The TiO_2-thickness of the rutile grains in Fig.2 (annealed at 800°C in Ar/O_2) are found to vary from about 40nm to 50nm compared to the perfectly smooth surface of the as-deposited layer. The grains are mainly oriented with their c-axis parallel to the wafer surface as indicated by the X-ray data.

Fig.2: Cross sectional TEM micrograph of the 50nm TiO_2-layer annealed at 800°C in Ar/O_2. The dielectric is covered by evaporated Al used as the gate electrode of the capacitors.

Properties of Deposited Dielectrics

The TiO$_2$-samples were also analysed by CV-measurements. Annealing the samples at 800°C in argon increases the capacitance due to an increase of the dielectric constant of the TiO$_2$-films. Therefore, the increase of the capacitance is directly correlated with the transformation of the film into the rutile structure. Annealing in argon/oxygen yields nearly the same results as the argon anneal for the 50nm TiO$_2$-films whereas for the 25nm TiO$_2$-films a decrease in capacitance is observed. This is due to oxidation of the silicon substrate during the Ar/O$_2$-anneal as confirmed by TEM-analysis. The CV-curves of all annealed TiO$_2$-samples looked like those of very thin SiO$_2$-layers and showed nearly no shift (less than 10mV) between curves recorded from accumulation to depletion and vice versa. This indicates that flatband voltage shifts due to trapping and detrapping of charges are not a dominant process for SiO$_2$ equivalent electric fields up to 10MV/cm. A typical CV-curve of a TiO$_2$-film is shown in Fig.3. Evaluation of the interface state density yields values below $10^{11}eV^{-1}cm^{-2}$ for samples annealed at 800°C; no forming gas anneal was performed for these samples.

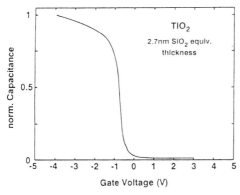

Fig.3:
CV-curve of a capacitor with a 50nm TiO$_2$-layer

The extraction of the dielectric constant of the TiO$_2$-layers from the CV-data cannot be done easily with our samples since the oxide equivalent thickness of the films was about 2.5nm. Therefore the uncertainty in the thickness of the native SiO$_2$-layer (1.8nm ± 0.3nm) causes large errors in the dielectric constant of the TiO$_2$-film. Furthermore these very thin TiO$_2$-films are more conductive than thicker films; this must be taken into account for the analysis of the CV-measurements. Unfortunately the measured capacitance values do not simply correlate with the dielectric constants of the native SiO$_2$- and TiO$_2$-layers. This is obvious by analysing the frequency dependence of the impedance. Up to now we have not been successful to simulate the measured impedance in consideration of the capacitance and conductance of the TiO$_2$- and SiO$_2$-layers as well as of the silicon substrate. Regarding the discussed limitations, our data reveal that the dielectric constant of the annealed TiO$_2$-films is at least 100 in accordance with the bulk values. For the calculation of a more accurate value, TiO$_2$-films with thicknesses above 100nm must be analysed.

Current densities versus oxide equivalent electric field for negative gate polarity are shown in Fig.4; data for positive gate are not plotted in Fig.4 but they are almost symmetric to the negative ones. Given are data for the 25nm and 50nm TiO$_2$-films before and after annealing at 800°C. A dramatic decrease in current density can be seen after annealing. There is nearly no difference in the current density between the 25nm and 50nm TiO$_2$-samples. The current density of the 25nm TiO$_2$-layer after annealing in Ar/O$_2$ is significantly lower since annealing in Ar/O$_2$ causes a thickness increase of the SiO$_2$-layer between the TiO$_2$-film and the silicon substrate to more than 5nm as discussed above. Compared to a 2.5nm SiO$_2$-dielectric the annealed TiO$_2$-films showed a reduction of the current density by two orders of magnitude at an electric field of 3MV/cm. Since for application in DRAMs the charge stored in the capacitor is the most interesting quantity the current density of the TiO$_2$-films is also given as a function of the charge storage density in the capacitor.

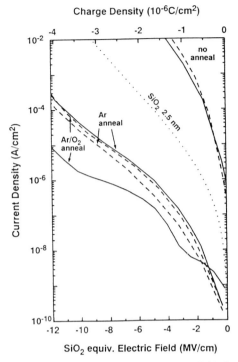

Fig.4:
Current density of the different TiO$_2$-samples vs. SiO$_2$ equivalent electric field for negative gate voltages. Data for the 25 and 50nm TiO$_2$-layers are given in full and broken lines respectively. For comparison the charge storage density is given on the upper scale.

Time dependent breakdown tests at various electric fields have been performed for the different TiO$_2$-samples. The 50nm TiO$_2$-layers annealed in Ar/O$_2$ showed the best results concerning time-to-breakdown. For capacitors with an area of $1.6*10^{-3}$cm^{-2} stressed at 13.5MV/cm (SiO$_2$ equiv. field) with negative gate polarity breakdown times of 5500sec are obtained at a cumulative failure of 63%. Extrapolation of the t$_{bd}$-values of 0.1cm^2 capacitors to lower electric fields yields lifetimes of about 10^{11}sec at a SiO$_2$ equivalent field of 5MV/cm and a cumulative failure of 63%. Defect densities were in the order of 0.1cm^{-2}. We have also investigated the cumulative failure distribution as a function of the areas of the capacitors. The dependence of the time-to-breakdown values on the areas indicate that a weak-spot mechanism is responsible for the observed failures.

4. Conclusions

Our data show that TiO$_2$ is well suited as a dielectric for future DRAM storage cells. Dielectric films with an oxide equivalent thickness of 2.5nm but compared to SiO$_2$ reduced current densities have been realized with sputtered TiO$_2$-layers. Annealing the films above 700°C is essential for low current densities and a high dielectric constant. Concerning charge transport and wear-out properties the current as well as breakdown characteristics seem to be determined by the weakest spots, i.e. the thinnest grains or grain boundaries in the TEM micrograph. Optimization of the processing conditions in order to get smoother TiO$_2$-films might further improve the electrical characteristics.

References

[1] B. Shen, I. Chen, S. Banerjee, G. Brown, J. Bohlman, P. Chang and R. Doering, Technical Digest, IEEE IEDM 87, 1987, p.582 and SISC 88, 1988, San Diego, California.
[2] F. Grant, Reviews of Modern Physics 31 (1959) 646.

Diffusion of hydrogen in and into LPCVD silicon oxynitride films

C.H.M. Marée, W.M. Arnold Bik, F.H.P.M. Habraken
Department of Atom and Interface Physics, University of Utrecht,
P.O. Box 80.000, 3508 TA Utrecht, The Netherlands

This paper describes results of experiments on the exchange of hydrogen and deuterium in LPCVD silicon nitride and oxynitride films at temperatures 700–1000°C. The situation in which both H and D are bound in the material is compared to the situation in which H in the material exchanges with deuterium from D_2 gas phase molecules. It appears that in both situations the H diffusion coefficient has a minimum value for O/N ≈ 0.3.

1 INTRODUCTION

For several years already it is known that relevant electrical properties of silicon nitride films are related to the presence and behaviour of hydrogen in this material (Topich and Turi 1982, Schols and Maes 1983, Stein 1988). In addition, it has been shown that hydrogen plays a decisive role in the thermal oxidation of silicon (oxy)nitride films (Kuiper et al. 1989, Habraken and Kuiper 1990). The latter process step is applied to build ONO triple layer structures. These observations motivate the detailed investigation of the hydrogen chemistry in LPCVD silicon nitride films. Since the controlled introduction of oxygen in silicon nitride during the deposition process of silicon oxynitride films (Kuiper et al. 1983) appears to improve certain properties of the insulator material (Maes 1991, Kapoor et al. 1990), the influence of oxygen incorporation on the behaviour of hydrogen is also studied. The purpose of the present work is to investigate if the mechanisms of incorporation of hydrogen into oxynitrides from the gas phase differs from that of diffusion of hydrogen, which is incorporated in the material during the deposition.

2 EXPERIMENTAL

Silicon oxynitrides were grown onto silicon substrates in a LPCVD process at 800°C. The reactant gases were SiH_2Cl_2, NH_3 (ND_3) and N_2O. The O/N incorporation ratio in the films were varied between 0 and 1 by adjusting the N_2O/NH_3 gas phase ratio (Kuiper et al. 1983). The as deposited films contain about 3 at% H for O/N < 0.5. For larger O/N ratios the hydrogen concentration is lower (Habraken et al. 1986). As analysis technique we applied Elastic Recoil Detection (Arnold Bik et al. 1990).

3 RESULTS AND DISCUSSION

3.1 Diffusion of hydrogen bound in the oxynitrides

In earlier work we have determined the diffusion coefficient of hydrogen (and deuterium) in LPCVD Si_3N_4 films and its temperature dependence (Arnold Bik et al. 1990). This work was carried out making use of layer structures consisting of a double layer of nitride in which one layer was grown using NH_3 and the other using ND_3. These samples were annealed in an inert gas ambient and subsequently the H and D profiles were determined using Elastic Recoil Detection. We have found that the value for the diffusion coefficient amounts to $10^{-17} - 5\times10^{-14}$ cm²/sec in the temperature range 700−1000°C and that it is characterized by an activation energy of about 3 eV.
In a similar study we have deduced that the

© 1991 IOP Publishing Ltd

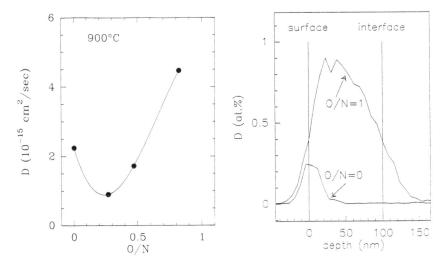

Figure 1: *Hydrogen diffusion coefficient at $900^\circ C$ in silicon oxynitrides as a function of O/N, as obtained from the double layer experiments.*

Figure 2: *D concentration depth profiles for oxynitrides with O/N = 0 and O/N = 1 after annealing at $800^\circ C$ as obtained by using ERD*

activation energy for hydrogen diffusion in LPCVD silicon oxynitride films is virtually not dependent on the O/N concentration ratio in the oxynitrides for O/N < 1. On the other hand, the diffusion coefficient shows a minimum value at O/N = 0.3 (fig. 1) (Arnold Bik et al. to be published). This observation explains the observed minimum in the rate of desorption of hydrogen during inert gas annealing from oxynitride films having an O/N ratio of about 0.3 (Arnold Bik et al. to be published).

The above mentioned work deals with the mechanistics of hydrogen, which is incorporated in the material in a =N-H bonding configuration during the deposition. The data are pertinent to the situations where the concentration of hydrogenic species (H and D) has not yet been changed as a result of the high temperature treatment.

3.2 Diffusion into the oxynitrides

In this section we report the study of the incorporation of deuterium into LPCVD, NH_3 grown, silicon oxynitride films from the gas phase. To this end the oxynitride films were annealed for 5 minutes in a 4% D_2/N_2 mixture at 800, 900 or 1000°C.

Figure 2 shows ERD deuterium depth profiles for the silicon nitride and the silicon oxynitride film with O/N = 1, after annealing the layers for 5 minutes at 800°C in the D_2/N_2 mixture. These profiles immediately show that the extent of D uptake in the nitride (O/N = 0) is much smaller than in the oxynitride with O/N = 1. Furthermore, in the nitride the presence of D is confined to a thin layer at the surface whereas in the sample with O/N = 1 deuterium appears to be distributed over the entire depth of the film. Figure 3 shows the total amount of D incorporated in the oxynitrides as a function of O/N for the various anneal temperatures. The amount of D incorporated has a minimum at O/N = 0.3 for all temperatures but increases strongly with O/N for O/N > 0.3. The extent of D incorporation shows a tendency to increase with increasing anneal temperature.

During the D_2/N_2 anneal hydrogen leaves the oxynitride. The hydrogen loss for the different temperatures is given in fig. 4 as a function of O/N. This amount of desorbed hydrogen has a minimum around O/N = 0.3 and it increases with increasing temperature for all O/N. This minimum in the amount of released hydrogen at O/N = 0.3 and the minimum in the amount

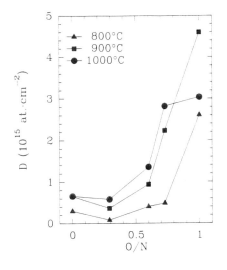

Figure 3: *Total amount of D incorporated in silicon oxynitride films during the 5 min. anneals at the indicated temperatures as a function of O/N*

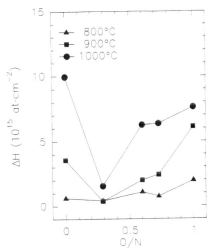

Figure 4: *The hydrogen loss during the 5 min. anneals at the indicated temperatures as a function of O/N*

of D taken up at this particular oxynitride composition is consistent with the minimum in the diffusion coefficient for hydrogen in the region of O/N = 0.3 (see figure 1).

D concentration profiles, which could be ascribed to a fast diffusion of deuterium molecules, not bound to the oxynitride network, have not been observed.

4 CONCLUSION

The process of incorporation of hydrogen from the gas phase into LPCVD silicon oxynitride during a hydrogen anneal seems not to differ largely from the diffusion of hydrogen, bound in the material. The diffusion coefficient shows a minimum for the oxynitrides having a O/N ratio of about 0.3. This deduction explains the earlier observed minimum in the hydrogen loss rate from silicon oxynitride films during high temperature treatment.

ACKNOWLEDGMENTS

The authors wish to thank A. Kuiper (Philips Research Labs, Eindhoven) for the oxynitride films and valuable discussions, E. Hengst and C. Denisse (ASM, Bilthoven) for help during the annealing treatments and the crew of the accelerator departement for running the tandem van de Graaff accelerator.

REFERENCES

Arnold Bik et al. 1990 — *W.M. Arnold Bik, R.N.H. Linssen, F.H.P.M. Habraken, W.F. van der Weg and A.E.T. Kuiper, Appl. Phys. Lett. 56 (1990) 2530.*

Arnold Bik et al. to be published — *W.M. Arnold Bik, C.H.M. Marée, M.J. van den Boogaard, F.H.P.M. Habraken and A.E.T. Kuiper, To be published.*

Habraken et al. 1986 — *F.H.P.M. Habraken, R.H.G. Tijhaar, W.F. van der Weg, A.E.T. Kuiper and M.F.C. Willemsen, J. Appl. Phys. 59 (1986) 447.*

Habraken and Kuiper 1990 — *F.H.P.M. Habraken and A.E.T. Kuiper, Thin Solid Films 193, 665 (1990).*

Kapoor et al. 1990 — *V.J. Kapoor, R.S. Bailey and R.A. Turi, J. Electrochem. Soc. 137, 3589 (1990).*

Kuiper et al. 1983 — *A.E.T. Kuiper, S.W. Koo, F.H.P.M. Habraken and Y. Tamminga, J. Vac. Sci. & Technol. B1, 62 (1983).*

Kuiper et al. 1989 — A.E.T. Kuiper, M.F.C. Willemsen, J.M.L Mulder, J.B. Oude Elferink, F.H.P.M. Habraken and W.F. van der Weg, J. Vac. Sci. & Technol. B7, 455 (1989).

Maes 1991 — H.E. Maes, in "LPCVD Silicon Nitride and Oxynitride Films, Material and Applications" Edited by F.H.P.M. Habraken, Research Reports ESPRIT (Springer, Heidelberg, 1991) p. 118.

Schols and Maes 1983 — G. Schols and H.E. Maes, in Silicon Nitride Thin Insulating Films, Proc. Vol. 83-8, (Edited by V.J. Kapoor and H.J. Stein) p. 94, Electrochem. Soc. Pennington, NJ (1983).

Stein 1988 — H.J. Stein, Proceedings Electrochemical Society 1988 Fall Meeting, Vol. on "Silicon Nitride and Silicon Dioxide Thin Insulating Films".

Topich and Turi 1982 — J.A. Topich and R.A. Turi, Appl. Phys. Lett. 41, 641 (1982).

Paper presented at INFOS '91, Liverpool, April 1991
Contributed Papers, Section 3

Characterization of SiO$_2$ films deposited by reactive excimer laser ablation of SiO target

A. Slaoui, E. Fogarassy, C. Fuchs and P. Siffert
Laboratoire PHASE, CRN, 23 rue du Lœss, F-67037 Strasbourg Cedex, France

> ABSTRACT : Silicon dioxide films have been prepared by ArF pulsed-laser ablation of a silicon monoxide (SiO) target in oxygen atmosphere, at various substrate temperature (20 – 450°C). The structural and electrical properties of the SiO$_2$ deposited films have been investigated. The effects of a rapid thermal annealing on the quality of the laser ablation SiO$_2$ films have been studied.

1. INTRODUCTION

There is considerable interest in the deposition at low temperature of silicon dioxide films. Several methods have been developed including ion beam methods (Todorov *et al.* 1986), plasma (Theil *et al.* 1990) and photon-assisted chemical vapor deposition (CVD) (Lan *et al.* 1990). Among these techniques, laser ablation deposition is thought to be a promising means for thin SiO$_2$ films fabrication as it offers some advantages such as minimized contamination from a heat source and a high deposition rate in addition to be very simple to implement (Hanabusa 1987, Sato *et al.* 1988, Fogarassy *et al.* 1989). A preliminary report of the present work was published previously (Fogarassy *et al.* 1990a). Here we investigate the properties of the laser ablation deposited SiO$_2$ films by measuring the infrared spectra (IR), deposition rate, etch rate, refractive index, breakdown voltage and annealing behavior.

2. EXPERIMENTAL

High-purity silicon monoxide bulk targets were irradiated in pure oxygen atmosphere (0.1 Torr), through a suprasil quartz window, with a pulsed ArF excimer laser (Lambda Physik EMG 201) working at 193 nm. The laser generated 300 mJ pulses of 20 ns duration at a repetition rate of 50 Hz. The beam was focused by a spherical lens onto the target under normal incidence to give energy densities ranging between 1 and 10 J/cm^2. During the laser irradiation, the target was rotated to avoid the formation of a deep crater able to modify material ejection and subsequent emission of a large number of macroscopic particles. Emitted species which consist mainly of silicon and oxygen were deposited onto < 100 > oriented Si substrates (resistivity of 1.5 Ω.cm) placed at a distance of 3 cm from the target. Substrate temperature was controlled by an electric heater and was varied in the range 20 – 450°C.

The deposited oxide films have been characterized by infrared (IR) absorption spectroscopy using a Perkin Elmer 938G double beam grating infrared spectrophotometer using a clean silicon wafer in the reference beam.

Film thicknesses were measured with a mechanical stylus and confirmed by IR absorption and their refractive indices were deduced from ellipsometry performed at 632.8 nm wavelength (Gaertner). Etch rates were measured in dilute hydrofluoric acid (100 : 1 H$_2$O : 49% HF) at 25°C. The electrical properties were measured by conventional high-frequency C-V technique, using MOS capacitors with 1 mm diameter aluminium dots as electrode. Some MOS structures were annealed at 450°C for 30 min

© 1991 IOP Publishing Ltd

in a forming gas (90% N_2, 10% H_2). The rapid thermal annealing (RTA) treatment was performed in a tungsten halogen lamp furnace (JIPELEC model FAV 4) in the range 400 – 1000°C for 30 sec.

3. RESULTS AND DISCUSSION

Figure 1 shows the spectra of thermally grown SiO_2 film (1100°C) and pulsed laser ablation SiO_x films (LA – SiO_2) deposited at various substrate temperatures with the laser fluence fixed at 3 J/cm². The absorptions around 1080 cm⁻¹ (Asymmetric SiOSi Stretching Mode ASM), 800 cm⁻¹ (SiOSi bending) and 450 cm⁻¹ (SiOSi rocking) are characteristics of silicon dioxide (Pliskin and Lehman 1965) and are observed for all the films. However, the laser-deposited oxide spectra exhibit additional features: (i) two absorptions at high frequencies, 3600 and 3350 cm⁻¹, usually attributed to O-H stretching modes, with the band at the higher frequency resulting from SiOH and the 3350 cm⁻¹ peak caused by loosely bonded SiOH or adsorbed water (Pliskin 1977) ; (ii) two small absorption bands are observed at 940 cm⁻¹ and 1630 cm⁻¹ attributed respectively to a Si-OH bending vibration with a non bridging oxygen atom (Theil et al. 1990, Pliskin 1977) and to absorbed water (H_2O) ; (iii) a pronounced shoulder on the high frequency side of the ASM band which extends up to 1300 cm⁻¹, which was already observed by Slaoui et al. (1989) in laser grown and irradiated thin film oxides. This feature was attributed by Bensch and Bergholz (1990) to disorder and/or defects within the SiO_2 thin film. In our case, the structural disorder can be caused by the very fast deposition rate (~ 1200 Å/min) as it can be attributed to energetic ions generated during the laser evaporation of SiO and which bombarded the substrate and the growing film during the deposition. A part of these energetic ions could lead to bond rupture and rearrangement of the ring structure. The IR results also show that hydrogen has reacted with the oxide to from SiOH. In our case, the hydrogen, OH or water molecules included in the target and in the deposition chamber are thought to be responsible for Si-OH groups incorporation.

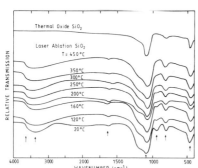

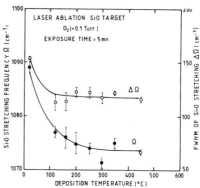

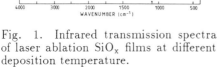

Fig. 1. Infrared transmission spectra of laser ablation SiO_x films at different deposition temperature.

Fig. 2. Plot of the change in the SiO stretching (●) peak frequency and (○) FWHM against deposition temperature.

The ASM peak position Ω and the full width at the half maximum absorption (FWHM) $\Delta\Omega$ of the LA – SiO_2 are reported in Figure 2 as a function of deposition temperature. For films deposited at room temperature, the Si-O stretching peak exists at 1090 cm⁻¹ and the peak width is of about 150 cm⁻¹. Comparable Ω but lower FWHM values were measured on thermal oxide films of equivalent thicknesses (5200 Å) (Sato 1970, Bensch and Bergholz 1990). These results suggest that stoichiometric oxide films without oxygen deficiency can be synthetized by laser ablation of SiO at 20°C. Rutherford backscattering spectroscopy (RBS) analysis also confirm that the films are stoichiometric (Fogarassy et al. 1990b). However, these films exhibit large half-width which are indicative of porous films. The porosity is evidenced by the presence of hydrogen bonded surface silanol groups (Si-OH) and water in the spectra of the deposited films. The porosity is futher substantiated by the very rapid HF-etch rate of 6000 Å/min and the very low refractive index of 1.26. The presence of bonding strain in the films probably induced by the very high deposition rate also contribute to the position and the half-width broadness of the Si-O stretching band. These stained-bonding films are expected to be more reactive than relaxed films with larger Si-O-Si bond angles, and in addition are more accessible to chemical attack because of localized increases in the bond-free volume (Theil et al. 1990). The strain in the films will also result in lower refractive indices and faster etch rates (Pliskin 1977). The broadening of the stretching vibration can also be due to the presence of the pronounced shoulder at the high-wavenumber side of the ASM band.

Hydrogen extraction from the LA-deposited SiO_2 films can also be achieved by using a rapid thermal annealing process. IR spectra of the LA $-$ SiO_2 films deposited at 20°C, before and after rapid thermal treatement are shown in Fig. 3. It can be seen that the Si-OH peak at about 940 cm^{-1} becomes weaker and disappears at 600°C while the absorption band at 3350 cm^{-1} has nearly vanished at 800°C. Another interesting feature in Fig. 3 is that the Si-O stretching peak at about 1090 cm^{-1} at 20°C tends to shift to a higher wavenumber as the annealing temperature is raised, which means that Si-O bondings become stronger and consequently the Si-Si distance increases. Furthermore, the FWHM of the ASM vibration is narrowed with increasing annealing temperature, reflecting a larger distribution of vibration energies after the heat treatment. On the other hand, the refractive index is found to increase with the annealing temperature whereas the etch rate strongly decreases. Deposited films treated at 800°C etch uniformly with rate of about 40 Å/min, very close to the value of thermally grown SiO_2. The change in IR stretching frequency, refractive index and etch rate with the annealing temperature give evidence for a rearrangement of Si-O bonds due to the outward diffusion of hydrogen, a relief in the bonding strain and/or the annealing of some structural defects.

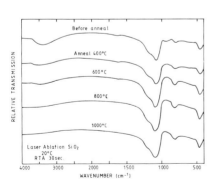
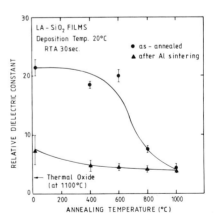

Fig. 3. Infrared transmission spectra of laser ablation SiO_x films deposited at 20°C after different rapid thermal annealing (RTA) temperatures.

Fig. 4. Relative dielectric constant vs. annealing temperature for LA-SiO_x films deposited at 20°C : (●) as RTA annealed and (▲) after RTA and post-metallization thermal treatment (450°C/15 min).

Figure 4 shows the annealing temperature dependence of the relative dielectric constant of the LA − SiO_2 films deposited at 20°C. The dielectric constant values are deduced from the capacitance-voltage measurements in the accumulation region. The dieletric constants decrease especially in the MOS structures sintered in forming gas, and approached to that of thermally oxidized Si, 3.8. These facts indicate that post-metallization annealing is effective for improving the electrical properties of the SiO_2 films as well as structural ones. The dielectric constant, i.e. polarizability is considered to be related to the bonds having a high-dielectric polarization, such as OH or Si-Si bonds. The decrease of the dielectric constant may be attributed to a decrease in the Si-OH and H_2O contents as well as to an increase in the Si-Si bond strength.

The minimum oxide charge density was $1.5 \times 10^{11} cm^{-2}$ in a SiO_2 film with a thickness of 6500 Å obtained by deposition at 20°C followed by a RTA treatment at 1000°C during 30 sec and a post-metallization annealing. Breakdown voltages have been measured for samples approximately 6000 Å thick. The samples deposited at 20°C have a maximum at 2-3 MV/cm probably due to the presence of a high concentration of hydroxyls in the film. Samples as-deposited and RTA treated breakdown at 4-8 MV/cm. Similar breakdown voltage values have been measured on chemically (Adams et al. 1980) and plasma (Adams et al. 1981) deposited silicon dioxide films.

4. SUMMARY

Silicon dioxide films have been deposited by pulsed laser ablation technique at 20° − 450°C using a silicon monoxide solid as a target. The refractive index increases and the etch rate and porosity decrease with an increase in the deposition temperature. The films obtained at 450°C give values comparable to that of a plasma deposited oxide. On the other hand, the relative dielectric constant was shown to be strongly affected by the annealing temperature. This fact might be related to a decrease in the Si-OH and H_2O contents in the films as well as to an increase in the Si-Si bond strength. The minimum value of the oxide charge density is about $1.5 \times 10^{11} cm^{-2}$ for films deposited at 20°C, RTA (1000°C/30 sec) treated and post-metallization annealed. Dielectric breakdown occurs at fields of 2-8 MV/cm. The low breakdown field makes these films unsuitable as a gate in some applications.

REFERENCES

Adams A C, Smith T E and Chang C C 1980 *J. Electrochem. Soc.* **127** 1787
Adams A C, Alexander F B, Capio C D and Smith T E 1981 *J. Electrochem. Soc.* **128** 1545
Bensch W and Bergholz W 1990 *Semicond. Sci. Technol.* **5** 421
Fogarassy E, Fuchs C, Stoquert J P, Siffert P, Perriere J and Rochet F 1989 *J. Less-Common Metals* **151** 249
Fogarassy E, Slaoui A, Fuchs C and Stoquert J P 1990 *Appl. Phys. Lett.* **57** 664
Fogarassy E, Slaoui A, Fuchs C and Siffert P 1990 *Appl. Surf. Sci.* **46** 195
Hanabusa H 1987 *Mat. Sci. Rep.* **2** 51
Lan W H, Tu S T, Yang S J and Huang K F 1990 *Jpn. Appl. Phys.* **29** 997
Pliskin W A and Lehman S 1965 *J. Electrochem. Soc.* **112** 1013
Pliskin W A 1977 *J. Vac. Sci. Technol.* **14** 1064
Sato K 1970 *J. Electrochem. Soc.* **117** 10
Sato T, Furuno S, Igushi S and Hanabusa M 1988 *Appl. Phys.* **A45** 355
Slaoui A, Fogarassy E, White C W and Siffert P 1988 *Appl. Phys. Lett.* **53** 1832
Theil J A, Tsu D V, Watkins M W, Kim S S and Lucovsky G 1990 *J. Vac. Sci. Technol.* **A8** 1374
Todorov S S, Shillinger S L and Fossum E R 1986 *IEEE Electr. Dev. Lett.* **7** 468

Low pressure MOCVD of tantalum oxide

E.P. Burte and N. Rausch
Fraunhofer Arbeitsgruppe für Integrierte Schaltungen, Artilleriestrasse 12, D-8520 Erlangen, FRG

ABSTRACT: Preparation and properties of tantalum oxide films have been studied for the application to ultra large scale integrated (ULSI) circuits as a new capacitor dielectric in low-power high-density DRAM's. Tantalum oxide films were deposited by a low pressure metal organic chemical vapor deposition (MOCVD) process in a hot wall-type vertical furnace at temperatures ranging from 375°C to 450°C. Stoichiometry, structure as well as electrical properties of the layers were examined before and after an annealing treatment in oxygen atmosphere at different temperatures.

1. INTRODUCTION

The application of conventional Oxide-Nitride-Oxide (ONO) triple layer structures as dielectric material in ULSI dynamic storage capacitors will reach its physical limits in terms of reduction of thickness in the generation of 64 Mb DRAM's because the direct tunnel current increases dramatically, if the effective silicon dioxide film thickness becomes lower than 5 nm. In order to maintain the required specific capacitance by selecting a film thickness, which is easy to produce with high reliability, and in order to obtain an effective SiO_2 film thickness showing sufficiently low leakage current densities for favorable memory operation, a new storage material exhibiting a much higher dielectric constant and a properly low leakage current is necessary. The most promising high-dielectric constant insulator for application to low-power high-density DRAM's beyond 64 Mb seems to be tantalum pentoxide (Ta_2O_5) which exhibits a dielectric constant of 22 and above, depending on the fabrication process.

Most of recent publications deal with Ta_2O_5 film produced by a reactive sputtering process of tantalum in oxygen ambient (Shinriki 1990). Sputtered amorphous Ta_2O_5 layers have shown to be suitable for memory cells. Very few studies report on tantalum oxide films deposited by low pressure chemical vapor deposition (Zaima 1990) which is the most appropriate method for application in trench-capacitors in contrast to the sputtering method exhibiting a very poor step coverage and shadow effects.

In this paper we deal with metal organic chemical vapor deposition (MOCVD) of tantalum oxide at low pressure performed in a hot wall-type vertical furnace which is a common method combining a high uniformity in radial as well as in axial direction of the tube with a high throughput. This paper reports selected details on the preparation and properties of tantalum oxide films produced by low pressure MOCVD.

2. EXPERIMENTAL

Tantalum oxide films were deposited on n-type (100) silicon substrates in a hot wall-type low pressure CVD system. Fig. 1 shows a schematic diagram of the vertical reactor used for the investigated MOCVD Ta_2O_5 process. This reactor provides down-stream introduction of the process gases oxygen, nitrogen, and tantalum ethylate $[Ta(OC_2O_5)_5]$ vapor. A mass flow controller (MFC) controls the flow of liquid tantalum ethylate. As the melting point of this metal organic compound is slightly above room temperature, storage tank, MFC, and tubes are kept at a temperature of 50 °C. In a chamber heated to 150 °C tantalum ethylate is vaporized. Nitrogen gas transports this vapor through a heated gas supply line (to prevent condensation) into the vertical reactor made of quartz. Ten 3-inch silicon wafers are placed horizontally in a vertical quartz boat. Typical deposition conditions are listed in table 1.

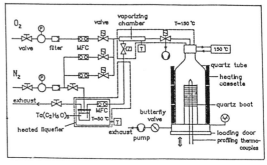

Fig. 1: Schematic diagram of the vertical LPCVD reactor

Table 1: Deposition Parameters

Gas flow rate:	N_2	50 - 150	sccm
	O_2	50 - 200	sccm
	$Ta(OC_2H_5)_5$	0.1 - 0.3	sccm
Temperature:		375 - 525	°C
Pressure:		150 - 500	mTorr
Period of deposition:		15 - 120	min

After the deposition most of the wafers were annealed for 30 minutes in dry oxygen at temperatures ranging from 500°C to 900°C.

The stoichiometric and structural properties of the tantalum oxide films were investigated by ellipsometry, secondary neutral mass spectroscopy (SNMS), Rutherford backscattering spectroscopy (RBS), X-ray photoelectron spectroscopy (XPS), high resolution transmission electron microscopy (HRTEM), and X-ray diffraction (XRD).

The tantalum oxide films were electrically characterized by applying capacitance-voltage and current-voltage measurements. Therefore, MIS capacitors with an Al gate electrode of 0.5 μm thickness prepared by thermal evaporation were fabricated. The Al gate metallization was patterned by a wet chemical process. The area of the generated circular Al dots was 8×10^{-4} cm^2.

3. RESULTS AND DISCUSSION

3.1 Composition and structure of the Ta_2O_5 layers

At constant temperature and gas flow deposition rate and thickness homogeneity depend on the chamber pressure. By evaluating the slope of the Arrhenius plot of the deposition rate versus 1000 K/T, the activation energy was determined to be 0.8 eV.

The values of film thickness were determined by ellipsometry. The optically measured values agree well with those determined by the high resolution transmission electron microscope (HRTEM). The as-deposited films show a refractive index n ranging from 2.17 to 2.20. After a heat treatment at temperatures higher than 700°C the refractive index tends to increase slightly. This observation can be understood by a densification of the material which was determined by RBS.

Using RBS data, the composition of the as-deposited film was calculated to be $Ta_{1.8}O_5$. The deviation from the stoichiometric ratio of tantalum to oxygen is within the limit of accuracy of RBS measurement. Also, XPS measurements did not show a chemical shift of the Ta 4f peak to a higher binding energy between tantalum and oxygen after a heat treatment at 1000°C in dry O_2 ambient. This means that tantalum is in the highest oxidation state. Therefore the as-deposited layers are assumed to be stoichiometric. The relatively low values of the dielectric constant (ε: 15-24) of as-deposited films correspond to a low specific density compared to annealed layers, which exhibit higher values of the dielectric constant (ε: 28-34).

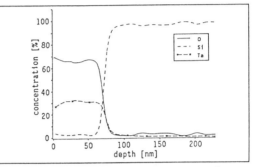

SNMS was applied to the samples in order to measure the atomic concentration versus depth of the elements oxygen (mass 16), silicon (mass 28) and tantalum (mass 181). A typical SNMS plot of a sample consisting of 70 nm tantalum oxide on silicon is shown in fig. 2. The SNMS analysis confirms the stoichiometric composition of these films. A SiO_2 interfacial layer formed during O_2 annealing could not be detected by SNMS.

Fig. 2: SNMS depth profile of Ta_2O_5 on silicon

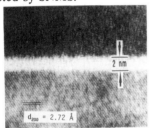

Fig. 3: HRTEM picture of Ta_2O5 on silicon. At the interface appears a SiO_2 layer of 2 nm thickness grown during the deposition.

XRD measurements show that the crystallization occured after a 30 min anneal at temperatures above 700 °C. Grains exhibiting an orthorhombic crystal structure were formed by annealing as revealed by HRTEM pictures and XRD spectra. HRTEM indicates that an interfacial silicon oxide layer of 2 nm thickness was grown on the silicon substrate during Ta_2O_5 deposition (see fig. 3). The thickness of this interfacial oxide layer increases with annealing temperature; but the growth rate is smaller than that on bare silicon in the same oxygen ambient. A 30 min heat treatment at 900°C results in an interfacial oxide of 4.1 nm thickness.

3.2 Electrical characterization

Fig. 4 shows a typical high frequency (1 MHz) capacitance-voltage plot of an as-deposited Al-Ta_2O_5-Si sample. Applying Terman's method, the interface trap density at midgap was estimated to be lower than $D_{it} = 3\times10^{11}$ $eV^{-1}cm^{-2}$. The capacitance value in accumulation was used to extract the dielectric constant of the corresponding Ta_2O_5 film. Annealing at 700°C improves the quality of the dielectric layer-silicon interface as the C-V curve of the annealed sample shows (see fig. 4). The increase in the slope of the curve near midgap ($C/C_I = 0.14$, $U_{Mid} = -0.4V$) and the shift of the curve along the voltage axis to the left indicates lower densities of fast interface traps and fixed interface charges, respectively. Fig. 5 shows the leakage current density versus applied electrical field of a sample before and after annealing in oxygen ambient at 900°C. Both samples exhibit rather high leakage current values at an applied field of 1 MV/cm in comparison to those of sputtered tantalum oxide films reported in literature (Hashimoto 1989). Nevertheless the leakage current could be reduced by annealing. Destructive breakdown occurs at 1.3 MV/cm. Assuming that the current transport through Ta_2O_5 is governed by Frenkel-Poole emission, j/E versus $E^{1/2}$ (j: current density, E: strength of electric field) was plotted (see fig. 6). Evaluating the slope of this curve, the optical dielectric constant was determined to be 1.3, which is far away from the theoretical value of 4.8 assuming a refractive index of 2.2. Therefore the Frenkel-Poole model is not appropriate to describe the current transport mechanism through Ta_2O_5.

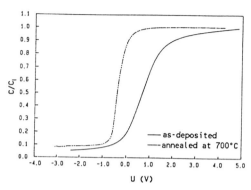

Fig. 4: C-V characteristics of an as-deposited and an annealed layer

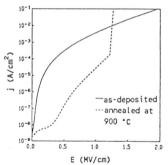

Fig. 5: I-V characteristics of an as-deposited and an annealed layer

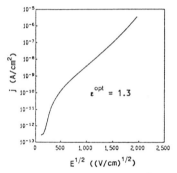

Fig. 6: Frenkel-Poole plot of an as-deposited layer

4. CONCLUSION

Tantalum oxide films have been successfully deposited onto silicon substrates by using a hot wall low pressure MOCVD process.

The atomic ratio of tantalum to oxygen was 2:5 revealing that stoichiometric Ta_2O_5 layers were grown.

The dielectric constant of as-deposited films ranged from 15 to 24. By annealing at 700°C it increased up to 30. Furthermore, annealing improved the quality of tantalum oxide films with respect to interface trap density and leakage current density.

In order to apply these films to future ULSI circuits further reduction of the leakage current through these films is necessary.

5. REFERENCES

Hashimoto C et al. 1989 *IEEE Trans. Electron Dev.* **36** 14
Shinriki H et al. 1990 *IEEE Trans. Electron Dev.* **37** 1939
Zaima S et al. 1990 *J. Electrochem. Soc.* **137** 1297

# Epitaxial fluoride films on semiconductors: MBE growth and photoluminescent characterization

N.S.Sokolov, S.V.Novikov, N.L.Yakovlev

A.F.Ioffe Physico-Technical Institute, 194021, Leningrad, USSR

ABSTRACT: Single crystalline Ca and Sr fluoride films have been grown on Si and GaAs by molecular beam epitaxy. The fluoride growth modes were investigated using dynamic RHEED studies. The films were doped with Eu^{2+} and Sm^{2+} ions to measure their strain and crystalline quality and the dependences of those on the growth temperature and film thickness by photoluminescence spectroscopy.

In the recent years single crystalline CaF_2, SrF_2 and BaF_2 films grown by molecular beam epitaxy (MBE) have attracted considerable attention (Schowalter and Fathauer 1986). These fluorides having cubic fluorite structure are good insulators with a band gap higher than 10 eV and lattice constants from 0.546 nm to 0.620 nm. There are various opportunities to apply these fluoride films for integration of elements based on different semiconductors as well as high temperature superconductors on common semiconductor substrate and fabrication of improved MIS structures and three dimensional (3-D) integral circuits.

Large binding energy of fluoride molecules in vapour phase results in molecular type of fluoride sublimation under heating and provides proper stoichiometric ratio during MBE growth. Schowalter and Fathauer (1986) showed that owing to the smallest free energy of (111)-face in fluorite structure the films can epitaxially grow on this face with atomically smooth surface. Such surface morphology is revealed from the streaked reflection high energy electron diffraction (RHEED) pattern.

The main parameters of our MBE-system are described by Gastev et al (1987). We carried out RHEED studies to investigate growth modes, epitaxial relations, and surface morphology of Ca, Sr and Ba fluoride layers on Si and GaAs. Novikov et al (1987) found that well pronounced specular beam (00-spot) intensity oscillations in RHEED patterns could be observed during the MBE growth of CaF_2 on Si(111) at 550 - 670°C. The period of these oscillations corresponded to the growth of one triple F-Ca-F layer (.315 nm). The observation of RHEED intensity oscillations is clear evidence for the two-dimensional growth mode of this fluoride on Si(111). Novikov et al (1991) succeeded to observe the RHEED intensity oscillations during epitaxial growth of SrF_2 and mixed $(Ca,Sr)F_2$ on Si and GaAs.

In this work we present the results of dynamic RHEED study during the epitaxial growth of $SrF_2/CaF_2/Si(111)$ structure (Figure 1). Part (I) of this figure shows the specular beam intensity behaviour during the deposition of the first two fluorite monolayers on a clean (7×7) reconstructed Si(111) surface at 700°C. The structure was then cooled to

© 1991 IOP Publishing Ltd

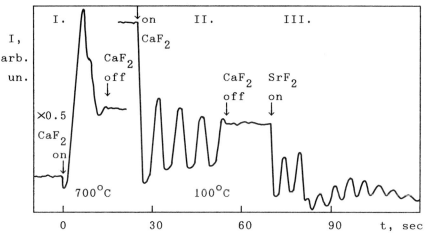

Figure 1. RHEED intensity oscillations during $SrF_2/CaF_2/Si(111)$ growth.

100°C and 4 fluorite layers were grown onto it (Part II). The presence of the oscillations (up to 30 in some experiments) at such low temperature evidences for a large surface migration length of fluoride molecules expected for ionic crystals (Yang and Flynn 1989). After the beginning of SrF_2 growth (Part III) two periods of usual shape are followed by an abrupt decrease of oscillation amplitude. This feature can be naturally explained by the breakdown of pseudomorphic growth mode when additional steps are induced on the growing surface. This information could be of value for the growth of insulating layers based on fluoride superlattices on GaAs.

As was shown in our previous papers (Sokolov et al 1989, 1990) Eu^{2+} and Sm^{2+} ions could be used as sensitive photoluminescent probes to characterize the crystallinity and to measure elastic strain in these films. In this paper we present new results in this field.

Liquid helium temperature photoluminescence (PL) spectra of $CaF_2:Sm^{2+}$ epitaxial layers of different thickness grown on Si(111) at 700°C are shown on the left side of Figure 2. Here and after the dashed curves are the PL spectra of Eu^{2+} or Sm^{2+} ions in bulk crystals and the upper scale is the strain in the plane of the interface. On the right side of the figure one can see the PL spectrum of 6 nm fluorite layer with the interface part (1 nm) grown at 700°C and the rest part grown at 100°C. As was shown before (Sokolov et al 1990) the compression of this layer indicates its pseudomorphic growth mode ($a_{CaF2} > a_{Si}$). The misfit accommodation occurs during the epitaxy of the other layers grown entirely at high temperature. Their tensile strain emerges during the postgrowth cooling because of the difference of the thermal expansion coefficients ($\eta_{CaF2} > \eta_{Si}$). The layers 30 nm thick have the largest tensile strain, the decrease of the strain in the thicker layers is due to the relaxation of the thermal stress. The main part of the strain (~.35%) in 1000 nm fluorite layer is due to the thermal strain arising as a result of cooling down from room temperature to 1.7 K. It evidences for an efficient stress relaxation in this film at room temperature. The remarkably narrow emission line of this film is comparable with that of the bulk crystal, which points out to high crystalline quality of the film.

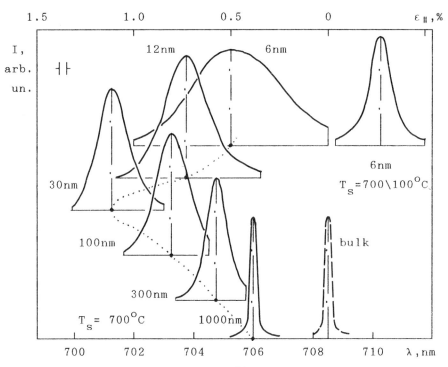

Figure 2. Photoluminescence of Sm^{2+} in CaF_2 layers on Si(111)

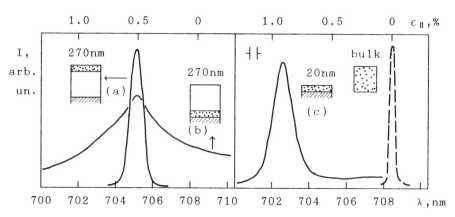

Figure 3. PL of selectively doped CaF_2/Si(111) structures

Selective doping by rare-earth ions appeared to be useful in measuring the strain and crystallinity through the depth of the film. The PL spectra of CaF_2/Si(111) films 270 nm thick doped with Sm^{2+} near the surface (a) and near the interface (b) are shown in Figure 3. The elastic component of the strain is the same, the interface part of the film is, however, more defective, which is revealed by the broader emission line (b). The comparison with the spectrum (c) of homogeneously doped 20 nm thick layer demonstrates that numerous defects near the interface appear during the overgrowth of the next layer and postgrowth cooling of the structure.

Our photoluminescent study revealed also that the planar elastic strain of fluorite layers on Si(100) is much lower than on (111) face. This can be explained by the three-dimensional growth mode of fluorite layers on Si(100) which results in more defective regions connecting growing clusters where thermal stress relaxation is easier.

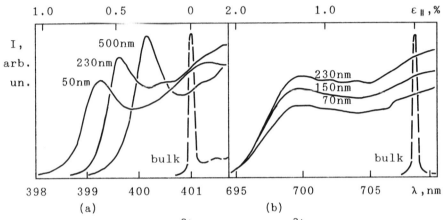

Figure 4. PL of $SrF_2:Eu^{2+}$ (a) and $CaF_2:Sm^{2+}$ (b) on GaAs(111)

Figure 4 shows impurity PL spectra of SrF_2 and CaF_2 layers on GaAs(111)B grown at $600^\circ C$ and $500^\circ C$ correspondingly. The substrates were preliminarily cleaned by thermal heating at $590^\circ C$ under As beam. A different dependence of the PL spectra on the film thickness suggests the difference in thermal stress relaxation in epitaxial SrF_2 and CaF_2 films. It could be due to the opposite signs of lattice mismatch in these two heterostructures ($a_{SrF2} > a_{GaAs} > a_{CaF2}$).

The authors are indebted to J.Alvares and E.Martynenko for their help in the fluoride MBE growth.

References

Gastev S V, Novikov S V, Sokolov N S, Yakovlev N L 1987 *Zh.Tekh.Fiz., Pisma* **13** 961 (1987 *Sov.Tech.Phys.Lett.* **13** 401)
Novikov S V, Sokolov N S, Yakovlev N L 1987 *Zh.Tekh.Fiz.,Pisma* **13** 1442 (*Sov.Tech.Phys.Lett.* **13** 603)
Novikov S V, Sokolov N S, Yakovlev N L 1991 *Abstracts of VI European Conference on Molecular Beam Epitaxy* (Tampere, Finland)
Schowalter L J, Fathauer R W 1986 *J.Vac.Sci.Technol.* **A4** 1026
Sokolov N S, Vigil E, Gastev S V, Novikov S V, Yakovlev N L 1989 *Fiz.tverd.Tela* **31**(2) 75 (*Sov.Phys. Solid State* **31**(2) 216)
Sokolov N S, Yakovlev N L, Almeida J 1990 *Solid State Commun.* **76** 883
Yang M H, Flynn C P 1989 *Phys.Rev.Letters* **62** 2476

Paper presented at INFOS '91, Liverpool, April 1991
Contributed Papers, Section 3

Improved surface treatments and passivation procedures of GaAs crystals controlled by photoluminescence measurements

S.K. KRAWCZYK, M. GENDRY, J. TARDY, F. KRAFFT and P.VIKTOROVITCH

Laboratoire d'Electronique (URA CNRS n° 848), Ecole Centrale de Lyon
36, avenue G. de Collongue, 69131 Ecully Cedex, France

P. ABRAHAM, A. BEKKAOUI and Y. MONTEIL

Laboratoire de Physico-Chimie Minérale (URA CNRS n° 116), Université Lyon I
43, Bd du 11 Novembre, 69622 Villeurbanne Cedex, France

R. SCHÜTZ, R. RIEMENSCHNEIDER, R.RICHTER and H.L. HARTNAGEL

Institute für Hochfrequenztechnik, Technische Hochschule Darmstadt
Merckstr. 25, 6100 Darmstadt, Germany

ABSTRACT: The objective of this work is to use photoluminescence measurements associated with other analytical techniques (XPS, STM and RHEED) to study GaAs surface treatments. We also propose a new surface passivation procedure, which is based on a thermal annealing of GaAs under a PH_3 atmosphere followed by a low temperature photon enhanced chemical vapour deposition (UV CVD) of Si_3N_4.

1. INTRODUCTION

Elaboration of GaAs/insulator interfaces with low surface recombination velocity is of great importance and has a direct impact on the threshold current of laser diodes, on the gain-bandwidth product of GaAs/GaAlAs bipolar transistors and on the reliability of the above devices.The assessment of the whole passivation process is usually achieved *a posteriori* by electrical characterization of the completed devices or test structures.

Previous works showed that room temperature photoluminescence intensity from InP crystals strongly depends on the surface recombination velocity and that in-process photoluminescence measurements (i.e. after the initial surface treatment, insulator deposition and annealing) can be used to precisely establish to what extent does each step in the process affect the final interface properties (Krawczyk et al 1984a, 1984b and 1990, Longère et al 1989).

The objective of this work is to extend the application of photoluminescence measurements to study and to optimise passivation procedures of GaAs crystals.

2. EXPERIMENTAL TECHNIQUES

Experiments were carried out with semi-insulating (S.I.) and n-type LEC GaAs wafers. Scanning photoluminescence (SPL) measurements were performed at room temperature using a SCAT IMAGEUR (SCANTEK, FRANCE). The photoluminescence signal is excited with an He-Ne laser and detected with a Si photodetector. SPL measurements consisted here of 200 x 200 data points. The highest spatial resolution is 1 μm. Complementary characterization techniques included XPS, RHEED and scanning tunneling microscpoy (STM).

© 1991 IOP Publishing Ltd

3. RESULTS AND DISCUSSION

We investigated the effect of chemical surface treatments (involving water solutions containing $H_2SO_4:H_2O_2$, $NH_4OH:H_2O_2$, HCl, HF and $(NH_4)_2S$) on the integrated photoluminescence intensity emitted by GaAs substrates. We have found that surface etching in basic solutions left the crystal with lower surface recombination velocity, in comparison with acid treatments. In particular, following the treatment in NH_4OH solutions, the PL signal increases by a factor of four, as compared with HCl and H_2SO_4 solutions (Fig. 1). This conclusion is almost independent on the relative concentration of the solution components. We also investigated the $(NH_4)_2S$ treatment proposed by Yablonovitch et al (1987). Although this treatment strongly increases the average PL signal (30-50 times), and the STM results indicated good short range (nanometer scale) spatial uniformity of the sample, the obtained SPL data show that the surface recombination velocity exhibits strong long range (tens of micrometer) non-uniformities (standard deviation of the PL signal close to 50%). In contrast to this, all other investigated wet treatments show excellent medium (micrometer range) and long range homogeneities (morphologies of photoluminescence images reveal typically only bulk non-uniformities and are unaffected by the treatments). This remains true even in case of the treatments in acidic solutions, which result in increased nanometer-range nonuniformities (as deduced from STM measurements).

Photoluminescence results are interpreted in terms of the physico-chemical properties of GaAs/oxide systems. It appears from XPS measurements that an increase of the surface recombination velocity (a decrease of the average integrated PL intensity) can be attributed to the presence of elemental As at the surface. The presence of elemental As at GaAs surfaces etched in acidic solutions was already demonstrated in previous works (e.g. Kohiki et al 1984, Solomun et al 1987) and may be attributed to a faster dissolution of Ga atoms than As atoms, which must be first transformed into an As-oxide (Stocker et al 1983). In contrast, the GaAs surface is more stable in basic solutions (Chang et al 1977) due to the chemisorption of hydroxyl groups on the unsaturated Ga orbitals (Aspnes et al 1985). Consequently, the removal of Ga atoms is slower. Although the native oxide obtained in this case is a mixture of both Ga-oxide and As-oxide, there is no creation of elemental As on the surface. The reversibility of the effect of acidic and basic treatments support the above model. However, with increased number of treatments, the PL intensity stabilizes at a lower value, which can be explained by increased surface recombination velocity due to the creation of surface microroughness. Creation of deep microroughness in strong acidic solutions was also reported by Aspnes et al (1985) on the basis of spectroellipsometry measurements.

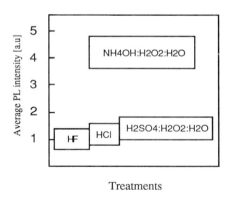

Fig. 1. Average room temperature PL intensities measured on semi-insulating GaAs substrates after various chemical treatments

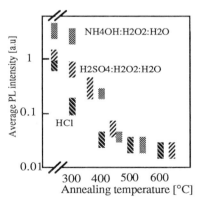

Fig. 2. Average room temperature PL intensities measured on semi-insulating GaAs substrates after various chemical treatments and annealings.

Since insulator deposition and epitaxial growth techniques usually require sample heating, we investigated how the photoluminescence intensity of variously treated GaAs substrates are modified by annnealings in the temperature range 200°C - 600°C. The annealing experiments were performed under N_2 flow, with constant time (10 mins) and temperature as parameter. The results shown in Fig.2 concern S.I. GaAs samples treated in HF, $H_2SO_4:H_2O_2:H_2O$ (5:1:1) and $NH_4OH:H_2O_2:H_2O$ (5:1:100) solutions.

We have found that the exposure of GaAs substrates with oxide free surface (HCl treatments) to temperatures 250°C - 300°C results in a dramatic decrease of the PL signal (factor of seven). This indicates that the surface recombination velocity increased at least by the same factor (Krawczyk et al, 1988). On the other hand, in case of the samples covered with thicker native oxides, the surface recombination velocity increases to the same level only after annealings in the range of 400°C-450°C. This holds true for the samples treated in both acidic or basic solutions.

Dramatic increase of the surface recombination velocity observed with oxide free GaAs surfaces subjected to low temperature (250°C - 300°C) annealings supports the model of Martin et al (1987), which supposes surface decomposition and migration of As atoms. This effect requires a low activation energy (0.3eV-0.5eV) and should be reduced by the presence of any type of oxide on the surface. The creation of free As atoms according to the reaction: $As_2O_3 + 2GaAs \rightarrow Ga_2O_3 + 4As$ (Schwartz et al 1979) can explain the increase of the surface recombination velocity observed in the case of oxidized samples in the range 300°C - 450°C. One can expect that annealing at higher temperatures (>450°C) would lead to As-free surfaces with lower surface recombination velocity, due to a thermal decomposition of As-oxides and to the evaporation of the elemental As. However, scanning photoluminescence depth profiling measurements (obtained after successive chemical etching steps) indicated that heat treatments at temperatures 400°C - 500°C induce non-radiative centers, which penetrate deeply into the semiconductor bulk (0.05μm- 0.5μm).This is attributed to the exodiffusion of As atoms resulting in the formation of As vacancies. According to theoretical predictions (e.g. Daw et al 1979), As-vacancies create a deep level in the upper part of the gap.

Finally, we investigated a new surface passivation procedure, which is based on a thermal treatment of GaAs under a PH_3 atmosphere (Viktorovitch et al 1991) followed by a low temperature (150°C) UV CVD of a Si_3N_4 layer. XPS analysis indicated that PH_3 treatments leave a thin superficial GaP layer (owing to As/P exchange), which prevents from the formation of an arsenic oxide on the surface (Fig.3). Photoluminescence measurements indicate a significant reduction of the surface recombination velocity for the samples treated in the temperature range 500°C-570°C (Fig.4). In addition, photoluminescence depth profiling characterisation indicates that PH_3 treatments prevents from the formation of non-radiative centers in the bulk of the crystal (Fig.5). RHEED diagrams indicated a well ordered surface structure for the samples treated under these conditions. After the insulator deposition, the samples treated with PH_3 show a higher PL intensity than those treated by classical wet solutions (for example a factor of three in comparison with HCl treated samples).

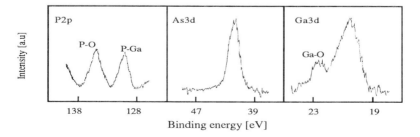

Fig. 3. P2p, As3d and Ga3d core-level spectra for GaAs surface treated at 550°C under PH3.

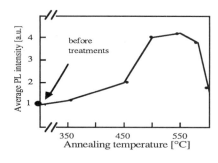
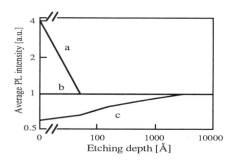

Fig.4. Average PL intensity measured on n-type GaAs substrates after PH_3 treatments carried out at various temperatures.

Fig.5. PL intensity measured on n-type GaAs substrates after successive chemical etching steps: a) sample annealed in PH_3 at 550°C; b) untreated reference sample; c) sample annealed at 550°C but without PH_3.

4. CONCLUSIONS

We have used SPL and SPL depth profiling measurements combined with other analytical techniques (XPS, RHEED, STM) to study chemical and thermal treatments of GaAs crystals. We have demonstrated that GaAs surface treatment in basic solutions leave lower surface recombination velocity (higher PL signal) in comparison with acid etchings. On the other hand, the stability of the electronic properties of GaAs surface during annealing depends on the oxidizing character of the treatments. Photoluminescence results were correlated with the physico-chemical properties of the semiconductor surface, in particular, with the formation of elemental arsenic on the surface. We propose a new surface passivation procedure, which is based on the annealing of GaAs crystals under PH_3 followed by the deposition of an insulating layer. We point out that the PH_3 treatment leaves the surface with no elemental As and no As-oxides. In addition, we have demonstrated that annealing of GaAs substrates at T>400°C results in the creation of non-radiative centers, which penetrate deeply into the bulk. This effect is suppressed, when annealings are carried out under PH_3 atmosphere.

REFERENCES

Aspnes D E and Studna A A 1985 *Appl. Phys. Lett.* **46** 1071
Chang C C, Citrin P H and Schwartz B 1977 *J. Vac. Sci. Technol.* **14** 943
Daw M S and Smith D L 1979 *Phys. Rev. B*. **20** 5150
Kohiki S, Oki K, Ohmura T, Tsuji H and Onuma T 1984 *Jap. J. Appl. Phys.* **23** L15
Krawczyk S K and Hollinger G 1984a *Appl. Phys. Lett.* **45** 870
Krawczyk S K, Bailly B, Sautreuil R, Blanchet R and Viktorovitch P 1984b *Electronics Lett.* **20** 255
Krawczyk S K, Garrigues M, Khoukh A, Lallemand C and Schohe K 1988 *Proc. 5th Conf. on Semi-Insulating III-V Materials* ed G Grossmann (Bristol: Adam Hilger) pp 555-560
Krawczyk S K, Schohe K, Klingelhöfer C and J. Tardy 1990 *Proc 2nd Int. Conf on InP and Related Materials* pp 265-270
Longère J Y, Schohe K, Krawczyk S K, Schütz R and Hartnagel H L 1989 *Appl. Surf. Sci.* **39** 151
Martin R and R. Braunstein R 1987 *J. Phys. Chem. Solids* **48** 1207
Schwartz G P, Schwartz B, DiStefano D, Gualtieri G J and Griffiths J E 1979 *Appl. Phys. Lett.* **34** 205
Solomun T, Richtering W and Gerischer H 1987 *Ber. Bunsenges. Phys. Chem.* **91** 412
Stocker H J and Aspnes D E 1983 *Appl. Phys. Lett.* **42** 85
Viktorovitch P, Gendry M, Krawczyk S K, Krafft F, Abraham P, Bekkaoui A and Monteil Y 1991 *Appl. Phys. Lett.* to be published.
Yablonovitch E, Sandroff C J, Bhat R and Gmitter T 1987 *Appl. Phys. Lett.* **51** 439

Paper presented at INFOS '91, Liverpool, April 1991
Contributed Papers, Section 3

Low temperature chemical vapor deposition of Si_3N_4 thin films assisted by electrical discharge

B Balland*, R Botton*, J C Bureau**, Z Sassi**, J Prudon*, M. Lemiti*

* Laboratoire de "Physique de la Matière" (URA-CNRS 358) Bât.502
** L.T.C.M. (URA-CNRS 116) Bât.401
I.N.S.A. Lyon, 20 avenue Albert Einstein, F - 69621 - Villeurbanne CX - FRANCE

ABSTRACT : An original process has been proposed which enables the fabrication of insulating (Si_3N_4, SiO_2,...) thin films, by means of an *in-situ* activation of the reactions at low temperature (T<400°C) and under low pressure (P = 1 to 2 torr). Our objective was to obtain a deposit in a DC discharge, the substrate being not used as electrode. The layers have been analyzed using SIMS and Infrared spectroscopy. M.I.S. structures have been realized and good electrical characteristics are obtained if the deposition temperature is less than 225°C and for small thicknesses.

1 - INTRODUCTION

Recent developments of microelectronics require dielectric thin films both showing good electrical properties and fabricated using a "cold technology", either by direct deposition as described by Hamano (1984), Morosanu and Segal(1980), Numasawa (1983), and Sze (1986), or by post-deposition treatment (Balland *et al* (1989), (1990)). In order to avoid secondary effects that could generate early damages in the devices, the deposition must be fast. Thus, in order to obtain a rapid film formation at low temperature (T<800°C), the rate of the deposition reactions must be enhanced. One process often used is UV photochemical activation. Enhancement is obtained by means of a mercury vapor lamp irradiating the sample through a quartz (for λ = 250 nm) or Suprasil (for λ = 185 nm) window. Under a pressure of about 1 torr and at a temperature less than 400°C a deposition velocity of 10 nm/min is obtained. HF electrical discharge (using a HF plasma, Numasawa (1983)) can be also used. Pressure and temperature conditions are respectively 0.5 to 3 torr and 100 to 400 °C. In this case, the electric field is very intense, and set perpendicular to the substrate surface, and favors contamination.

In our present work, an original process is proposed which enables the fabrication of polysilicon or insulating (Si_3N_4, SiO_2,...) thin films, by means of an *in-situ* activation of the reactions at low temperature (T < 400 °C) and at low pressure (P = 1 to 2 torr).

2 - EXPERIMENTS

The gases (both reaction and vector gases) were used to perform a DC electric discharge lamp. The "pseudo" lamp thus realized was set to work in the normal glow discharge domain. The samples were not used as electrodes, as was the case in the methods described in the introduction. The samples were placed parallel to the discharge current, between the anode and the cathode. The discharge current was mainly due to electrons (typically 1 mA/cm^2 at 1 torr). A low intensity (several volts per cm) electric field was oriented parallel to the sample surface. The penetration probability of the ions accelerated by the electric field was

© 1991 IOP Publishing Ltd

thus close to zero. This configuration minimized the contamination of the films during their formation.

During the deposition process, several activation mechanisms can interfere : UV activation (UV photons generated by the reaction gases), photo-sensibilization, direct impact activation (as it is observed during thermal activation). In order to simplify we shall limit the presentation of the technique to the case of Si_3N_4 deposition (from SiH_4 and NH_3) on silicon substrates. The "pseudo" lamp (no wall between the lamp and the reaction gases) was a small horizontal tunnel furnace placed in a vacuum vessel. The electrodes were placed at the extremities of the furnace, the substrate being positioned horizontally on the lower part of the furnace. The reaction gases were introduced through the anode. To enable a good distribution of gases in the furnace, sets of small holes were drilled in the anode and the cathode. The gases were evacuated through the cathode. The pressure inside the furnace (typically 1.5 torr) depended on the diameter of the holes drilled in the cathode and of the gas flows (60 cm^3/min). The substrates were nitrided using a mixture of argon (vector gas) containing typically 0.08 % SiH_4 and 1.7 % NH_3.

The deposition parameters that have been studied are : (i)- deposition duration (from 0 to 20 minutes) ; (ii)- temperature (320, 225, 160, 100 and 65 °C) ; (iii)- pressure (from 0.25 to 4 torr) ; (iv)- composition of the reaction gases (NH_3/SiH_4 ratio from 5 to 60) ; (v)- film position in the discharge ;(vi) - current intensity (from 0.5 to 3 mA). It has been observed that the substrate temperature was the major sensitive parameter. Owing to the gas composition, the intensity of the visible discharge was too low to be observed. In a DC discharge, several activation levels can be found and they have been investigated. The layer formed on the substrate varied in thickness valong the path of the glow discharge long as shown Figure 1. At 225°C (which is taken as the reference temperature) no layer was observed during the first

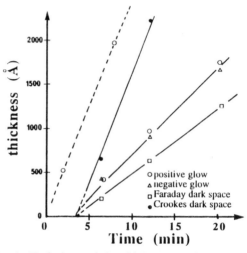

Fig. 1. Variations of the thicknesses of the deposited layers with the deposition time in different domains of the discharge.

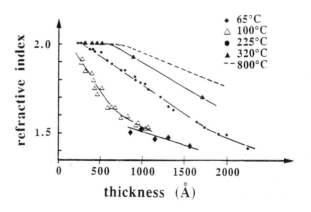

Fig. 2. Variations of the refractive index of the deposited layers with their thicknesses at different temperatures.

4 min, then deposition rate increased. With respect to the different domains of the discharge, it increased with the following speeds : 10 nm/min in the positive light domain (close to the anode) ; 6 nm / min inthe Faraday dark space ; 10 nm/min in the negative light domain, and more than 20 nm/min in the Crookes dark zone (near the cathode). The most favorable domain of deposition is the domain close to the anode for it can be easily extended by simply modifying the distance between the electrodes and needs no increase of the discharge potential. This domain of the discharge is the most valuable one to obtain quality depositions. The deposition velocities slowly decreased with temperature and were not sensitive to pressure variations, but they increased with intensity. Finally, they slightly increased as the NH_3 / SiH_4 ratio decreased.

3 - RESULTS AND DISCUSSION

Physico-chemical and electrical analyses (ellipsometry, SIMS, Infrared spectroscopy, electrical characteristics C-V and G-V) have been used to investigate the quality of the films thus deposited. The refractive index has been determined using ellipsometry and Figure 2 shows the variation of the index with the layer thickness. The refractive index varies very slowly with respect to the discharge domain. For layers having widths less than 40 nm, it has a quasi constant value $1.9 < n < 2$ (n= 2 for the nitride) but decreases if temperature is lowered under 200°C. The refractive index is only slightly sensitive to the other checked parameters. The relative dielectric permittivity is almost constant, ranging from 5 to 7, with a mean value of 6, which is close to the nitride one. We have observed too that introducing oxygen in the reaction gases favored a

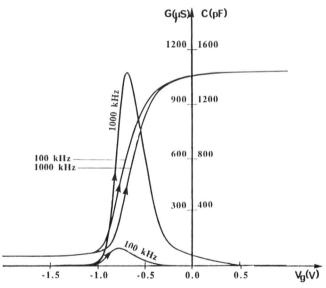

Fig. 3. C-V and G-V electrical characteristics. Deposition temperature : 100°C ; area = 1mm^2 ; thickness = 40 nm.

decreasing of the refractive index versus temperature but not versus the layer width. Knolle (1989) and Ling (1986) have shown that the presence of oxygen had a complex influence on the refractive index $(1.4 < n < 2.6)$. The composition of layers having various thicknesses have been analyzed using SIMS and Infrared spectroscopy. With temperatures higher than 200 °C, Infrared investigations showed the presence of great amounts of both the Si_3N_4 nitride and hydrogenated species (NH, NH_2, SiH), and no Si-O bonds. For lower temperatures, all the Infrared lines assigned to bonds with nitrogen decreased and Si-O lines increased in intensity. This can be interpreted in terms of a degenaracy of Si_3N_4 towards oxinitride as stated by Knolle and Osenbach (1988), Nguyen (1983) and Rand and Roberts (1973), and even silicon oxide, the influence of accidentally introduced oxygen beeing more sensitive when temperature is lowered. The formation of a "pure" nitride layer at low temperature and low pressure, without formation of oxinitride implies that the residual atmosphere should be free of oxygen. On the contrary, introducing oxygen in the reaction gases can lead to the controlled formation of oxinitrides. SIMS analyses showed the presence of a nitride layer, the width of which varied with the deposition temperature. At 225 °C, the layer extends over some 10 nm. As the analysis penetrates the layer up to the layer/substrate interface, a depletion in silicon and an enrichment in nitrogen are observed.

M.I.S. structures have been realized using these nitrided layers. The flat-band shift ΔV_{FB} and the interface state density N_{it} have been investigated through high and low frequency C-V characteristics as shows Figure 3. The values of the flat-band shift ΔV_{FB} depend on the discharge domain and decrease with temperature. Good electrical characteristics are obtained with T ≤ 225 °C; thus ΔV_{FB} is about -1V, the width of the hysteresis of the C-V cycle is less than 0.1 V and N_{it} is less than 10^{11} states/cm^2. the electrical characteristics are better for thinner films and lower deposition temperatures. Figure 4 shows that the flat-band shift strongly increases (for constant thickness) when the deposition temperatures increases from 100°C to 320°C. Over 320°C and up to 800°C, a saturation effect is observed on the flat-band shift.

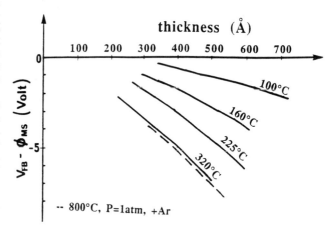

Fig. 4. ΔV_{FB} versus the insulating film thicknesses and for different temperatures.

4 - CONCLUSION

Our initial objective was to test the possibility to obtain a deposit in the positive light domain of a DC electrical discharge, the substrate being maintained at a floating potential and not being used as an electrode. Our results have thus been obtained with a very simple experimental setup. Nevertheless, they show the great interest that can be found in developing this original process which enables the formation, at low temperature, of insulating thin films showing satisfying electrical properties.

REFERENCES

B Balland B, Bureau J C and Glachant A 1989 *Appl. Surf. Sci.* **89** 210
B Balland B, Glachant A, Bureau J C and Plossu C 1990 *Thin Solid Films* **190** 103
Hamano K 1984 *Jap. J. of Appl. Phys.* **23** 1209
Knolle W R 1989 *Thin Solid Films* **168** 123
Ling C H 1986 *Jap. J. of Appl. Phys.* **25** 1490
Morosanu C E , Segal E 1980 *Rev. Roumaine de Chimie* **25** 315
Nguyen V S 1983 , *Proc. of the Symp. on Silicon Nitride* The Electrochem. Soc. **83** 453
Numasawa Y 1983 *Jap. J. of Appl Phys.* **22** 2792
Rand M J and Roberts J F 1973 *J. Electrochem. Soc.* **120** 446
Sze S M 1986 *Semiconductor devices - Physics and Technology* (N-York: J Whiley & Sons)

Depassivation of P_b sites by heat and electric field

E H Poindexter,[1] G J Gerardi,[2] F C Rong,[1] W R Buchwald,[1] and D J Keeble[3]

[1] Electronics Technology and Devices Laboratory, Fort Monmouth, NJ 07703 USA
[2] Department of Chemistry, William Paterson College, Wayne, NJ 07470 USA
[3] Department of Physics, Michigan Technological University, Houghton, MI 49931 USA

ABSTRACT: Depassivation of P_b sites in Si/SiO_2 structures by high negative bias field or moderate thermal anneal is maximized with damp oxides, but is nearly undetectable with dry or wet oxides. Thermal activation energy for the process is a minimum with damp oxides. The decline in P_b generation with too much H_2O may be modeled by a single rate-limiting reaction, resembling the volcano effect on electrodes, or the Marcus inversion regime in other charge transfer situations.

1. INTRODUCTION

It has recently been shown (Blat et al 1991) that the negative-bias-temperature instability (NBTI) of metal-oxide-silicon (MOS) structures can be modeled by a single electrochemical reaction of the form

$$\equiv SiH + A + hole^+ \longrightarrow \equiv Si\cdot + AH^+ . \tag{1}$$

The detailed kinetics of the NBTI revealed that the reaction is first order, with no evidence of any sequential or reverse reaction. A similar reaction has also been proposed for the related phenomena of heat- or field-induced depassivation of P_b centers, as directly observed by electron spin resonance (ESR) (Gerardi et al 1989). In the latter study, it was found that the reaction was initially favored by H_2O in the oxide, but was ultimately reduced and terminated by too much H_2O. This perhaps peculiar behavior was ascribed to a concerted second reaction which was blocked when the interface region was saturated with H_2O.

The single-reaction kinetics of the NBTI make it desirable to reconsider models for the P_b depassivation, and to seek a single-reaction model. In this paper, we report additional data on P_b generation by bias field or heat. The better picture of the reaction thereby provided is re-examined and compared with analogous reactions in other, more thoroughly explored charge-transfer systems, in hopes of developing an improved model.

2. EXPERIMENTAL APPROACH

Silicon wafers with (111) face were used to avoid the complicated chemistry of the (100) face. They were variously oxidized in dry O_2, steam, or O_2 moistened with a measured $[H_2O]$, followed by selected anneals, to complement elsewhere-reported samples (Reed and Plummer 1988, Brower 1989).

© 1991 IOP Publishing Ltd

The entire sample set provided a wide concentration range of hydrogenous species [hyd] in the oxide. Samples are designated by the measured oxidant [H_2O], or by an equivalent [H_2O] estimated from the extensive oxidation study of Razouk and Deal (1979). The resultant ordering is at best only semiquantitative; however, the conclusions here rely on very conspicuous qualitative trends. Oxides were usually grown at 1000C to about 100nm thickness; experimental results were insensitive to either parameter. Electric field was applied by the corona method, and reached 12-13MV/cm regardless of [hyd]. Thermal anneal was performed in Ar or N_2 ambient. Activation energy E_a for thermal depassivation was determined from simple 4-point anneals, judged adequate for large differences and strong trends. [P_b] was observed by ESR.

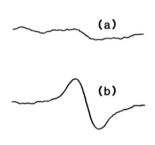

Fig. 1. ESR signals from P_b centers in damp-oxide Si wafers (a) unstressed (b) field-stressed at 300K

3. OBSERVED RESULTS

A typical example of P_b generation by negative electric field at 300K is shown in Fig. 1; ESR signals for thermal depassivation without field at about 700K look the same. This sample had 1000ppm [H_2O] in the oxidant, which gives about the optimum [hyd] for easiest and strongest generation of P_b. The depassivation effect declines rapidly with increase or decrease of [hyd] from this value. In the 1000ppm region, P_b generation by electric field typically required 15-20s; a few samples were so unstable that measurements were not feasible. In striking contrast, very dry or very wet samples showed no depassivation after minutes or hours of exposure to the field. Field-less thermal depassivation also proceeded very readily in samples of this intermediate [hyd], requiring about 20-40s, but requiring hours with dry oxides, and an undetermined longer time (if ever) with wet. Wet oxides were also tested with rapid thermal anneal, and temperatures >1000K were needed for depassivation in the 10-20s range. The thermal E_a was about 0.6eV at 1000ppm [H_2O], and increased to ≥1.5eV with dry or wet oxides. The ultimate field-generated [P_{be}] and thermal E_a are plotted versus [H_2O] in Fig. 2.

The good correlation between [P_{be}], thermal E_a, and depassivation time provide strong mutual confirmation of these 3 aspects of the process, which were not determined individually with precision. The data in Fig. 2 comprise a reaction fingerprint which may well be unique in Si/SiO_2 chemistry.

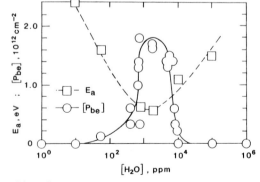

Fig. 2. Depassivation of P_b centers: thermal E_a and field-induced [P_{be}] vs. [H_2O] in oxidizing O_2 ambient

4. DISCUSSION

The complex fingerprint of the P_b generation reaction narrows the scope of models which might be proposed. The decline in [P_{be}] toward the dry region of [hyd] could at least partly be due to lack of reactant A needed in the depassivation. However, the decline in [P_{be}] with too much [hyd], and the variation in E_a throughout, are not ascribable in any simple way to [hyd].

The electrochemical nature of Eq. 1 suggests comparison with analogous charge-transfer phenomena on electrode surfaces, well studied for many years. A pertinent feature often noted in such reactions is the so-called "volcano" effect (Morrison 1977). This effect occurs when a parameter such as binding energy of a key moiety in a reactant or product molecule may be monotonically varied by systematic chemical substitution or adjustable conditions to cause a monotonic increase in the free energy of the reaction. It might be expected that a series of increasingly exothermic reactions would show increasing reaction rates. Yet, as first predicted by Marcus (1957), this expectation is not necessarily valid; the reaction may actually be retarded if the free-energy change ΔG is excessive. This principle is illustrated in Fig. 3, showing reaction coordinates for a system which becomes more exothermic from left to right. The corresponding E_a at first declines with increasing ΔG, goes to zero, and then increases. If reaction rate constant K is defined by an Arrhenius expression

$$K = K_0 \exp(-E_a/kT) , \qquad (2)$$

then the reaction rate would be a maximum at the E_a minimum (as observed for the depassivation here). The region in which ΔG increases, but K decreases, has been termed the Marcus inversion regime; it was quite controversial until recent decisive experiments (Closs and Miller 1988).

We now seek a specific embodiment of Eq. 1 which gives a suitable variation in ΔG as [H_2O] increases. The straightforward assumption that reactant A is simply H_2O is viable. It has been computed theoretically and verified experimentally that ΔG of H_3O^+ formation from H^+ and H_2O increases greatly as local [H_2O] increases (Conway 1981). Specifically,

$$H^+ + H_2O \longrightarrow H_3O^+ , \quad \Delta G = 8eV \qquad (3)$$

$$H^+ + 4H_2O \longrightarrow H_3O^+ \cdot 3H_2O \text{ or } H_9O_4^+ , \quad \Delta G = 13eV \qquad (4)$$

Still more H_2O would further increase the exothermicity, but only slightly. These ΔG values do not apply to the complete reaction, since the bond-

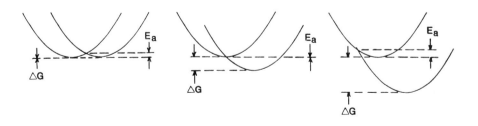

Fig. 3. Reaction coordinates showing initial decrease, vanishing, and eventual increase of activation energy E_a in a series which has a monotonic increase in free-energy change ΔG.

breaking energy of

$$\equiv SiH \longrightarrow \equiv Si^{\cdot} + H \tag{5}$$

and the ionization energy of

$$H + hole^+ \longrightarrow H^+ \tag{6}$$

must also be considered; molecular orbital calculations indicate that the overall single-H_2O reaction (Eq. 1 with A = H_2O) is slightly endothermic with $E_a = \sim 2eV$ (Edwards 1990). The E_a is in agreement with the observed E_a for dry oxides reported here. The large increase in ΔG as H_2O is added would more than overcome these unfavorable values.

If H_2O is indeed the attacking reactant, then the negative field bias would assist the reaction in several ways: (a) attracting holes to the interface; (b) orienting the polar H_2O molecule so that the negative O end approaches the $\equiv SiH$; (c) lowering of the energy of the charged H_3O^+ product; and (d) promptly removing the charged product from the reaction site, to inhibit any reverse reaction. In contrast, models assuming A to be H or OH^- are not so much helped by the field. Furthermore, there is no plausible source of H or OH^- in thermal SiO_2 at 300K, though either might be expected with radiation. For depassivation at 700K without electric field, the situation is less clear-cut. Thus, although H_2O seems on balance to be the most likely attacking reactant, it should remain as a working hypothesis at this time.

In conclusion, the depassivation of P_b centers by negative electric field or moderate thermal anneal has been found to be consistent with well-established single-reaction rate-limited models. The most likely depassivating reactant is H_2O. This tentative model also suggests a role for H_2O and H_3O^+ in radiation-induced P_b generation, and as factors in the oxide charge inevitably found near the Si/SiO_2 interface after thermal oxidation.

REFERENCES

Blat C E, Nicollian E H, and Poindexter E H 1991 J. Appl. Phys. 69 1712
Gerardi G J, Poindexter E H, Caplan P J, Harmatz M, Buchwald W R, and Johnson N M 1989 J. Electrochem. Soc. 136 2609
Reed M L and Plummer J D 1988 J. Appl. Phys. 63 5776
Brower K L 1989 Semiconductor Sci. Technol. 4 970
Razouk R R and Deal B E 1979 J. Electrochem. Soc. 126 1573
Morrison S R 1977 The Chemical Physics of Surfaces (Plenum: New York) pp 291, 329
Marcus R A 1957 J. Chem. Phys. 26 867
Closs G L and Miller J R 1988 Science 240 440
Conway B E 1981 Ionic Hydration in Chemistry and Biophysics (Elsevier: Amsterdam) pp 394, 404
Edwards A H 1990 preliminary results

Paper presented at INFOS '91, Liverpool, April 1991
Contributed Papers, Section 4

Information on the spatial distribution of P_b defects at the (111)Si/SiO$_2$ interface revealed by ESR observation of dipolar interactions

A. Stesmans and G. Van Gorp

Physics Department, Katholieke Universiteit Leuven, 3001 Leuven, Belgium

ABSTRACT: K-band electron spin resonance at 4.3 K has unveiled the dipolar interaction effects between the $[111]{*}Si{\equiv}Si_3$ (P_b) defects at the (111)Si/SiO$_2$ interface. Analysis of this 2 dimensional dipolar structure indicates that the P_b's are randomly spread over all Si lattice sites in the (111) plane of the interfacial terraces. In addition, a ^{29}Si super-hyperfine doublet of splitting $a_\parallel^{SHF}$ = 14.8±0.2 G has been identified.

1. INTRODUCTION

The Si/SiO$_2$ structure, still being an essential building block in semiconductor device technology, remains a subject of intense research. A key factor in this matter regards the interface defects, which, as generally assumed, are naturally generated to relieve strain.

The dominant defect (see, e.g., Poindexter and Caplan 1983) at the (111)Si/SiO$_2$ interface is the $[111]{*}Si{\equiv}Si_3$ center, which is simply an unpaired sp^3 hybrid on an interfacial Si atom perpendicular to the (111) interface. These defects, termed $[111]P_b$ centers as measured by electron spin resonance (ESR), account for 50-100% of all interface traps and are thus the main cause of the less than ideal behaviour of the Si/SiO$_2$ interface in devices. Hence the continuous interest in the full characterization of these defects. After conventional (~ 1 atm $\cdot$O$_2$; 950°C) thermal oxidation, typically a fraction $f \equiv [P_b]/N_a \approx 0.5\%$ of the $N_a = 7.830 \times 10^{14}$ Si atoms cm^{-2} in the (111) Si surface are the site of a paramagnetic P_b center. A noteworthy observation is that its thermochemical properties are dominated by interaction with H; at temperatures $T \geq 225°$C, the defect readily bonds to H resulting in a diamagnetic entity symbolized as HP_b. This process, however, may be reversed by high vacuum annealing at $T \geq 550°$C.

Mainly from ESR observations, the individual atomic model of this defect has been well established. Yet, in a somewhat broader perspective, some aspects remain unknown. One such property is the precise *distribution* of the defects at the interface. Or put differently, with what particular interface feature(s) or sites are the P_b defects linked? That distribution is of significant interest to understand the physical process(es) by which they are formed. Furthermore, it is clear that, if ascribing a somewhat co-establishing role to the P_b defects as relaxation nuclei, this insight may add significantly to the understanding of the inherent interface structure, which ultimately may provide a clue to the oxidation mechanism.

The present work reports on the *dipole-dipole* (DD) interaction within the essentially 2 dimensional (2D) P_b spin system. In contrast with the 3D case, the nature of the DD interaction, that is, resolved structure and spectral

© 1991 IOP Publishing Ltd

broadening, as revealed in the ESR spectrum of a 2D spin system, relates closely to the areal distribution of the spins. This provides a —so far lone standing— means to disclose information about the in-plane location of the P_b's. The combination of a systematic analysis of the reversible H-passivation of P_b defects with low-T ESR has enabled the first successful identification of the dipolar structure (Stesmans and Van Gorp 1990). The salient DD features complemented with correct theoretical-computational signal simulation lead to the first details about the areal P_b distribution.

2. EXPERIMENTAL DETAILS

Slices measuring 1x9x0.117 mm^3, were cut from a p-type Si wafer (Cz-grown, 10 Ωcm) polished to optical flatness on both sides. These slices were generally submitted to tree types of thermal treatments; (1) Dry thermal oxidation at 920-956°C in 99.999% pure O_2 at 0.2-1.1 atm for 84-130 min. (2) Hydrogenation (passivation) in H_2 (99.9999%;~1.1 atm) for 11-79 min at 253-353°C. Sequential H passivations are applied to decrease $[P_b]$ stepwise on one and the same sample. (3) Degassing (dissociation) at 752-835°C in vacuum ($\leq 2 \times 10^{-7}$ Torr) for 752-835 min. An important observation is that repeated application of steps (2) and (3) in an arbitrary sequence does not affect the total defect density $[P_b^*] \equiv [P_b] + [HP_b]$ (comprising both ESR-active and passivated defects) grown in after the initial oxidation step. Steps (2) and (3) provide thus a reproducible way to monitor $[P_b]$.

20.2-GHz ESR observations were routinely carried out at 4.3K with the applied magnetic induction $B \| [111]$ (perpendicular to the interface) to obtain the most favourable signal-to-noise ratio. Modulation of B at ~120 kHz resulted in the detection of absorption-derivative spectra $dP_{\mu a}/dB$.

3. EXPERIMENTAL RESULTS AND DISCUSSION

A comparison of undistorted ESR signals as observed on one sample for various $[P_b]$ values clearly unveils the dipolar interaction as shown in Fig. 1. In this figure, the low-$[P_b]$ signal (f=0.05%) of peak-to-peak linewidth $\Delta B_{pp} = 1.30 \pm 0.03$ G is seen to broaden to 1.90 ± 0.03 G for f increasing to 1.5% ($[P_b] = 1.2 \times 10^{13}$ cm^{-2}). Interesting is the attendant substantial change in line shape; doublets, labelled B, C, and D, become clearly resolved. This particular fine structure depends on the in-plane arrangement of the spins, which may thus provide information on the areal P_b distribution.

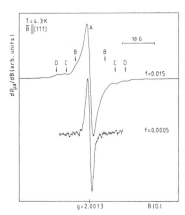

Fig. 1. K-band ESR spectra of $[111]P_b$ defects at the (111)Si/SiO$_2$ interface

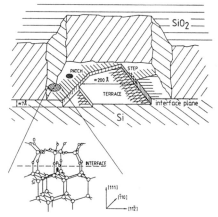

Fig. 2. Sketch of the (111)Si/SiO$_2$ interface

For $f \to 0$, the residual signal is observed, that is, void of DD effects. It is characterized by the residual natural width $\Delta B_{pp}^R = 1.29 \pm 0.03$ G and a width at half height of the absorption spectrum $\Delta B^R = 1.98 \pm 0.03$ G. While the residual signal shape and width allow an in-depth evaluation of the residual line broadening mechanisms operative, correct knowledge of that residual signal, used to convolute dipolar histograms, is essential to obtain reliable signal simulations.

Having resolved the DD structure, the main task left is to develop correct theoretical calculations of the spectra starting from a supposed P_b distribution model. By scanning various distributions, the fitting quality with the experiment will then select the correct model. Accepting, as well evidenced, that the P_b's are mainly distributed over one (111) Si plane, it is important to mention that it is not necessarily so that all Si sites in that plane are allowed as P_b sites; perhaps it is only a subset of the Si lattice sites, henceforth referred to as a *net* or *array*. Tracing that array is the main purpose of this work.

It is clear that one may envisage an immense number of arrays. And as the correct calculation of a dipolar spectrum for each such model implies a formidable computational effort, it is important to select only those models which are deemed realistic. This selection is probably best guided by the present knowledge of the Si/SiO_2 interface as inferred from various research methods. The evolving interface picture then, as sketched in Fig. 2, is one which is atomically abrupt, consisting of flat terraces separated by ledges, typically 1-3 atoms high and 110-220 Å apart (see, e.g., Stesmans and Van Gorp 1991). Along one report, the otherwise undistorted flat terraces comprise small domains of inhomogeneities —called *patches*— measuring 3-8 atomic distances across in average. Thus, in an idealized view, there are 3 salient interface features, i.e., terraces, steps, and patches, to which the P_b's could preferably be linked. This led us to consider 5 possible arrays. The P_b's could only reside at steps (1), patches (2), and terraces. In the latter case, of course, various arrays may be envisaged. In the extreme case that all Si sites at the terrace [or, equivalently, the (111) plane, in view of the terrace extension] are allowed —the most general array—, it is referred to as the *random distribution* (RD) *model* (3) (cf. all dots in Fig. 3). An alternative array (4) at terraces may be obtained by accepting the ditrigonal ring capping of a P_b defect (Cook and White 1988), as illustrated in the insert of Fig. 2. If incorporating such capping into a kind of crystalline SiO_2 interface transition layer, one may imagine the six (SiO_4) membered rings to cover the Si surface in a close-packed way. The central Si atoms then, representing the possible P_b sites, again make up a regular 2D

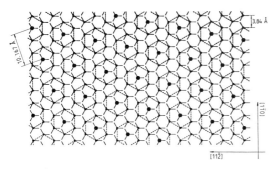

Fig. 3. Schematic top view of an unreconstructed (111)Si surface.

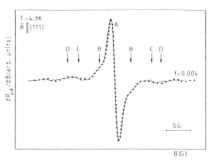

Fig. 4. Fitting of theory (dots) to the [111]P_b ESR spectrum.

triangular net (cf. large dots in Fig. 3) of lattice constant $\sqrt{7}$ times larger than for the RD array. It is clear that this model incorporates a significant coestablishing nature of the P_b's. Finally, model (5), like model (3), also assumes the RD array, but with the additional boundary condition that for a particular P_b concentration, the P_b's tend to be self-avoiding, i.e., distributed as far apart as possible.

Dipolar spectra have been calculated for each model for various P_b densities, assuming a *random distribution* of the P_b's within each model. And as there is so far no analytical line shape expression available for a dilute 2D spin system, we had to resort to a computational approach. We upgraded a reliable, though computationally burdensome, ab initio quantum mechanical computational approach (Brower and Headley 1986), by extending to sufficiently large (111) Si regions. Fittings have been carried out for $f \approx 4 \times 10^{12}$ cm^{-2} because of practical constraints on computer time and memory.

The best, almost perfect fit, is obtained for the RD model, as shown by the dotted curve in Fig. 4. As convoluting line shape, the experimental residual P_b signal, as simulated by a sum of Gaussian curves, has been used. It has also been incorporated in this model that, the 1st, 2nd, 3rd, and 4th nearest neighbour interactions do not appear in the fine structure, in agreement with experiment. The "step" and "patch" models are excluded because they predict a too peaky spectrum, while the ditrigonal ring model (4) is disfavoured because it exhibits a too strong B doublet.

Within the RD model the various structural features observed are well explained. Doublet B is the added result of the 8th, 9th, and higher nearest neighbour interactions. The structure C and D, in fact, is a mixture of various doublets, i.e., 5th, 6th, and 7th nearest neighbour fine structure, and a ^{29}Si superhyperfine (SHF) doublet, characterized by a splitting $a_{\parallel}^{SHF}$ = 14.8 ± 0.2 G. The latter doublet, arising from hyperfine interaction of the unpaired P_b electron with the nearest Si neighbours in the Si substrate, had to be incorporated in the simulation. Since the fine structure intensity may vary relatively to the central signal A, the structure at positions C and D may alter with increasing f, as observed (cf. Fig. 1).

4. CONCLUSIONS

Sensitive K-band ESR at 4.3K in combination with a H-based tuning of $[P_b]$ has enabled the first successful separation of the dipolar structure in the $[111]P_b$ signal for $B\|[111]$. The dipolar structure indicates that the P_b's are predominantly randomly distributed at the interfacial terraces over the array of all Si lattice sites. This suggests that these defects are generated at the interface to release strain in a random manner.

The fine structure due to the 1st, 2nd, 3rd, and 4th nearest neighbour interaction is not observed, which may result from antiferromagnetic spin pairing or the self-avoiding nature of the P_b's. In addition, the existence of a SHF doublet of 14.8 G splitting has been demonstrated.

REFERENCES

Brower K L and Headley T J 1986 *Phys. Rev.* **B34** 3610
Cook M and White C T 1988 *Phys. Rev.* **B38** 9674
Stesmans A and Van Gorp G 1990 *Phys. Rev.* **B42** 3765
Stesmans A and Van Gorp G 1991 *Appl. Phys. Lett.* **57** 2663 and refs therein
Poindexter E H and Caplan P J 1983 *Prog. Surf. Sci.* **14** 201

Radiation damage in SiO$_2$/Si structures induced by high energy heavy ions

M.C. Busch[1], A. Slaoui[1], E. Dooryhee[2], M. Toulemonde[2] and P. Siffert[1]
[1]Laboratoire PHASE, CRN, B.P. 20, 67037 Strasbourg Cedex, France
[2]CIRIL, rue Claude Bloch, B.P. 5133, 14040 Caen, France.

ABSTRACT : We have investigated the structural and electrical properties of SiO$_2$/Si structures irradiated by high energy (> 0.5 GeV) Xe and Ni ions. Structural analysis of the irradiated SiO$_2$ films, performed with infrared spectroscopy, indicates atomic displacements, broken and strained Si-O bonds induced by the irradiation. Electrical measurements of irradiated SiO$_2$/Si structures show an increase of the interface state density D_{it} and of the oxide fixed charge density N_f with the ion fluence. These results are compared to defects induced by heavy ions irradiation in silica.

1. INTRODUCTION

Several studies have investigated the defects induced by γ/x rays, electrons or protons irradiation in SiO$_2$/Si structures (Ma and Dressendorfer (1989), Di Maria (1978), Powell (1975), McLean (1980)). The interaction of these light particles with the target is dominated by electronic energy losses (Pfeffer (1985), Chengru et al. (1988)). In contrast, few work has concerned the effects of irradiation by heavy ions especially in the high-energy range (> 1 MeV/amu). At low energies the collisions cascades result directly in displacements of target atoms whereas at high energies the production of defects is controlled by the electronic energy loss (Dooryhee et al. (1988)) causing in insulators some anisotropic permanent damage.

Very few studies have so far dealts with the response of SiO$_2$/Si systems to high energy ion irradiation. In this paper, we present experimental results on radiation damage in thermal SiO$_2$ films grown on silicon induced by MeV Ni and Xe ion bombardment investigated with infrared spectroscopy and electrical characteristics.

2. EXPERIMENTAL DESCRIPTION

The substrates used in this study were $4-6$ Ωcm, $<100>$, $600\,\mu$m n-type silicon wafers. The oxidation was performed in dry O$_2$ (3% HCl) at 1000° C and followed by nitrogen annealing at the oxidation temperature. The oxide thickness is of about 800 Å. The SiO$_2$/Si structures have been irradiated at room temperature by 0.8 GeV Xe or by 0.6 GeV Ni with fluences in the range of 10^8 to 5×10^{12} ions/cm^2 ($\pm$ 20%). Irradiation was performed at GANIL (Caen, France) with instantaneous fluxes limited to $\sim 5\times 10^8$ ions/cm^2/s to avoid overheating. The projected ranges of the ions are 50 and $100\,\mu$m for Xe and Ni, respectively, so they were stopped in the silicon substrate.

The SiO$_2$ films were characterized using ellipsometric measurements (Gaertner HeNe at 6328 Å) and infrared (IR) spectroscopy (Perkin Elmer G983). Their electrical characteristics were analyzed by using the admittance-frequency measurements (HP 4192A).

3. RESULTS AND DISCUSSION

Figure 1 shows infrared absorption spectra before and after Xe and Ni ions irradiation at a fluence of 5×10^{12} cm^{-2}. The irradiation induces important changes in the position and the width of the Si-O stretching band, as well as in the area of this peak. This absorption band normally found at 1080 cm^{-1} by Pliskin (1965), before irradiation is shifted to lower wave numbers as the fluence increases, and reached 1045 cm^{-1} at a fluence of 5×10^{12} Xe/cm^2. Further effects of implantation are a decrease of the amplitude of the Si-O peak and an increase of the full width at half maximum (FWHM) as the ion fluence increases. These results are summarized in table I. On the other hand, the weak band at 810 cm^{-1} corresponding to the Si-O bending band (Pliskin (1965)) is slightly shifted to higher wave numbers.

Any vibration of hydrogen-bond silanol groups (SiOH) nor adsorbed water were detected by IR analysis of irradiated SiO$_2$ films. Moreover, Rutherford backscattering analysis, in agreement with ellipsometric measurements, shows that the irradiated SiO$_2$ films are stoechiometric. Thus, porosity and oxygen deficiency can not be responsible for the shift and the broadening of the Si-O stretching band (Pliskin and Lehman (1965)). Fritzsche et al. (1972) have observed similar behaviour of the Si-O peaks and width but also an increase in the film thickness of ion implanted thermal SiO$_2$ at energies of about 100 keV and doses up to 10^{16} ions/cm^2. They explain their results in terms of broken bonds and displacements of target atoms due to elastic nuclear collisions. At high energies, the production of defects is in part controlled by the electronic energy loss (Langevin et al. (1986)). Dooryhee et al. (1988) have shown that energetic ions in SiO$_2$ produces paramagnetic defects (oxygen vacancies (E') and peroxy radicals (OHC)) as γ/x-rays do with however specific features. On the other hand, Duraud et al. (1988) have shown that energetic heavy ions in fused silica not only induce oxygen vacancies but also distorsion of the bond angle between adjacent SiO$_4$ tetrahedra. Chengru et al. (1988) have suggested that the production of defects in vitreous SiO$_2$ by MeV heavy ion implantation should be related to the generation process of ion tracks. Simon (1960) has shown that neutron radiation in vitreous SiO$_2$ produces a decrease in the IR frequency of about 15 cm^{-1} and a Si-O-Si bond angle decreases of about 4° without a change in the nearest neighbor Si-O distance. Using the IR data, we have calculated the variation in the bond angle Si-O-Si (Lucovsky et al. (1987)) as a function of the ion fluence. The results are reported in table I for the ion Xe. We find that the 5×10^{12} Xe/cm^2 irradiation induces a decrease of about 10° in the angle. This change in the bond angle is equivalent to a decrease from about 3.04 to 2.95 Å in the Si-Si distance.

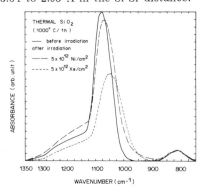

Fig. 1. IR absorption spectra of thermal SiO$_2$ films before and after irradiation with Ni and Xe at 5 10^{12} ions/cm^2.

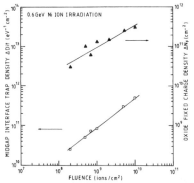

Fig. 2. The change of midgap interface state density, ΔD_{it} and oxide fixed charge number, ΔN_f, vs. the fluence.

Fluence (Xe/cm^2)	$A_{Si-O}^{Str.}$	$\Omega_{Si-O}^{Str.}$ (cm^{-1})	$\Delta\Omega_{Si-O}^{Str.}$ (cm^{-1})	Si-O-Si angle	d_{Si-Si} (Å)	Etch rate (Å/min)
0	0.1646	1080	73.3	144.5	3.05	120
10^{11}	0.1646	1080	73.5	144.5	3.05	167
2 10^{11}	0.1486	1075	76.8	142.85	3.03	220
10^{12}	0.1312	1069	82	141	3.017	440
5 10^{12}	0.097	1045	100	134.3	2.95	640

Table 1 : IR characteristics and etch rate for Xe irradiated SiO$_2$/Si samples.

All these observations mean that in our case the structural damage mechanism involved here may be different from the simple electronic and nuclear process. The changes in our IR characteristics and Si-O-Si bond angles result probably from an increase of strained and/or broken bonds and from atomic displacements in the SiO$_2$ films when the ion fluence increases. This interpretation is consistent with etch rate experiment which is very sensitive to Si-O bond arrangement (Pliskin (1971)). The data are reported in table I. The etch rate increases with the dose and is found to be five times faster for a 5×10^{12} Xe/cm^2 irradiated oxide than for unirradiated one. On the other hand, the loss in Si-O band area after irradiation, can be interpreted as a loss of absorbing oscillators (Fritzsche et al. (1972)). Since the area and the Si-O peak frequency reached their initial values after thermal annealing at 900° C during 30 min, the loss of oscillators is not due to sputtering of the oxide and may be understood as a consequence of broken bonds. The etch rate of the irradiated and annealed oxide sample is the same than for an unirradiated oxide suggesting a complete rearrangement of the SiO$_4$ tetrahedrons.

After Ni irradiation, similar IR results were observed. However, the decrease of the Si-O-Si bond angle is only about 4° for a fluence of 5×10^{12} Ni/cm^2 corresponding to a decrease from 3.04 to 3.01 Å in the Si-Si distance. This result shows how much important is the electronic stopping power of the incident ion (Ni : 22 MeV cm^{-2} mg^{-1}, Xe : 77 MeV cm^{-2} mg^{-1}) on the radiation induced damage in SiO$_2$. At the same ion fluence and energy, heavier ions lose more energy and produce more defects during an ion's passage through the SiO$_2$ (Dooryhee et al. (1988)).

We define the normalized fraction of damage F_d as $F_d = (S - S_r)/(S_d - S_r)$ where S, S_r, S_d are the areas of Si-O stretching band of, respectively, the irradiated, unirradiated and completely damaged samples. We have took as S_d the value corresponding to the 5×10^{12} Xe/cm^2 irradiated sample since for much higher fluences the position of the Si-O peak does not change. The evolution of F_d as a function of fluence ϕ is found to follow the expression $F_d = 1 - \exp(-A\phi)$, where A is the initial damage rate. The measured values of the damage rate A for Xe and Ni irradiations are 8×10^{-13} cm^2 and 6×10^{-14} cm^2, respectively. These values are in very good agreement with those reported by Toulemonde et al. (1990) for α-quartz substrate.

The electrical measurements were performed on SiO$_2$/Si structures by using a mercury probe. Midgap interface state (ΔD_{it}) and oxide charge density (ΔN_f) generations due to radiation are plotted in figure 2 as a function of Ni ion fluence. We observe a nearly linear relationship between the electrical defects and the ion fluence. Many works has concerned the effects of ionizing radiations on the electrical properties of thermally grown oxide as induced by light particles. A strong correlation has been found between the interface trap state density and trivalent silicon (Pb centers) defects (Poindexter et al. (1981)), caused by the rupture of strain bonds at SiO$_2$/Si interface, and between the fixed charges density in the oxide and the E' defects (Lenahan et al. (1982)) caused by the rupture of the Si-O bond. In the case of energetic heavy ions,

Chengru et al. (1988) have shown that the E' defects in $v - SiO_2$ induced by 17 MeV Cl and F ions at low ion fluences ($< 5 \times 10^{12}$ part./cm^2) are similar to those created by γ-ray or proton irradiation. They have also observed a linear relation between the E' defects density and the ion fluence before saturation. Langevin et al. (1986) which have studied the effects of 0.8 GeV oxygen and 3.6 GeV krypton irradiation on fused silica have also found that the E' defect concentration increase has a linear relation to ion fluence. Therefore, we can suggest that the 0.6 GeV Ni irradiation of our SiO_2/Si structure has induced the formation of E' centers in SiO_2 and Pb centers at the interface, which are responsible of the increase of D_{it} and N_f, respectively.

4. CONCLUSION

SiO_2/Si structures have been irradiated by 0.8 GeV Xe and 0.6 GeV Ni ions at fluences up to 5×10^{12} ions/cm^2. The structural analysis of irradiated SiO_2 layers as performed by infrared spectroscopy shows that the energetic heavy ions induce strains and variations in bond force constants as well as displacements of atoms in the target. The production of structural damage is found to be strongly related to the ion fluence and to the electronic energy loss of the ion in the oxide.

The electrical analysis of the irradiated structures gives evidence for generation of interface state traps (D_{it}) at the SiO_2/Si interface as well as trapped positive charges in the oxide (N_f) which are correlated to Pb and E' centers, respectively. The measured D_{it} and N_f are found to increase linearly with ion fluence.

REFERENCES

Chengru S, Mangi T and Tombrello T A 1988 *J. non-Cryst. Sol.* **104** 85
Di Maria D J 1978 *The Physics of* SiO_2 *and its Interfaces* ed. S.T. Pantelides, (Pergamon, New-York) pp 160
Dooryhee E, Langevin Y, Borg J, Duraud J P and Balanzat E 1988 *Nucl. Instr. and Meth.* **B32** 264
Duraud J P, Jollet F, Langevin Y and Dooryhee E 1988 *Nucl. Instr. and Meth. Phys. Res.* **B32** 248
Fritzsche C R and Rothermund W 1972 *J. Electrochem. Soc.* **119** 1243
Griscom D L 1978 *The Physics of SiO_2 and its interfaces* ed. S.T. Pantelides (Pergamon, Elmsford, NJ) pp 232
Langevin Y, Dooryhee E, Borg J, Duraud J P, Lecomte C and Balanzat E 1986 *Appl. Phys. Lett.* **49** 1699
Lenahan P M and Dressendorfer P V 1982 *IEEE Trans. Nucl. Sci.* **NS-29** 1459
Lucovsky G, Manitini M J, Scivastava J K and Irene E A 1987 *J. Vac. Sci. and Technol.* **B5** 530
Ma T P and Dressendorfer P V 1989 *Ionizing Radiation Effects in MOS devices and Circuits, New-York* Wiley Interscience
Mclean F B 1980 *IEEE Trans. Nucl. Sci.* **NS-27** 1651
Pfeffer R L 1985 *J. Appl. Phys.* **57** 5176
Pliskin W A and Lehman H S 1965 *J. Electrochem. Soc.* **112** 1013
Pliskin W A 1971 *J. Vac. Sci. and Technol.* **8** 512
Poindexter E H, Caplan P J, Deal B E and Razouk R R 1981 *J. Appl. Phys.* **52** 879
Powell R J *J. Appl. Phys.* **46** 4557
Simon I 1960 *Modern Aspects of the Vitreous State* ed. J.D. Mackenzie (Butterworths, London) pp 120.
Toulemonde M, Balanzat E, Bouffard S, Grob J J, Hage-Ali M and Stoquert J P 1990 *Nucl. Instr. and Meth.* **B46** 64

Paper presented at INFOS '91, Liverpool, April 1991
Contributed Papers, Section 4

The effect of zinc in silicon dioxide gate dielectrics

T. Brozek [a], V. Y. Kiblik [b], O. I. Logush [b], G. F. Romanova [b]

[a] Institute of Microelectronics and Optoelectronics,
Warsaw University of Technology, Koszykowa 75, 00-662 Warsaw, Poland

[b] Institute of Semiconductors, Ukrainian Academy of Sciences,
Prospect Nauki 45, 252650 Kiev-28, USSR

ABSTRACT: Electrophysical and radiation properties of silicon-silicon dioxide (Si-SiO2) systems doped with zinc were investigated. It has been found that zinc slightly improves parameters of MOS structures. Zinc-related centres in the oxide exhibit donor properties and during irradiation effectively capture positive charge. A small amount of zinc in an oxidizing ambient (0.001-0.002%) improves radiation hardness of the Si-SiO2 interface and the near-surface silicon region.

1. INTRODUCTION

Electrophysical properties of thermal SiO2 and the Si-SiO2 interface depend, among other factors, on the presence of foreign atoms (other than silicon and oxygen) in the MOS system (Snel and Varwey 1989). For this reason many attempts have been made to improve properties of MOS systems by means of intentionally introduced impurities. Frequently, however, such a treatment while improving some parameters degrades others, e.g. radiation hardness (Dressendorfer 1989).
In our investigations we have studied electrophysical and radiation properties of MOS structures with thermal oxides doped with zinc. The choice of zinc was based on its good solubility in silicon and its healing action on dislocation loops in the near-surface silicon region. Chang and Tsao (1969) showed that zinc can affect characteristics of MOS structures by creation of negatively charged states in SiO2. Nevertheless, we may expect that zinc, as a transition metal, creates donor states in the oxide and at the interface. According to our approximate calculations (Kiblik et al. 1987) zinc centres can introduce donor levels (as interstitials) in SiO2 and the transition SiOx region. Fuller and Morin (1957) considered such a case for silicon, but no experimental evidence for this has been found until now. On the other hand, isolated zinc atoms in the silicon lattice act as double acceptors, susceptible to the influence of ionizing radiation (Lebedev et al. 1987) and hydrogen passivation (Stolz et al. 1989).

2. EXPERIMENT

Experiments were carried out on silicon structures which had oxides grown in the atmosphere containing zinc. Silicon wafers (n-type, 4,5 Ωcm, <111> oriented) were oxidized in wet oxygen at 950°C to thicknesses 70-80 nm. Zn atoms were introduced into the hot zone in the form of aqueous

© 1991 IOP Publishing Ltd

solution of $ZnCl_2$. The zinc content in the oxidizing ambient varied from 0 to 0.0025%. The higher concentractions resulted in poor quality oxides.

Our SIMS data confirmed the presence of zinc in the doped oxides (a linear decrease of its concentration from the outer oxide surface toward the interface) and showed that there was no pile-up of the impurity near the silicon surface. We observed also a decrease of the refractive index of SiO_2 (determined ellipsometrically) when zinc was present in the oxidizing atmosphere (see Table 1). Changes in the refractive index may be caused by changes of morphology (microstructure and stoichiometry disorder) and/or changes of the mechanical stress level in the oxide and at the interface (Landsberger and Tiller 1986).

MOS capacitors were formed by means of thermal evaporation of aluminium and were subsequently annealed in dry N_2. For evaluation of radiation sensitivity some samples were γ-irradiated with a Co-60 source up to the dose of 10 Mrads. In order to determine basic electrophysical parameters (such as oxide charge Q_{ot}, surface states density D_{it} and minority carrier lifetime τ_g) high-frequency C-V and C-t characteristics were measured at every stage of the experiment.

3. RESULTS AND DISCUSSION

Electrophysical parameters of MOS structures with zinc-doped oxides and reference oxide are collected in Table 1. It can be seen that the moderate doping of the Si-SiO_2 system with zinc does not lead to significant changes - the values of Q_{ot} and D_{it} are comparable or even slightly lower for zinc containing samples than for the reference ones.

TABLE 1. The zinc contents in oxidizing atmosphere and corresponding parameters of investigated structures

Zinc contents, %	Q_{ot}, cm^{-2}	D_{it}, $eV^{-1}cm^{-2}$	τ_g, μs	n
0	4.8×10^{11}	5.7×10^{11}	45	1.48
0.0005	4.2×10^{11}	5.7×10^{11}	40	1.46
0.001	4.3×10^{11}	5.0×10^{11}	86	1.45
0.0025	3.9×10^{11}	5.2×10^{11}	55	1.46

3.1. Oxide charge

A zinc atom, located in the interstitial position (i.e not chemically bonded) of the oxide matrix can create a donor level in the forbidden gap of SiO_2. Our calculations (Kiblik et al. 1987) show that its energy corresponds to about E_c-6.0eV and E_c-3.8eV for neutral and ionized state respectively. Hence, it is highly probable that at the thermodynamic equilibrium zinc centres may be neutral, but during excitation these levels can effectively trap holes. Differences in Q_{ot} for non-irradiated samples may be caused by a small quantity of substitutional zinc atoms acting as acceptors, which was observed by Chang and Tsao (1969). The possible reason for our not having observed such a strong negative charge action is that their zinc concentration (as we estimated) was one thousand times higher.

Radiation-induced changes of Q_{ot} are presented in Fig.1. where one can see that positive charge build-up is faster for oxides containing zinc. Their radiation sensitivity is about 1.5-2 times higher than that of reference one. This effect is more visible in Fig. 2, where changes of flat-band voltage of biased MOS samples under the influence of γ-rays are shown.

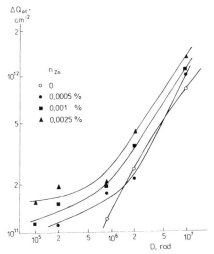

Fig. 1. Radiation-induced changes of Q_{ox} for oxides grown with various zinc contents

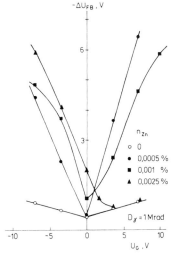

Fig. 2. Changes of U_{FB} vs gate voltage during irradiation for structures under investigation

3.2. Interface states

Zinc atoms in silicon act as double donors with corresponding deep levels inside the energy gap. One can then expect that the presence of zinc near the Si-SiO$_2$ interface results in generation of zinc-related interface states. Contrary to this, in the doped MOS samples the density of surface states (its mean value) is comparable and even lower than for the reference structures. It may mean that other mechanisms are also involved. We can consider here the decorating of dislocations and other defects (e.g. oxidation stacking faults) with zinc atoms resulting in an improvement of the interface quality and/or lowering of the mechanical stress level in the near-surface region as a result of zinc addition (the behaviour of the refractive index suggests loosening of the oxide matrix). In this case the number of strained Si-O bonds is also lower, which according to Sakurai and Sugano (1981) should lead to lower density of surface states.

In spite of the lowered radiation

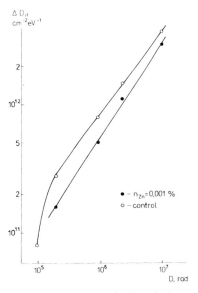

Fig. 3. Radiation-induced changes of D_{it} vs dose for zinc doped and reference samples

hardness of the bulk zinc doped oxides, the interface of doped structures was less sensitive to the influence of ionizing radiation than that of reference samples (Fig. 3). The dependence of this improvement effect on the zinc contents in oxidizing ambient is shown in Fig. 4.

There are two possible mechanisms responsible for the observed effect. Firstly, the lower level of the mechanical compressive stress in the oxide near the interface leads to lower strain gradient, which according to Grunthaner et al. (1982) results in weaker interface states generation during irradiation. Secondly, hydrogen species released under γ-rays in the oxide bulk, after reaching the interface can passivate zinc-related centres and turn them into electrically inactive states (Stolz et al. 1989) which reduces the density of initially present interface traps. This mechanism can also explain the data of Brożek et al. (1990), where irradiation followed by thermal treatment (which anneals radiation-induced states) leads to even lower density of interface states than the initial one.

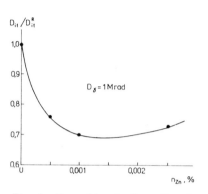

Fig. 4. Normalized D_{it} for irradiated samples with various Zn contents (D_{it}^R refers to MOS structures with undoped oxide)

3.3. The near-surface silicon region

The defectivness of the near-surface silicon region can be characterized by the minority carrier lifetime τ_g, since the reciprocal lifetime is proportional to the defect density. Fig. 5 shows the dependence of the defectivness level on the zinc concentration. The lowest defectivness was found for the sample with 0.001% zinc additive. This sample is also the most stable against radiation-defect generation. One can see that there is only a narrow range of zinc concentration when its "healing" action is observed. Above this range the precipitation of zinc probably takes place.

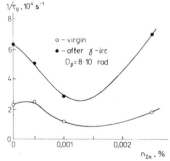

Fig. 5. $1/\tau_g$ vs zinc contents for non-irradiated and gamma irradiated MOS structures

References

Brożek T, Kiblik V Y, Litovchenko V G, Lisovskii I P 1990 *Proc. 18th Yugoslav Conf. on Microelectronics MIEL'90 (Ljubljana)* pp 399-402
Chang C Y and Tsao K Y 1969 *Sol. St. Electr.* **12** 411
Dressendorfer P V *Ionizing Radiation Effects in MOS Devices and Circuits* ed T P Ma and P V Dressendorfer (New York: J Wiley and Sons) pp 354-364
Fuller C S and Morin C J 1957 *Phys. Rev.* **105** 379
Grunthaner F J, Grunthaner P J and Maserjian J 1982 *IEEE Trans. Nucl. Sci.* **29** 1462
Kiblik V Y, Litovchenko V G, Plotnikova L G and Brozek T 1987 *Proc. 2nd All-Union Sci. Conf. "Physics of oxide films" Petrozavodsk USSR)* p 84
Landsberger L M and Tiller W A 1986 *Appl. Phys. Lett.* **40** 143
Lebedev A A, Sultanov N A and Ecke W 1987 *Sov. Phys. Semicond.* **21** 10
Sakurai T and Sugano T 1981 *J. Appl. Phys.* **52** 2889
Snel J and Varwey J F 1989 *Instabilities in Silicon Devices* ed G Barbottin and A Vapaille (Amsterdam: North Holland) vol. 2 pp 381-402
Stolz P, Pensl G, Grunebaum D and Stolwijk N 1987 *Mat. Sci. & Eng.* **B4** 31

Paper presented at INFOS '91, Liverpool, April 1991
Contributed Papers, Section 5

Drain bias instability in polycrystalline silicon thin film transistors

N D Young, S D Brotherton, and A Gill

Philips Research Laboratories, Redhill, Surrey, UK.

Abstract. The characteristics of poly-Si TFTs are degraded by operation at high drain bias, due to the creation of near conduction band edge acceptors, and near mid gap donors, by hot carriers generated at the drain junction. The drain current at high biases is most likely enhanced by high gain parasitic bipolar action, and this can be reduced by increasing gate length. Improved stability is also achieved using an LDD structure.

1. Introduction

There is currently interest in both Poly-Si TFTs and αSi:H TFTs for electronics on glass applications. Poly-Si TFTs can be used for a wider range of circuit appications than αSi:H TFTs, because they demonstrate a higher field effect mobility, with a wide range of values from 5 to 150cm^{-2}V^{-1}s^{-1} having been reported. However, the higher the mobility, the more likely it is that hot carrier degradation effects will become important: It has been reported recently that such degradation effects can be seen in poly-Si TFTs, (L in the range 2-6µm), with mobilities of only 5-10cm^{-2}/V-s, (Young and Gill 1990a and 1990b). The present work examines the device degradation more fully, and discusses the mechanism for the enhanced drain current in which the hot carriers are generated.

2. Device Fabrication.

Poly-Si TFTs were made using three different processes, all of which enable the use of low cost glass substrates without glass distortion problems. In the first two processes self aligned structures were formed using phosphorus ion implantation for the source, drain and gate regions, and activation at 600°C, (as has been described previously by Brotherton, Ayres and Young (1991)). In one process the active layer was formed using columnar LPCVD poly-Si, and in the other it was formed using amorphous LPCVD silicon or αSi:H which was subsequently crystallised at 600°C. The third process involved formation of coplanar poly-Si TFTs using αSi:H and excimer laser crystallisation.

3. Results and Discussion

Typical transfer characteristics for devices fabricated by the three different technologies are shown in figure 1(a). In all three cases the leakage current, (I_L= 1.6-4.0x10^{-14}A/µm), subthreshold slope, (S=0.53-0.66 V/dec per 0.1µm of SiO$_2$), and threshold voltage, (V_T=3.5-5.5V per 0.1µm of gate oxide), are quite similar, whereas the field effect mobility varies due to a variation in both grain size and structure. The mobility for the columnar poly-Si devices lies in the

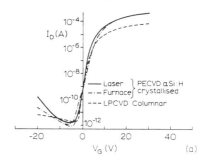

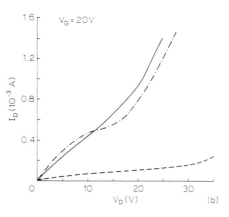

Figure 1   Device Characteristics

© 1991 IOP Publishing Ltd

range 5-10cm^2/V-s, whereas that for the crystallised devices is 20-30cm^2/V-s.

The output characteristics for the same devices are shown in figure 1(b). It can be seen that the drain current does not saturate in a standard MOSFET manner, but instead continues to rises quite steeply above pinch off. This will be referred to as drain current enhancement. If devices are operated in the region of drain current enhancement, then degradations to the basic device characteristics occur. It is the purpose of this work to characterise this degradation, and to determine the mechanism of the current enhancement.

The degradation of the transfer characteristic caused by passing a high drain current is shown for an n channel columnar poly-Si device in figure 2(a). Qualitatively similar degradation is also seen for crystallised Si devices. The reduction in the on state current, and the increase in the off state current, have been caused just by making the output characteristic measurement shown in figure 2(b). Such effects are consistent with the formation of interface states by hot carriers, as has been suggested previously by Young and Gill, (1990a). The on current degradation is due to the formation of acceper levels near the conduction band edge, (which become negatively charged when $V_G > V_T$), and the off current degradation to the formation of near midgap donors: Figure 2(c) shows the leakage current in more detail, both before and after the stressing. The slope at low drain biases of 1/2 suggests that the leakage current mechanism is generation in the space charge region at the drain, (Brotherton, Ayres and Young, (1991)), and it can be seen that the generation current has increased during the stress. It is proposed that these states are donors, (rather than acceptors), since no detectable threshold voltage shift is seen in the subthreshold part of the transfer characteristic, (ie the states must be neutral for $V_G > 0V$). Careful consideration of the hysteresis in the output characteristic, (figure 2(b)), shows that it is consistent with this state creation model.

The results for similarly stressed p channel devices are shown in figure 3. These devices show a hysteresis in their transfer characteristics, even before stressing. Fortunately, this saturates very rapidly, and it is the saturated loop which is shown in the figure. On stressing, a negative threshold shift occurs for the negative limb of the characteristic, and additionally

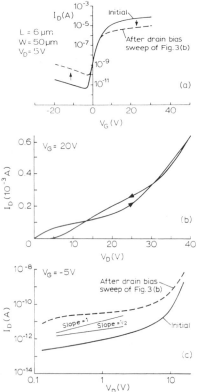

Figure 2 Drain Bias-Stress for n Channel Devices

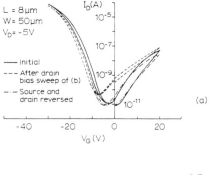

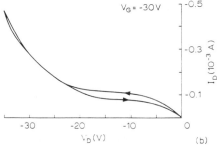

Figure 3 Drain Bias-Stress for p Channel Devices

leakage current degradation results. These effects are again consistent with the formation of near mid gap donor levels. If the source and drain contacts are reversed, the leakage current degradation is reversed, (since the defects are no longer in the drain junction region), though there is a small difference for V_G=10-20V which is most likely associated with the change in the occupancy of the near conduction band edge acceptors. The output characteristic sweep responsible for the degradation, (shown in figure 3(c)), shows a hysteresis consistent with the negative threshold voltage shift. Thus, when both n and p channel devices are stressed at high drain biases seemingly different, but consistent, degradation effects occur.

It is now necessary to examine the output characteristics in more detail to ascertain the mechanisms for the current enhancement, and hot carrier degradation. Figure 4 shows the output characterisics for devices from all three technolgies as a function of gate length. The enhanced current is evidently L dependent in all cases, suggesting that it cannot be due solely to a simple junction phenomena at the drain, such as impact ionization or field enhanced generation. However, it is possible that such mechanisms are occurring in conjunction with either punch through, or a bipolar feedback effect. If we look at the destructive avalanche breakdown voltage of TFTs in the off state, (essentially operating them as simple pn junctions), we see an L dependence for $L \lesssim 6\mu m$, and an L independence for $L \gtrsim 6\mu m$, indicating that punch through occurs only for $L \lesssim 6\mu m$ in this case. However, punch through will, (in principle), occur at lower drain bias for the on state, (since the gate bias assists in the lowering of the barrier at the source), and therefore we might expect, (if anything), punch through to occur for slightly larger L devices in the on state. In the cases of both punch through, and bipolar feedback, we might an L^2 to the phenomena: In the former the breakdown voltage will be proportional to L^2, and in the latter the current gain will be proportional to $1/L^2$: Figure 5 shows output characteristics for L=3,6 and 20μm, measured in more detail. This clearly shows the effect of gate length on the current enhancement. The current gain, (defined to be the increase in drain current above the saturation value,

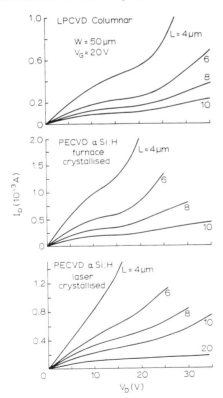

Figure 4 The L Dependence of Output Characteristics

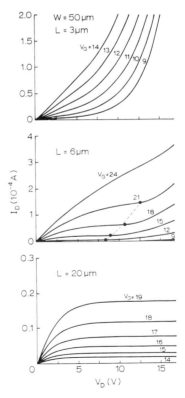

Figure 5 The Gate Bias Dependence of Output Characteristics

divided by the saturation value), for L=4,6,8 and 10μm devices measured just above threshold, (to allow a reasonable saturation before enhancement occurs), is shown in figure 6. A good $1/L^2$ dependence is confirmed. Furthermore, with regards to figure 5(b) it can be seen that the initiation of the current enhancement, (solid points), occurs at higher V_D as V_G is increased: This is inconsistent with punch through. Thus we believe that the enhanced current in the on state is due to the bipolar feedback effect for L≳4μm. If this is indeed the case, the electron-hole pair generation in the drain junction must be quite weakly dependent on drain bias, and the bipolar gain must be high, to produce curves such as figure 5(b) and 5(c).

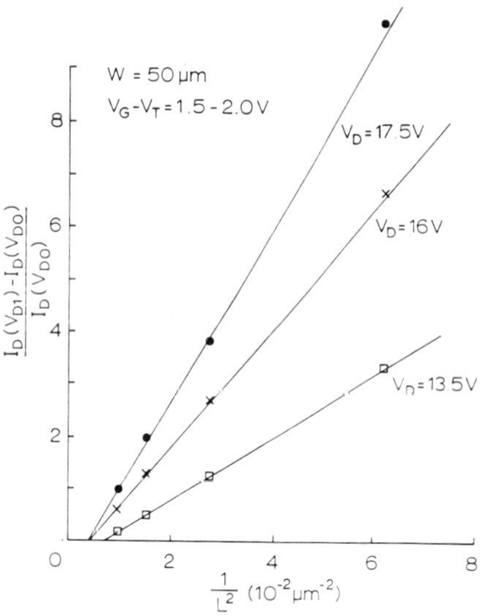

Since all of the drain current passes through the high field junction region at the drain, it might be expected that hot carrier generation will be proportional to I_D, for a fixed drain field and mobility. Thus reducing the enhancement by increasing L is of use in increasing stability, (lowering the bipolar gain). However, reducing the drain field has an equally important effect, (even though the generation is quite weak), as demonstrated in figure 7 where an LDD architecture, (similar to that given by Seki et al), has been employed.

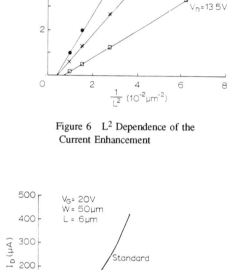

Figure 6 L^2 Dependence of the Current Enhancement

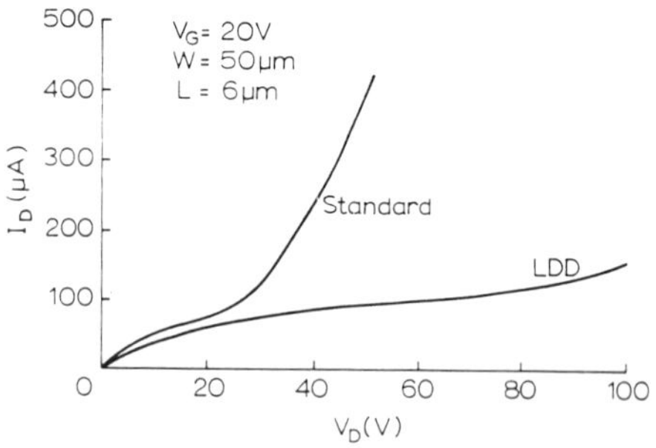

Figure 7 Enhanced Current Reduction Using LDD

5. Conclusions

Poly-Si TFTs fabricated using three processes demonstrate similar current enhancements at high drain biases, and similar degradations to device characteristics result after being driven at these high levels. The degradations in both p and n channel devices are due to the formation of acceptor states near the conduction band edge, and near mid gap donor states, by hot carrier injection. The enhanced current shows a gate length dependence which most likely arises from a high gain parasitic bipolar effect for L≳6μm. Stability can be improved by using larger L devices, and LDD architectures.

5. Acknowledgement

We are grateful to D McCulloch for the laser annealing of poly-Si films.

6. References

Brotherton S D, Ayres J R, and Young N D, (1991), accepted for publication in Solid State Electron.
Seki S, Kogure O, and Bunjino T, (1987), IEEE Electron Device Letts, EDL 8, 9, 434.
Young N D and Gill A, (1990a), Semiconductor Sci and Technol, 5, p728.
Young N D and Gill A, (1990b), ESSDERC90, edited by W Eccleston and P J Rosser, (Bristol: Adam Hilger), p303

Paper presented at INFOS '91, Liverpool, April 1991
Contributed Papers, Section 5

Low temperature plasma oxidation of polycrystalline silicon

P.K.Hurley, J.F.Zhang, W.Eccleston and P.Coxon *
Department of Electrical Engineering and Electronics. Liverpool University P.O. Box 147 Liverpool L69 3BX, U.K.
* *GEC Research Limited, Hirst Research Centre, East Lane, Wembley, Middlesex HA9 7PP, UK.*

The electrical properties of oxides grown on intrinsic polycrystalline silicon (polysilicon) by plasma anodisation are reported. The breakdown field of the oxide is found to increase with oxide thickness, which is in contrast to thermal oxidation of polysilicon where a continual decrease in breakdown field with oxide thickness is observed. From consideration of this result, and the current/voltage characteristics of the oxides, it is proposed that the growth of oxide by plasma anodisation reduces the surface roughness of the polysilicon. It is also demonstrated that the technique can be used to produce the gate oxide of polysilicon thin-film transistors on glass.

1. Introduction

Thermally grown oxide on polysilicon has been widely studied for its applications in floating gate memory elements, insulators in multilevel systems and passivating layers. In certain applications, such as large area active matrix displays, high temperature (> $850°C$) thermal oxidation can not be used, as all fabrication steps are limited to the softening point of the glass substrate, typically $630°C$. The method most commonly employed to produce the gate oxide of polysilicon thin film transistors (TFT's) for large area displays is plasma enhanced chemical vapour deposition. An alternative method which is capable of producing oxide films on semiconductors at temperatures below $600°C$ is plasma anodisation[1]. The objective of this paper is to report the electrical properties of the plasma grown oxide films on polysilicon, and to demonstrate a polysilicon TFT where the gate oxide has been fabricated using the technique.

2. Oxide Breakdown

The polysilicon films were deposited on n type single crystal silicon substrates by low pressure chemical vapour deposition (LPCVD), at $630°C$ and approximately 40mT. The polysilicon films were not intentionally doped. Using the oxidation conditions: temperature = $400°C$, current density = $4.4mA/cm^2$, pressure = 0.1 Torr, and total power = 915 W, the oxidation time was varied to obtain oxide thicknesses from 140 to 2000 Å. The refractive index of the films varied from 1.44 to 1.47, with an average value of 1.45, which indicates that the oxide is close to stiochiometric SiO_2. Test capacitors of diameter 0.38-1.0 mm were formed by evaporation of high purity aluminium through a shadow mask onto the oxide surface. To obtain the variation of breakdown field with oxide thickness, the applied gate voltage was positive (electron injection from the polysilicon), with a voltage ramp rate of 0.5V/s. A slow ramp rate is necessary for intrinsic polysilicon as no majority carrier type exists. Consequently, the accumulation layer of electrons has to be formed by electron/hole pair generation in the polysilicon. From measurements of the generation lifetime in the polysilicon,[2] the value of 0.5V/s was selected to ensure that the generation rate could follow the increase in gate charge.

A graph showing the variation of the average breakdown field strength with the oxide thickness is shown in figure 1. The dotted line represents the variation exhibited by thermally grown oxide on untextured polysilicon.[3] The sample size is 25 capacitors, and electrical breakdown is said to occur when a current of 10^{-6} amps ($J > 0.88 mA/cm^2$) flows through the oxide. No capacitors passed this current without experiencing irreversible electrical breakdown. The result indicates that either: the intrinsic properties of the oxide are dissimilar to thermally grown oxide, or that the oxide growth has a different effect on the polysilicon surface. As previous results have demonstrated that the electrical properties

© 1991 IOP Publishing Ltd

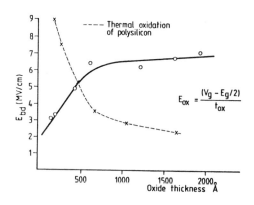

Figure 1 Variation of breakdown field strength with oxide thickness

of plasma grown oxides are similar to thermally grown oxide[4], the most probable explanation is that the growth of the oxide has a different effect on the polysilicon surface.

In the case of thermally grown oxide, the decrease in breakdown field strength with increasing oxide thickness has been attributed to an enhancement of the polysilicon surface roughness.[3,5] This occurs as the rate of thermal oxidation is dependent on the orientation of the silicon surface, and for polysilicon the grains have a random orientation. This results in differential oxidation rates across the polysilicon surface. Furthermore, preferential oxidation occurs in the vicinity of the grain boundaries. In the case of plasma anodisation the rate of oxidation is not dependent on the orientation of the silicon surface. Consequently, an enhancement of the roughness of the polysilicon surface due to differential oxidation rates will not occur. Moreover, the rate of oxide growth is proportional to the electric field at the oxide silicon interface. As the electric field is enhanced in the vicinity of protuberances on the polysilicon surface, more silicon is consumed in these regions. Consequently, the roughness of the polysilicon surface would reduce with increasing oxide thickness, explaining the variation of breakdown field with oxide thickness in figure 1.

To check this idea it is necessary to see if the model for the oxide growth makes any predictions which can be tested experimentally. If the surface is being smoothed, then the electric field intensification at protuberances on the polysilicon surface will decrease continually with oxide thickness. In this case the electric field E^+, necessary to create a leakage current density J_o, for electron injection from the polysilicon, will increase with oxide thickness. Furthermore, the current/voltage characteristics should become asymmetric with increasing oxide thickness, with a higher electron injection from the gate. For thermal oxidation of polysilicon electron injection is always higher from the polysilicon.[3,5,7]

Table 1 demonstrates that, for electron injection from the polysilicon, the leakage field E^+ does increase with the oxide thickness. The current/voltage characteristics for oxide thicknesses of 175 Å and 1200 Å, and both polarities of gate voltage, are shown in figure 2. For the thin oxide (175Å) the plots are symmetric, indicating that the oxide is uniform, and conforming to the oxide surface. For the thicker oxide (1200Å), the plots are polarity dependent, with a higher injection current from the aluminium/oxide interface. These results are both consistent with a reduction in surface roughness due to the oxide growth. Cross sectional transmission electron micrographs before and after plasma oxidation support the implications of the electrical measurements, showing that the oxide is thicker at sites of surface roughness.

t_{ox} (Å)	E^+ (MV/cm)
175	4.2
330	4.7
590	5.6
1330	7.4
1600	7.3

Table 1 Variation of leakage field with oxide thickness

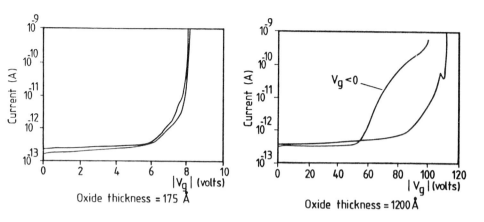

Figure 2 Current/voltage characteristics for oxide thicknesses 175 Å and 1200 Å

3. Oxide Conduction

In the previous section it has been assumed that the mechanism of high field conduction is Fowler-Nordheim tunnelling,[8] which has been demonstrated to be the case for thermally grown oxide on polysilicon.[9] An alternative mechanism of high field conduction is Poole-Frenkel[10], which is controlled by thermionic emission of electrons from traps in the oxide. This mechanism of conduction differs fundamentally from Fowler-Nordheim in that it is limited by the bulk of the oxide, as opposed to the electrodes. For this mechanism of conduction the current/voltage characteristics are insensitive of the polarity of the applied voltage, and the current density J is temperature dependent, with a plot of $\log_e J$ versus $1/kT$ being linear. From the previous section it was observed that for thicker oxides the conduction is polarity dependent, which is inconsistent with a bulk limited conduction. As a further check the variation of the current density with temperature was determined. A plot of $\log_e J$ versus $1/kT$ is shown in figure 3. The current is relatively insensitive to the temperature, and the plot is not linear. The broken line is the theoretical variation of the current density with temperature[8] for an effective mass of $m^x = 0.6 m_o$, and a barrier height of 3.3eV. It is noted that the variation of current density with temperature conforms to the theoretical prediction for tunnelling.

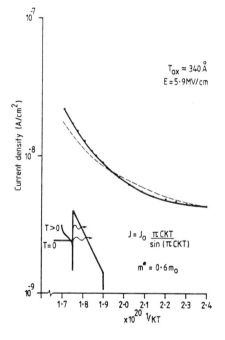

Figure 3 Variation of current with temperature

From this result, and the asymmetry in the conduction, it can be concluded that the mechanism of high field conduction for plasma grown oxides on polysilicon is Fowler-Nordheim.

4. Polysilicon thin film transistors

In attempting to produce a polysilicon TFT on glass by plasma anodisation a number of issues were being addressed:
1. Is it possible to oxidised polysilicon on an insulating substrate ?
2. Does exposure to the plasma result in significant incorporation of oxygen into the grain boundaries ? This could result in an increase in the density of states between the valence and conduction bands.[11]
3. How do the field effect mobilities compare with TFT's with deposited gate oxides ?

The plasma anodisation of polysilicon on glass was achieved using a special quartz mask and a peripheral ring contact to the surface of the polysilicon. The special anode arrangement allows the current during oxidation to flow laterally through the polysilicon film. The subthreshold characteristic of a polysilicon TFT fabricated by this technique is shown in figure 4. Analysis of the transfer characteristic gives a threshold voltage of 5.5 volts and a field effect mobility of approximately 40 cm^2/Vs. The field effect mobility of a polysilicon TFT deposited under identical conditions, with a deposited gate oxide, was 16 cm^2/Vs. The higher value of field effect mobility, and the scaling of the threshold voltages with oxide thickness, indicates that no significant incorporation of oxygen into the grain boundaries occurs during oxide growth. It has been found that the field effect mobilities of plasma grown gate oxide TFT's are either higher or similar to TFT's with deposited gate oxides. The higher values can be explained as a reduction in scattering at the oxide/polysilicon interface, due to a reduction in surface roughness.

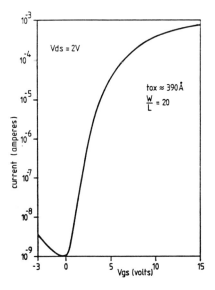

Figure 4 Subthreshold characteristics: Gate oxide grown by plasma anodisation

5. Conclusion

The variation of breakdown field strength with oxide thickness, and the current/voltage characteristics indicate that the growth of an oxide by plasma anodisation reduces the roughness of the polysilicon surface. These findings suggest that the technique could be used to smooth the polysilicon surface prior to device fabrication. It has also been demonstrated that, by using plasma anodisation a polysilicon TFT with a grown gate oxide can be fabricated, whilst keeping all the process temperatures below 600°C.

References
1. S.Taylor, W.Eccleston and K.J.Barlow, *J.Appl.Phys.* 64, 6516 (1988).
2. P.K.Hurley, S.Taylor and W.Eccleston, *Appl. Phys Lett.* 54, 1525 (1989).
3. L.Farone, *IEEE Trans. Electron Devices* 32, 1785 (1986).
4. J.F.Zhang, P.Watkinson, S.Taylor, W.Eccleston and N.D.Young, *Proc. Conf. ESSDREC 90, Nottingham* 265 (1990).
5. E.A.Irene, E.Tierney and D.W.Dong, *J.Electrochem. Soc.* 127, 2321 (1982).
6. K.Barlow, *Ph.D Thesis, University of Liverpool* (1987)
7. D.J.DiMaria and D.R.Kerr, *Appl. Phys. Lett.* 27, 505 (1975)
8. M.Lenzlinger and E.H.Snow, *J.Appl.Phys.* 40, 278 (1969).
9. P.A.Heimann, S.P.Muraka and T.T.Sheng, *J.Appl.Phys.* 39, 6240 (1982).
10. J.Frenkel, *Phys. Rev.* 152, 785 (1938)
11. C.R.M.Grovenor, *J.Phys.C.* 18, 4079 (1985)

Paper presented at INFOS '91, Liverpool, April 1991
Contributed Papers, Section 5

The effect of phosphorus implantation dose on the gap-state density in polycrystalline silicon

P.Vassilev, R.Paneva*, V.K.Gueorguiev, L.I.Popova

Institute of Solid State Physics, Sofia 1784, blvd. Lenin 72, Bulgaria
* Institute of Microelectronics, Sofia

ABSTRACT: The field-effect conductance in thin-film polysilicon $MOSFET's$ is used to determine the effect of phosphorus implantation dose on the gap-state density distribution in thin polycrystalline silicon films. The results for phosphorus implanted $MOSFET's$ with doses varying from 5×10^{11} to 5×10^{12} cm^{-2} are presented. The flat-band voltage and the position of the Fermi level in the polysilicon film are determined by combining the incremental method and the temperature dependence of the field-effect conductance. The dependence of the density of gap states and of the Fermi level position on the implantation dose is discussed.

1. INTRODUCTION

The wide use of polycrystalline silicon (polysilicon) in large-area electronics applications and 3-D integration requires adequate understanding of the dependence of the material properties on the preparation conditions. The electronic states associated with the defects, present at the grain boundaries and at the polysilicon-SiO_2 interface, control the performance of polysilicon MOS field effect transistors. The effect of the conventional fabrication conditions, e.g. the effect of the channel implantation dose, and the effect of various grain-boundary passivation processings (RF hydrogen plasma passivation, hydrogen ion-implantation, etc.) on the electrical properties of thin poly-Si films and the polysilicon-SiO_2 interface is of particular interest. Recently, it has been demonstrated that the field-effect conductance (FEC) method (Fortunato et al 1986, 1988) may be used to determine the effective density of localised states (DOS) in undoped and lightly doped unpassivated and passivated polysilicon thin films. The DOS distribution, which strongly influences the devices characteristics of small-grain $MOSFET's$, has already been shown to be dependent on the $LPCVD$ preparation conditions (viz., the pressure during deposition). The plasma hydrogenation effect in polysilicon films has also been assessed in terms of DOS (Migliorato et al 1987). The present paper aims at revealing the dependence of the DOS distribution on the phosphorus implantation dose in the case of thin $LPCVD$ polysilicon films.

2. EXPERIMENTAL

Thin film transistors ($W = 50\mu m$, $L = 50\mu m$) were fabricated on 4-in $\langle 100 \rangle$ $15\Omega.cm$ p-type silicon wafers, which were oxidised in wet O_2 ambient at $1000°C$ to oxide thickness of 1 μm. Low-pressure chemical deposition ($LPCVD$) at $620°C$ and pressure 390 mTorr was used to deposit 220 nm of undoped polysilicon. A conventional $LOCOS$ self-aligned $CMOS$ technology was used with 50 nm pad-oxide and 100 nm $LPCVD$ Si_3N_4. The wafers were doped by phosphorus ion-implantation through 35 nm of dry SiO_2 at an energy of 85 keV. The implantation dose ranged from 5×10^{11} to 1×10^{13} cm^{-2}. The gate oxide was grown at $1000°C$ in dry O_2.

© 1991 IOP Publishing Ltd

The resulting oxide thickness was 45 nm. $LPCVD$ polysilicon gates were deposited at $620°C$ and phosphorus doped at $900°C$, and source and drain regions were doped by phosphorus and boron ion-implantation. A multilayer metalisation and $PECVD$ $Si_xN_yH_z$ encapsulation with annealing in H_2 ambient at $450°C$ for 30 min were carried out. The $MOSFET$ structures are isolated from each other by thermal oxide, which is illustrated in Fig. 1. No other processings, which are usually used to improve the electrical properties of the polysilicon TFT's (recrystallisation, H-plasma passivation, hydrogen ion-implantation, etc.) have been carried out. Table 1 presents the SUPREM II calculated P concentrations.

The random distribution of the potential barrier heights at the grain boundaries, the random distribution of the grains dimensions and the extended states conductivity has led to the assumption of "effective" band bending and spatially uniform distribution of the gap states (Fortunato et al 1986). Following the analysis of FEC and reducing the Poisson equation to one dimension, the gap-state density $N_g(E)$ is given by:

$$N_g(E_f + \Psi_s) = \frac{\epsilon_{Si}}{2q} \frac{\partial^2}{\partial \Psi_s^2} \left(\frac{d\Psi}{dx}\bigg|_{x=0} \right)^2 \tag{1}$$

where: Ψ_s is the band bending at the surface ($x = 0$), and the x-axis is taken perpendicular to the surface; E_f is the Fermi level; ϵ_{Si} is the polysilicon dielectric constant; and q is the electron charge.

Using the incremental method (Suzuki 1982), from this equation one may calculate the DOS in a comparatively wide range in the forbidden gap. The availability of both N^+ and P^+ source-drain regions allows the determination of $N_g(E)$ below and above the Fermi level E_f. The flat-band voltage V_{fb}, needed to initiate the calculation procedure, and the position of the zero-temperature Fermi level E_{fo} with respect to the conduction band in the bulk E_{cb} may be determined from the temperature dependence of the field-effect conductance, as demonstrated by Fortunato et al (1988).

3. RESULTS AND DISCUSSION

The similarities between the transfer characteristics of the n-channel $TFTs$ (Fig. 2) with different phosphorus implantation doses in the range of $5 \times 10^{11} - 1 \times 10^{13}$ cm^{-2} are indications of the comparatively weak effect of the implantation dose on the on-state and off-state characteristics. As also seen in Fig. 2, the transfer characteristics

Table 1: The phosphorus concentration as calculated with SUPREM II.

Implantation dose $[cm^{-2}]$	Phosphorus concentration $[cm^{-3}]$
5×10^{11}	3.06×10^{16}
1×10^{12}	6.12×10^{16}
5×10^{12}	3.06×10^{17}
1×10^{13}	6.12×10^{17}

Table 2: The dependence of the flat-band voltage V_{fb} and the position the Fermi level ($E_{cb} - E_{fo}$) on the implantation dose.

Impl. Dose $[cm^{-2}]$	0	5.10^{11}	1.10^{12}	5.10^{12}
$E_{cb} - Efo$ [eV]	.56	.55	.53	.52
V_{fb} [V]	.45	.70	.40	-.30

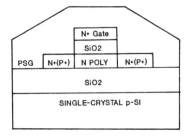

FIG. 1 Cross section of the polysilicon MOSFET

of the undoped $TFTs$ are quite similar to those of $TFTs$ with implantation doses of 5×10^{11} and 1×10^{12} cm^{-2}. This feature in Fig. 2 may be related to the density of filled trapping levels in the upper and lower halves of the bandgap, which appears to be high enough to control the space charge density in the $TFT's$ channel (and which is one of the assumptions leading to the simple expression for $N_g(E)$ in Eq. 1). Moreover, the ohmic behaviour of the $I_d(V_{ds})$ characteristics, illustrated in Fig. 3, indicate that the drain-channel junction is non-rectifying for V_g at the minimum of the transfer characteristics in Fig. 2 (Brotherton et al 1988). The comparatively high density of states in the forbidden gap, which may be ascribed to the preparation procedure, explains for the low on-state currents I_{on} of the $MOSFET's$.

The comparison between the DOS distributions (Fig. 4a and Fig. 4b) shows that the implantation of phosphorus with doses as low as 5×10^{11} cm^{-2} and the conventional technological processings associated with the implantation slightly increase the gap-state density near the Fermi level only, thus suggesting the creation of dangling-bond states as a result from the implantation (Fortunato et al 1986). A further increase of the implantation dose, however, results in a decrease of the gap-state density well above the Fermi level, although the slope remains unchanged, while $N_g(E)$ near E_f continues to increase. In the field-effect conductance measurements, the latter is manifested in the lower flat-band currents I_{fb} of the $MOSFETs$ in this case, resulting in a higher $\frac{I_{on}}{I_{fb}}$ ratio.

The slight increase of the gap-state density below E_f in the case of higher

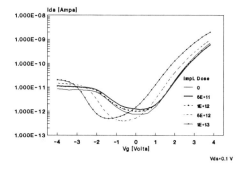

FIG. 2 The TFT transfer characteristics for various implantation doses

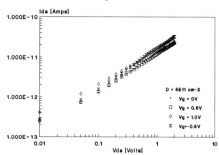

FIG. 3 Ids(Vds) characteristics for values of Vg close to Vfb

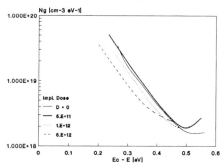

Fig. 4a Density of states Ng(E) above Ef

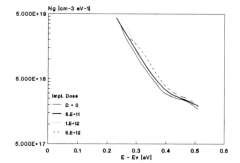

Fig. 4b Density of states Ng(E) below Ef

implantation doses may be related to the comparatively higher leakage currents, illustrated in Fig. 2. Once again it should be noted, that the FEC method, which is used to determine the gap-state density, does not distinguish between bulk and interface states. Nevertheless, the above discussion is accurate enough as far as it is based on the comparison between samples which have undergone similar preparation procedures.

Table 2 presents the calculated V_{fb} and $E_{cb}-E_{fo}$ values of the phosphorus implanted polysilicon films. As can be seen from this table, the implantation with doses $> 5 \times 10^{12}$ cm^{-2} is effective enough to overcome the pinning of the Fermi level at midgap, which is typical of unpassivated polysilicon and is attributed to the high density of traps at the grain boundaries. It can be seen from both Fig. 2 and Table 2 that the flat-band voltage V_{fb} decreases with increasing the phosphorus implantation dose.

4. CONCLUSIONS

The analysis of gap-state density in thin polysilicon films shows that even low phosphorus implantation doses, rendering phosphorus concentrations in the order of 10^{16} cm^{-3}, noticably increase the density of states in the vicinity of the Fermi level, while the slope of the $N_g(E)$ dependence above the Fermi level remains unchanged. The increase of the implantation dose results in a slight increase of the DOS below the Fermi level and in a lowering of the flat-band voltage of polysilicon $MOSFETs$. In terms of device performance, the increase of the DOS near the Fermi level in the case of higher implantation doses results in a higher $\frac{I_{on}}{I_{fb}}$ ratio.

REFERENCES

Brotherton S D, Young N D and Gill A 1988 *Solid State Devices* p 135
  (Eds. Soncini G and Calzolari P U, Elsevier Science Publishers)
Fortunato G, Migliorato P 1986 *Appl.Phys.Lett.* **49**, No.16 pp 1025-1027
Fortunato G, Meakin D B, Migliorato P, and LeComber P G 1988
  Phil.Mag. B **57** pp 573-586
Migliorato P, Meakin D B 1987 *Appl.Surface Sci.* **30** pp 353-371
Suzuki T, Osaka Y, Hirose M 1982 it Jap.J.Appl.Phys. **21** L159

Purely electronic dielectric breakdown of thin SiO$_2$ films

J. Suñé [2], E. Farrés, M. Nafría, and X. Aymerich
Departament de Física, Universitat Autònoma de Barcelona, 08193-Bellaterra, SPAIN.

ABSTRACT: The dielectric breakdown of thin-oxide silicon-gate MOS capacitors is shown to be essentially a reversible electronic phenomenon. Thermal effects are consequence of the breakdown and not their cause and can be limited by limiting the current which flows through the breakdown spots.

1. INTRODUCTION

It is widely recognized that there is a direct relation between the dielectric breakdown of thin SiO$_2$ films and their previous degradation by electron injection. A lot of work has been dedicated to investigate the degradation of these films when subjected to electrical stress (generation of bulk and interface electron and hole traps, slow trapping instabilities, generation of weak spots of reduced injection barrier,...), and there is still much interest for the study of these phenomena (Heyns 1989, DiMaria 1990, Porod 1990, Olivo 1988). However, much less is known about the physics of the actual breakdown mechanism. That is, very little is known about the mechanism which causes the sudden failure of the oxide (detected as an abrupt increase of the DC conductivity), and about the role that the generated defects (electron traps, interface states,...) play in the change of the properties of the insulator film. There is also a general tendency to believe that thermal effects are intrinsic to the breakdown and, as a consequence, that this is an irreversible phenomenon (contrarily to what happens with the breakdown of semiconductors). In this paper, however, we experimentally demonstrate that, at least in some cases, the dielectric breakdown is a reversible electronic phenomenon (a switching between two conduction states of very different conductivities). Thus, thermal effects are demonstrated to be a consequence of the breakdown and not their cause.

When a MOS capacitor is subjected to a constant-voltage stress which is not stopped after the occurrence of the first breakdown, many other breakdown events can be detected (as sudden current increments) because the electrical stress is maintained in the rest of the capacitor area in spite of the local increase of the current at the breakdown spots. If the stress gate voltage is not changed during the whole experiment, a small substrate series resistance is required for the observation of several local breakdowns. Otherwise, due to the high current which flows through one breakdown spot, almost all the applied voltage would drop in the substrate, and the degradation of the "non-broken" area of the

---

[1] Work partially supported by the Comisión Interministerial de Ciencia y Tecnología (CICYT) under project number MIC88-0340-C02
[2] Present adress: Università di Bologna, DEIS, Viale Risorgimento 2, 40136-Bologna (Italy)

capacitor would be drastically slowed down because of the reduction of the effective oxide field. In the case of thin oxide films, it was shown that these multiple breakdown events are sometimes reversible (Suñé et al. 1989,1990), but some doubts concerning the role of thermal effects still remained because of possible self-healing disconnections. In spite of the use of chromium or poly-Si gate electrodes whose melting points are much higher ($\sim 1400°$ C) than that of aluminum ($\sim 700°$ C), and of the observation by SEM of an apparently unaltered electrode surface after the electrical detection of many breakdown events, self-healing effects could not be completely ruled out. It is the main purpose of this paper to show that, if thermal effects are limited to a minimum (by choosing the appropriate samples and by using external protective circuits), the dielectric breakdown can be shown to be a reversible electronic phenomenon.

2. RESULTS AND DISCUSSION

As shown in a previous work (Suñé et al. 1990), using moderately doped n-type substrates, it is possible to provoke very "soft" breakdown (which provoke limited increase of the local current) when injecting from the poly-Si gate. This is due to the fact that the current which flows through the breakdown spot is severely limited by the spreading resistance (the spot area is very small) in the depleted n-type substrate. It was shown that the breakdown produced by electron injection from the gate to the substrate is much softer than that caused by injection from the substrate into the degenerate poly-Si gate. Since the local current is limited by the spreading resistance under the breakdown spot, one expects the thermal effects to be more efficiently limited by lightly doped n-Si substrates. However, it is convenient to use moderately doped substrates to avoid the problem of a larger series resistance. A doping level of approximately 10^{18} cm^{-3} warrants a sufficient limitation of the local breakdown current and, at the same time, a tolerable series resistance.

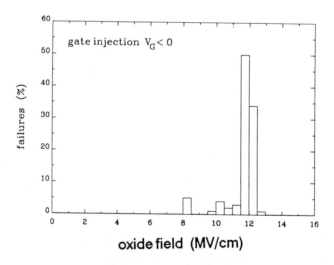

Fig. 1. Experimental distribution of breakdown fields of n$^+$ poly-Si/SiO$_2$/Si(n) capacitors of area $S = 1.25 \times 10^{-3}$ cm^{-3}. The substrate doping is $N_D = 10^{18}$ cm^{-3}, and the oxide thickness $t_{ox} = 13.5$ nm

We have subjected our n⁺ poly-Si/SiO$_2$/Si capacitors with $N_D \approx 10^{18}$ cm^{-3} and $t_{ox} =$ 13.5 nm to both static (constant-voltage and constant-current) and ramped-voltage measurements. The breakdown field distribution (corresponding to 100 capacitors of area $S = 1.25 \times 10^{-3}$ cm^2) is shown in figure 1. All samples broke down at fields larger than 8 MV/cm and 90% above 11 MV/cm. This indicates that the samples are of good quality as also shown by the low mid-gap interface state densities measured from the combination of high-frequency and quasi-static C-V (N$_{ss} \sim 10^{10}$ cm^{-2}). In constant-current experiments only one breakdown event can be observed because the applied voltage suddenly drops when the first breakdown occurs, and the degradation of the rest of the capacitor area is practically stopped. On the contrary, in constant-voltage tests, many breakdown events can be provoked as shown in figure 2, where a typical evolution of the current during a constant-voltage test at $V_G = -19$ V is depicted.

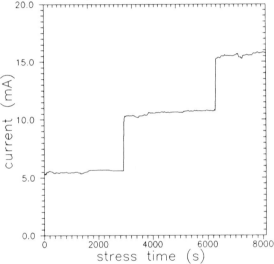

Fig. 2. Evolution of the current during a constant-voltage stress experiment at $V_G = -19$ V.

Three breakdown events appear as current steps of approximately 5 mA (the first one is not well resolved because it occurred after 3 seconds of stress, $Q_{bd} \approx 40$ C/cm^2, after showing the typical turn-around evolution evolution). The second and first breakdowns appear after much longer times because of the effective reduction of the oxide field in the rest of the capacitor (due to the substrate series resistance). Contrarily to previously published results (Suñé et al. 1989,1990) no reverse events were observed in these experiments. This can be attributed to both the occurrence of thermal effects and to the particular combination of applied voltage and load resistance (the sum of the spreading resistance under the spots and the substrate series resistance). Nevertheless, we have succeed in demonstrating that the observed breakdown events are intrinsically reversible by externally limiting the current during the application of voltage ramps. Using an HP4145B Semiconductor Parameter Analyzer with a compliance current limit of 10 mA, we have been able to observe an hysteresis cycle in the I-V curve as shown in figure 3 (curve a). If the compliance limit is removed, the hysteresis disappears and the whole I-V is changed (curve b). Curve a is the post-breakdown I-V after one breakdown event and curve b is the post-breakdown I-V after two events (note that the current at -19 V is approximately the same than in the first and second flat current terraces shown in figure 2). The compliance limit eliminates the thermal effects and allows to show that the breakdown is initially a purely electronic phenomenon. In these particular samples, however, thermal effects substantially alter the oxide structure when the breakdown occurs without the protective compliance limit. This can be the reason for not having observed reverse events during constant-voltage stress experiments. In our opinion, the occurrence of thermal effects at the very moment of the breakdown depends on the stress conditions, on the injection barrier height (the larger the barrier, the more limited the thermal effects) and on the load resistance (i.e. on the substrate doping). In any case, the results shown in figure 3 clearly demonstrate that the breakdown has to be understood as a fast local switching between two well-defined conduction states of drastically different conductivities. The nature of the high-conductivity state has not been determined yet (and can depend on the samples characteristics), but we believe that trap-assisted resonant tunneling is a reasonable hypothesis (Riccó et al. 1983, Suñé et al. 1989,1990)

To our knowledge, these are the first experimental results which demonstrate that the breakdown of thin SiO_2 films is essentially a reversible electronic phenomenon.

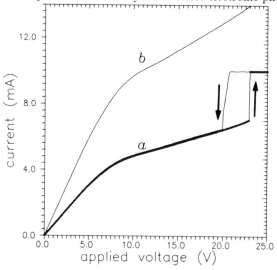

Fig. 3. Post-breakdown I-V characteristics after one breakdown event and protected (after the occurrence of the 1st breakdown) by a current compliance limit of 1 mA (curve a), and after two breakdown events (curve b). The transition between curve (a) and (b) is provoked by removing the external current limitation.

DiMaria D J, 1990 J. Appl. Phys. 68 5234
Heyns M M, Krishna Rao D, and De Keersmaecker R F, 1989 Proc of INFOS-89, Appl. Surf. Sci. 39 327.
Olivo P, Nguyen T N, and Riccó B, 1988 IEEE Trans. Electron Devices ED-35 2259.
Porod W, and Kamocsai R L, 1990 Appl. Phys. Lett. 57 2318.
Riccó B, Ya Azbel B, and Brodsky M H, 1983 Phys. Rev. Lett. 51 1795.
Suñé J, Farrés E, Barniol N, Martín F, and Aymerich X, 1989 Appl. Phys. Lett. 55, 128.
Suñé J, Aymerich X, and Heyns M M, 1990 Proc. of the 1st ESREF conf., Bari (Italy) pp. 313-320.

Paper presented at INFOS '91, Liverpool, April 1991
Contributed Papers, Section 6

Spatially-resolved measurements of hot-carrier generated defects at the Si-SiO$_2$ interface

A Asenov, J Berger, P Speckbacher, F Koch, and W Weber[*]
Physik Department E16, TU München, D-8046 Garching, Germany
[*]Corporate Res. & Develop., Siemens AG, D-8000 München 83, Germany

ABSTRACT: We report on a new approach for spatial profiling of hot-carrier-induced interface traps in MOSFETs. The method is based on forward bias measurements of the drain to substrate current as a function of gate voltage. In combination with 2-D numerical simulation it provides detailed information for the stress-generated defects which act as recombination centers.

1. INTRODUCTION

In small MOSFETs hot carriers from the channel are injected into the gate oxide. The injected carriers can lead to both charge trapping in the oxide and formation of interface traps (Hofmann et al 1985, Asenov et al 1987, Schwerin et al 1987). The hot-carrier degradation of MOSFETs is most frequently evaluated by monitoring changes in transistor characteristics $I_D(V_G)$ and/or $I_D(V_G)$, threshold voltage V_T and transconductance g_m. Also, reverse-bias (Giebel et al 1989) and forward-bias (Acovic et al 1990) measurements of the current of the drain contact operated as a gated diode have been successfully used for monitoring the stress created defects. These defects are located in the high field region of the channel near the drain. Their spatial distribution is a subject of some interest. Initially charge pumping measurements (Ancona et al 1988) have been used to evaluate the spatial profile of hot-carrier-induced interface traps in MOSFETs. Recently it has been shown (Speckbacher et al 1990) that measurement of reverse bias generation current between the drain and substrate of MOSFET as a function of gate voltage V_G and in combination with 2-D numerical simulation can provide a detailed measure of the spatial distribution of the interface states.

In this paper we present a new approach for spatially resolved identification of the stress-generated interface states. It is based on the measurement before and after stress of the forward drain-to-substrate current as a function of gate voltage. The method is sensitive, reliable and easy for implementation. The interpretation of the measured data, which also needs 2-D numerical simulation, is simpler than in the case of reverse current measurements. Several examples of the spatial distribution of stress generated defects in n- and p-channel transistors are presented in order to demonstrate the capability of the method.

2. DESCRIPTION OF THE METHOD

Let us consider an n-channel MOSFET. Under forward-bias conditions, for gate voltages V_G smaller than the threshold voltage V_T, the depletion layer surrounds the drain (Figure 1 a). The drain-to-substrate forward-bias current

© 1991 IOP Publishing Ltd

I_F consists of two components, the ideal diode diffusion current I_{dif} and a bulk or surface recombination current I_{rec} in the depletion region (Sze 1981). For small forward bias applied, the recombination current dominates and can be used as a monitor for the concentration of the recombination centers. According to Shockley-Read-Hall statistics mid-gap levels are the exponentially most effective centers for recombination. The recombination through such levels exhibits a sharp peak at the position where the injected electron and hole concentrations coincide and drops exponentially away from this point. The magnitude of the recombination peak is proportional to the carrier concentration at the crossing point. The shape of the surface recombination rate along the interface for $V_D=-0.2V$ is compared in Figure 1,b with the generation rate at reverse-bias conditions ($V_D=0.2V$). For this case the generation rate is almost constant when the mid-gap energy lies between the electron and hole quasi-Fermi levels and it is proportional to the intrinsic concentration.

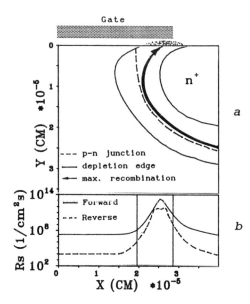

Fig. 1 a) Depletion region and maximum recombination pointer; b) Surface recombination ($V_D=-0.2V$) and generation ($V_D=0.2V$) rates for a constant density of defects along the surface.

When V_G is swept from V_T into accumulation, the recombination maximum scans like a pointer over interface toward the drain. The difference $\Delta I_F(V_G)$ between the measured currents after and before stress provides information for the spatial distribution of the stress-generated defects. Under the assumptions that 1) the stress generated defects are mainly interface states, 2) the energy distribution of these states near mid-gap is smooth, and 3) the excess electron and hole surface recombination rates are equal ($\Delta S_n = \Delta S_p = \Delta S$), the excess recombination current ΔI_F after stress can be expressed as followed

$$\Delta I_F(V_G) = qW \int_{x_1}^{x_2} \Delta S(x) f(V_G,x) dx, \qquad f(V_G,x) = \frac{n(V_G,x)p(V_G,x) - n_i^2}{n(V_G,x) + p(V_G,x) - 2n_i}, \qquad (1)$$

where x_1 and x_2 are the boundaries of the depletion region along the surface, q is the unit charge, n and p are carrier concentrations, W is the transistor width. At a given V_G the function f is sharply peaked at the point $x_m(V_G)$ where $n(V_G,x_m)=p(V_G,x_m)$. If $\Delta S(x)$ varies slow in the vicinity of x_m we obtain from (1) the simple relation

$$\Delta S(x_m) = \frac{\Delta I_F(V_G)}{qWi}, \qquad i = \int_{x_1}^{x_2} f(V_G,x)dx \qquad (2)$$

A 2-D numerical simulation provides the shape and the peak position of the

Hot Carrier Phenomena

function f for the gate voltages corresponding to the measurement conditions. Thus the position of the peak in combination with (2) determines the spatial distribution of the excess surface recombination velocity $\Delta S(x)$.

We note that the magnitude of the excess surface recombination velocity $\Delta S(x)$ is proportional to the density $\Delta N_{SR}(x)$ of those interface defects which act as recombination centers. In terms of elemental quantities $\Delta S(x) = v_T \sigma_{SR} \Delta N_{SR}$, where v_T is the thermal velocity and σ_{SR} is the capture cross section.

3. RESULTS AND DISCUSSION

The measured devices are n- and p-channel MOSFETs with gate oxide 20nm, channel length 1.5μm and channel width 10μm. The p-channel transistors have 0.43μm deep B-implanted junctions. The n-channel transistors are with LDD implantation (dosis $3*10^{13}$ cm^{-2}). The highly doped part of the junction is an As-implantation, spaced 0.2 μm from the gate.

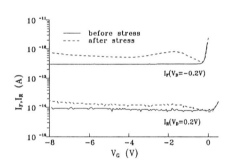
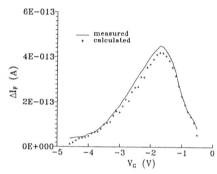

Fig. 2 Forward- and reverse-bias drain-to-substrate current as a function of V_G

Fig. 3 Difference ΔI_F between forward bias currents after and before stress as a function of V_G

The forward-bias drain-substrate current I_F as a function of gate voltage for an n-channel MOSFET before and after stress is compared in Figure 2 with the reverse-bias current I_R. The stress conditions were $V_G = 3V$ and $V_D = 8V$, stress time was 720 s. This figure highlights the main advantages of the forward bias measurements. First, the measured forward current is about two orders of magnitude higher than the reverse one. Secondly due to the sharpening of the pointer in the forward-mode measurement (see Figure 1 b) the spatial resolution is much higher. Finally, at forward-bias the relative change of the current corresponding to the same amount of the stress-created defects is higher than in reverse bias.

The difference ΔI_F between the forward bias currents is plotted in Figure 3 (solid line). The spatial distribution of the excess surface recombination velocity and stress-generated defect density (under the assumption that capture cross section is $\sigma_{SR} = 10^{-15}$ cm^2) is given in Figure 4. The typical value of the defect density is in agreement with that previously published by Haddara (1986) from charge pumping measurements. It also agrees with the numerical simulation results of Asenov (1987).

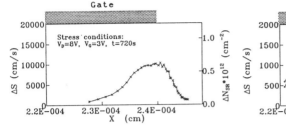
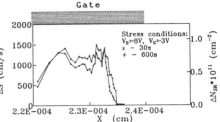

Fig. 4 Spatial distribution of the excess surface recombination velocity ΔS and recombination center density ΔN_{SR} - n-channel LDD MOSFET

Fig. 5 Spatial distribution of the excess surface recombination velocity ΔS and recombination centers density ΔN_{SR} - p-channel MOSFET

To verify the assumption that the spatial variation of ΔS (x) is smoother in comparison with **f** we recalculate the excess recombination current ΔI_F using already determined ΔS(x) and check for consistency.

Results for the V_G=-3V, V_D=-8V stress-cycle defects in p-channel MOSFET are presented in Figure 5 for the two stress times 30s and 600s. The defects are found to be located inside the drain near the metallurgical p-n junction.

There are two main error sources typical for all methods combining experimental measurements with numerical simulation results. First, the precise description of **f** shape provided by 2-D numerical simulation, is dependent on the detailed knowledge of the actual doping profile distribution. Second, charge trapping during the stress cycle will modify the potential distribution and consequently the shape of the **f(x)** function.

4. CONCLUSIONS

We have presented a profiling method for interface deep level defects which is easy to implement with greater sensitivity and precision than other methods. The new method was applied to n- and p- channel MOSFETs to study the spatial distribution of the stress generated interface states. The defect states were found to be localized in a small region near the drain. Under similar stress conditions in p-channel transistors a relatively lower density of recombination centers was observed.

REFERENCES

Acovic A Dutoit M and Ilegems M 1990 IEEE Trans. Electron Dev. **37** 1467
Ancona M G Saks N S and McCarthy D 1988 IEEE Trans. Electron Dev. **35** 2221
Asenov A Bollu M Koch F and Scholz J 1987 Appl. Surface Sci. **30** 319
Giebel T and Goser K 1989 IEEE Electron Dev. Lett. **10** 76
Haddara H and Cristoloveanu S 1986 Solid-State Electron. **29** 767
Hofmann K R Werner C Weber W and Dorda G 1985 IEEE Trans. Electron Dev. **ED-32** 691
Schwerin A Hänsch W and Weber W 1987 IEEE Trans. Electron Dev. **ED-34** 2493
Speckbacher P Asenov A Bollu M Koch F and Weber W 1990 IEEE Electron Dev. Lett. **11** 95
Sze S M 1981 Physics of Semiconductor Devices (New York: John Wiley & Sons pp 92-3

Paper presented at INFOS '91, Liverpool, April 1991
Contributed Papers, Section 6

Modeling and characterization of submicron P-channel MOSFETs locally degraded by hot carrier injection

A Hassein-Bey and S Cristoloveanu

Laboratoire de Physique des Composants à Semiconducteurs (UA-CNRS)
Institut National Polytechnique, ENSERG, BP 257, 38016 Grenoble Cedex, France.

Abstract. The influence of the localized defective channel region formed by hot carrier injection on the basic characteristics of P-channel transistors is systematically investigated and modeled. A parameter extraction method in stressed MOSFET's is proposed, based on the comparison of $I_D(V_G)$ characteristics before and after stress.

1. Introduction

Considerable work has been done to study the aging of N-channel MOS transistors. For the integration of CMOS structures to be optimized, the degradation of P-channel MOSFET's must also be carefully investigated. We demonstrate that P-channel transistors are not merely mirror images of their N-channel counterparts. Two-dimensional numerical simulations, used before for N-channel transistors (Haddara and Cristoloveanu 1987), are now performed with the aim of analyzing the influence of the parameters of the defective region (length, type and density of defects) on the ohmic region characteristics of P–MOSFET's. Based on these simulations, practical methods for the parameters extraction in degraded P–channel transistors have been achieved and tested.

2. Model presentation

Two-dimensional simulations are performed to solve the Poisson equation in the semiconductor and the Gauss equation at the semiconductor-oxide interface. This results in a nonlinear system of equations in the semiconductor, oxide and at the interface. The main program parameters are channel length L, localization and length ΔL of the degraded zone, type and concentration of defects. This program gives the surface potential profile along the channel $\Psi_s(x)$, the inversion layer concentration and the current $I_D(V_G)$ and transconductance $g_m(V_G)$ characteristics in the linear region. It is found that the acceptor interface traps effect is very different in P- and N- channels, whereas the donor trap effects in P-MOSFET's can be analyzed in much the same way as for acceptors in N-MOSFET's.

3. Simulation results

The effect of interface traps can be decomposed into two distinct components : charge effect and mobility reduction effect. The interface traps affect directly the surface potential Ψ_s: according to the presence of donor or acceptor traps, Ψ_s decreases or increases, respectively. In the case of a localized generation of *acceptor traps*, there is a barrier lowering in the degraded zone near the drain (Fig.1(a)). The resulting "extension" of the drain junction induces a channel length reduction. The transconductance $g_m = dI_D/dV_G$ tends to increase, in particular for low densities of traps since it is inversely dependant on the length L. However, an opposite effect that reduces the transconductance is the mobility degradation caused by the interface traps. For large densities of traps ($N_{it} \geq 5 \times 10^{11} eV^{-1} cm^{-2}$ in Fig.1(b)), the mobility degradation effect dominates and offsets the channel length reduction, making the transconductance to drop.

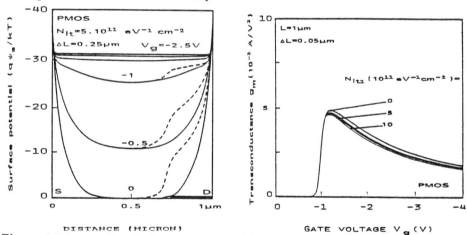

Figure 1 Influence of localized acceptor states : (a) longitudinal surface potential profile before (—) and after (- - -) stress and (b) transconductance degradation.

The case of *donor* interface traps is similar with that of acceptor traps in N-channel MOSFET's. Since the surface potential is lowered during the trap generation, the resistance of the localized region is increased. The coupling between the defective and undamaged portion of the channel causes a transconductance *overshoot* (for $N_{it} = 6 \times 10^{11} eV^{-1} cm^{-2}$ in Fig.2(a)). It is due to the narrow degraded region which can control the whole channel (Haddara and Cristoloveanu 1987).

The trapping of a *fixed* negative charge in the oxide leads to a very clear transconductance increase. Unlike the case of interface states, the fixed oxide charge is not modulated by the surface potential and therefore does not depend on gate voltage. In addition, there is almost no mobility reduction related to fixed charges. As a consequence, the effect of channel shortening is more visible after electron trapping in the oxide than after generation of acceptor interface traps. Since the degraded transistor has an apparent channel length $L - \Delta L$, it is easy to determine ΔL from the transconductance increase.

4. Extraction of the defective region parameters

The electrical stress can create degradation in a very narrow interface region near the drain junction. We compare the simulated characteristics of a virgin MOSFET (before stress) with a degraded MOSFET (after the stress), in order to extract the major parameters of this degraded region : length ΔL and interface traps density N_{it}. The value ΔL is evaluated by comparing transconductance curves $g_{b,a}(V_G)$ before (b) and after (a) aging, so that we have :

$$\frac{\Delta L}{L} \simeq \left(1 - \frac{g_b}{g_a}\right) \tag{1}$$

We now apply the two-piece model that consists in dividing the degraded MOSFET in two serial transistors (Haddara and Cristoloveanu 1987). One, formed by the rest of the channel with a length $L - \Delta L$ (Fig.2(a)), is virgin whereas the second is uniformly degraded along its length ΔL. In weak inversion, the transistor resistance before stress, R_b, is given by :

$$R_b = A \frac{L}{\mu_1} \exp\left(-\beta \frac{V_G - V_{th1}}{N_1}\right) \tag{2}$$

where A is a constant, μ_1 is the effective hole mobility, V_{th1} is the threshold voltage and $N_1 = (C_{ox} + C_D + qN_{it})/C_{ox}$. The oxide C_{ox} and depletion C_D capacitances are known from the process parameters, whereas N_{it} can be measured by charge pumping. After the stress, the total resistance R_a becomes :

$$R_a = A \frac{L - \Delta L}{\mu_1} \exp\left(-\beta \frac{V_G - V_{th1}}{N_1}\right) + A \frac{\Delta L}{\mu_2} \exp\left(-\beta \frac{V_G - V_{th2}}{N_2}\right) \tag{3}$$

where the first term represents the contribution of the undamaged region and the second term accounts for the defective region; N_2 has the same expression as N_1 but after the stress; μ_2, V_{th2} and N_{it2} are, respectively, the effective mobility, threshold voltage and density of states in the damaged region. Combining R_a and R_b yields the function $F(V_G) = (R_a - R_b)/R_b$ (Haddara and Cristoloveanu 1988) :

$$F(V_G) = \ln\left(\frac{\mu_1 \Delta L}{\mu_2 L}\right) + \beta V_G \left(\frac{1}{N_1} - \frac{1}{N_2}\right) - \beta \left(\frac{V_{th1}}{N_1} - \frac{V_{th2}}{N_2}\right) \tag{4}$$

The slope of $F(V_G)$ gives the density N_{it2}. In N-channel MOSFET's, the ratio $\Delta L/L$ can be neglected with respect to $\Delta R/R_b$. In P-channel MOSFET's, these two ratios are on the same order of magnitude and $\Delta L/L$ is no longer negligible. We now understand the interest of the previous step of the extraction procedure which gives $\Delta L/L$. By injecting the value $\Delta L/L$ in Eq.(4) yields N_2, hence N_{it2}. The plot of dF/dV_G, shown in Fig.2(b), is typical and gives $V_{th1} = -1.1V$, $V_{th2} = -0.4V$ and $N_{it2} = 5.7 \times 10^{11} eV^{-1} cm^{-2}$. By considering the simulated $I_D(V_G)$ curves as though they were experimental characteristics, we compare the parameters of the degraded region, obtained with the extraction method, with the true values injected in the program. A systematical investigation, conducted for different values of ΔL and N_{it2}, has demonstrated that the accuracy of the extraction method is in general better than 30 %, in spite of the rough two-piece model used.

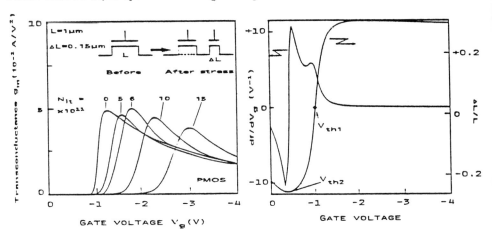

Figure 2 (a) Transconductance degradation for various densities of donor traps and (b) illustration of the extraction methods corresponding to Eqs.(1) and (4).

5. Conclusion

The effect of the localization of acceptor interface traps and negative oxide charges on the ohmic region characteristics of P-channel MOSFET's has been analyzed and modeled. The channel length shortening is a typical feature of P-channel MOSFET's. The transistor characteristics depend closely on the degraded region length and defect density. The major differences between N- and P- channels have been emphasized. Appropriate extraction methods have been developed, which provide accurate values for both ΔL and N_{it2} and allow a more detailed insight into the physical mechanisms involved in the aging P-channel MOSFET's.

References

Haddara H and Cristoloveanu S 1987 *IEEE Trans. Electron Dev.* **ED-34** 378
Haddara H and Cristoloveanu S 1988 *Sol. St. Electron.* **31** 1573
Hassein-Bey A and Cristoloveanu S 1991 *Proc. MRS'91*

Gate-controlled electroluminescence from reverse-biased Si-MOSFET drain contacts

A Kux, M Schels, F Koch, and W Weber*

Physik–Department, TU München, D–8046 Garching, Germany
* Corporate Res. & Develop., Siemens AG, D–8000 München 83, Germany

ABSTRACT : We report on electroluminescence from MOSFETs operated as a reverse–biased gated diode. The resulting electric field configuration causes the light emission to be sensitive to the Si–SiO$_2$ interface. We observe an interface–defect related luminescence signal. Degradation of the device due to hole trapping in the oxide with time reduces the emission intensity.

1. INTRODUCTION

Light emission from Si–devices is of considerable interest even if their emission efficiency does not reach that of the direct–gap III–V semiconductors. For interchip communication, transmitting high bitrates over short distances, using only Si–technology is a real advantage. A recent report (Noël 1990 a,b) claims that Si–Ge superlattices show efficient light emission at a photon energy of $0.825 eV$. We show that light of the same energy can be obtained from a MOSFET operating as a reverse–biased gated diode.

2. EXPERIMENTAL NOTES

The measurements were performed on n–channel MOSFETs with a gate oxide thickness of $40 nm$, a channel width of $10 \mu m$ and an effective channel length of $1.7 \mu m$. The source and drain contacts were abruptly doped with As–concentrations up to $5 \times 10^{20} cm^{-3}$. Fourteen of these transistors were operated in parallel to get a higher light intensity.

The samples were mounted in a cold finger cryostat and kept at liquid nitrogen temperature. The emission spectra were recorded by a BOMEM DA3 Fourier transform spectrometer using a North Coast Ge–diode for detecting the light. The spectra are normalized to account for the spectral sensitivity of the detector and the spectrometer.

3. RESULTS AND DISCUSSION

In Figure 1 we show emission spectra obtained from the MOSFET when the drain contact is reverse–biased with respect to the substrate ($V_{DSub} = 18.5V$) and an additional negative gate–voltage ($V_{GSub} = -10V$) is applied. Source and drain contacts are connected in parallel to actually double the emission intensity. We observe two separate

© 1991 IOP Publishing Ltd

emission peaks, one at $0.82eV$ and the other at the Si band gap energy of $1.10eV$. The $0.82eV$ peak becomes considerably weaker and the $1.10eV$ signal completely vanishes when the measurement is repeated after keeping the MOSFET at the same voltages for about half an hour, indicating degradation of the device.

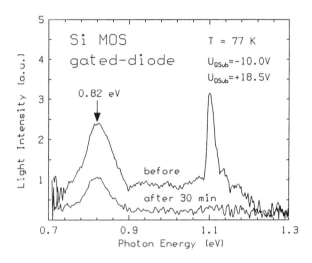

Fig. 1. Light emission spectra of the reverse–biased gated diode. The second spectrum is recorded after operating the diode with the indicated voltages for 30 minutes.

To understand the mechanisms for light emission and degradation we must consider the potential and electric field for the gated diode. The negative gate–voltage effectively squeezes down the depletion layer width at the Si–SiO$_2$ interface and thus provides very high electric fields (Figure 2). Avalanche generation creating the carriers necessary for light emission is most effective in this region.

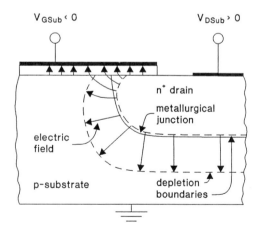

Fig. 2. Schematic view of potential and electric field for the reverse–biased gated diode with negative gate–voltage.

Under reverse bias band–to–band recombination giving rise to the $1.10eV$ luminescence occurs only where the carriers are generated in the avalanche region. The reason is that the avalanche–generated carriers are driven to their respective majority carrier region by the electric field. There they do not contribute to radiative recombination .

Hot Carrier Phenomena

The recombination process leading to light emission at $0.82eV$ is due to holes arriving at the Si–SiO$_2$ interface. The electric field has a component pulling the holes towards the Si–SiO$_2$ interface where they recombine with electrons occupying an interface defect. A possible such defect is the dangling bond. Its density of states peaks at $\approx 0.3eV$ and $\approx 0.8eV$ above the Si valence band (Poindexter 1989). In a very narrow region near the drain contact we find negatively charged dangling bond states. There the capture of holes can be expected to lead to light emission at an energy of about $0.8eV$.

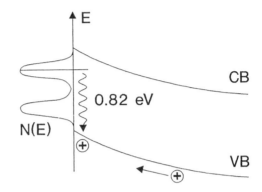

Fig. 3. Holes driven to the surface by the electric field recombine with electrons in negatively charged dangling bond states to generate the $0.82eV$ luminescence (N(E) indicates the density of dangling bond states).

The $0.82eV$ luminescence is not visible in the forward biased diode. Under forward bias conditions we have no surface sensitivity. Minority carriers are injected over the whole depletion layer region. The great diffusion length of minority carriers in Si ($\approx 10\mu m$ to $100\mu m$) gives a large generation volume in the bulk and causes band gap radiation to be dominant. This is also the case for photoluminescence investigations of MBE Si layers (Lightowlers et al 1989).

A very similar $0.82eV$ luminescence peak is also present under MOSFET operating conditions in the same device (Herzog et al 1989).

The holes driven to the Si–SiO$_2$ interface are also responsible for the degradation effect. Some of them can be trapped in the oxide. This leads to a modification of the potential in the interface region. The result is a lower electric field and a correspondingly lower avalanche generation rate. Indeed we observe a decrease in reverse current from $0.73mA$ for the first measurement in Figure 1 to $0.24mA$ for the measurement recorded 30 minutes later. The smaller number of avalanche generated carriers is responsible for the lower emission intensity.

4. CONCLUSIONS

Light emission spectra of Si MOSFETs operated as a reverse–biased gated diode show Si band-to-band recombination at $1.10eV$ and an additional strong emission peak at an energy of $0.82eV$. The avalanche generation of carriers near the drain contact close to the Si–SiO$_2$ interface makes the luminescence sensitive to this region. The recombination line at $0.82eV$ involves recombination at dangling bond states at the interface. Some of the holes arriving at the interface can be trapped in the oxide leading to device degradation and a decrease in luminescence intensity under constant applied voltages because the surface field is reduced.

REFERENCES

Herzog M, Schels M, Koch F, Moglestue C and Rosenzweig J 1989 *Electromagnetic Radiation from Hot Carriers in FET Devices, Proceedings of the International Conference on Hot Carriers, Tempe, Arizona*
Noël J P, Rowell N L, Houghton D C and Perovic D D 1990a *Appl. Phys. Lett.* **57** 1037
Noël J P, Rowell N L and Houghton D C 1990b *37*[th] *National Symposium of the American Vacuum Society, Toronto*
Poindexter E H 1989 *Semicond Sci Technol* **4** 970
Lightowlers E C, Higgs V, Davies G, Schäffler F and Kasper E 1989 *Thin Solid Films* **183** 235

Paper presented at INFOS '91, Liverpool, April 1991
Contributed Papers, Section 6

Electron spin resonance study of trapping centers in SIMOX buried oxides

John F. Conley and P. M. Lenahan
The Pennsylvania State University, University Park, PA 16802

P. Roitman
National Institute of Standards and Technology, Gaithersburg, MD

ABSTRACT: We combine electron spin resonance and capacitance versus voltage measurements to study E' centers in a variety of SIMOX buried oxides. The oxides had all been annealed above 1300°C. Our results clearly show that E' centers play an important role in the trapping behavior of these oxides.

1. INTRODUCTION

Silicon-on-insulator (SOI) promises many advantages over conventional silicon technology. These advantages include total isolation, high speed, high packing density, low power consumption, and radiation hardness.

Among the many SOI technologies, separation by implanted oxygen (SIMOX) has emerged as the leading method for the formation of the buried oxide (BOX). In SIMOX, a high dose of oxygen ions ($>10^{18}$ cm^2) at a high energy (>100 keV) is implanted deep into silicon to form the BOX layer. Initially, the process resulted in a BOX with many oxide precipitates near its surface (1). However, high temperature (T ≥ 1300°C) annealing allows for the dissolution of these oxide "islands" into the BOX by Ostwald ripening (2). Since the SIMOX process is so different from conventional thermal growth, the resulting oxides may exhibit radically different charge trapping behavior. Indeed, much study of the BOX is needed before the potential advantages of SIMOX technologies can be fully exploited with regard to radiation hardness.

Charge trapping in SIMOX due to irradiation was recently studied by Boesch et al. (3). Boesch et al. concluded that efficient trapping of radiation induced holes exists in the bulk of the oxide and that initial trapping and subsequent thermal detrapping of electrons also takes place in the BOX. However, they did not deal with the structure of these charge trapping centers.

During the past four years, several groups have investigated SIMOX oxides with ESR.(4-7). Nearly all of these investigators have dealt with paramagnetic centers at precipitate surfaces. (These precipitates can be eliminated by a high temperature anneal (4)). However, in SIMOX oxides subjected to x-irradiation, Stahlbush, et al. (6) found an ESR signal similar to that of an E' center. A very recent study of Stesmans et al. (7) also involved gamma irradiation; they observed several ESR defects, including E' centers. In these earlier studies the authors did not directly correlate their ESR results to charge trapping in the oxides.

In this study, we use electron spin resonance (ESR) and capacitance versus voltage (CV) measurements along with vacuum ultraviolet (hc/λ = 10.2 ev) and ultraviolet (hc/λ = 5 ev) irradiation sequences to explore the role of E' trapping centers in SIMOX buried oxides. (The E' center is a silicon back-bonded to three oxygens (8-11). In thermally grown oxide films, it is the dominant deep hole trap and is a hole trapped in an oxygen vacancy (10-12). It has a zero crossing g-value of 2.0005.) There is experimental evidence for neutral E'

© 1991 IOP Publishing Ltd

defects in PECVD oxides (13). In this study, we present experimental evidence which links E' centers to charge trapping in SIMOX oxides.

2. EXPERIMENTAL PROCEDURE

The SIMOX buried oxides studied were approximately 4000Å thick and received an anneal at 1300°C or 1325°C for five hours in 99.5% argon 0.5% oxygen or eight hours in 99.5% nitrogen 0.5% oxygen. Before any measurements were made, a surface oxide and the top layer of silicon were removed by etches in HF and KOH (HF attacks only SiO_2 and KOH attacks only Si).

Our electron spin resonance (ESR) measurements were conducted at room temperature on an IBM ER-200 X-band spectrometer. A TE104 "double" resonant cavity was used with a calibrated "weak-pitch" standard so that accurate determinations of both relative and absolute number of spins could be obtained. (Relative spin-concentration measurements are estimated to be accurate to ±10% while absolute spin-concentration measurements are accurate to a factor of two. Capacitance versus voltage (CV) measurements were taken at room temperature using a 1-MHz Boonton capacitance bridge and a mercury probe. The density of space charge was determined from shifts in the CV curve.

3. IRRADIATION

E' centers were generated by exposing (bare) BOX to vacuum ultraviolet light (VUV) from a deuterium lamp. In some cases, an LiF filter was used, and the oxides were illuminated briefly while positively biased. The filter passes only photons with $hc/\lambda < 10.2 eV$. Most of these photons are absorbed in the top 100Å of the oxide where they create electron hole pairs. The positive bias drives holes across the oxide while electrons are swept out. In other cases, the oxides were VUV illuminated without the filter ($hc/\lambda \lesssim 10.2 eV$) and unbiased for an extended period (~ 20 hours).

Ultraviolet illumination (UV) from a sub-SiO_2 bandgap ($hc/\lambda < 5.5 eV$) mercury-xenon lamp was also used in combination with positive bias. The UV illumination results in the internal photoemission of electrons from the Si. The positive bias drives electrons across the oxide.

Biasing was performed by depositing low-energy ions created by corona discharge (14) onto the samples. (These ions have essentially thermal kinetic energy.) Corona charging allowed the generation of a uniform electric field over the large surface area samples (~1 cm^2) required for ESR measurements. The surface potential was measured with a Kelvin probe electrostatic voltmeter.

4. RESULTS

We exposed a variety of SIMOX buried oxides to 20 hours of vacuum ultraviolet (VUV) light from a 50 watt deuterium lamp. The behavior of all samples was qualitatively the same. A strong E' signal is generated by the VUV illumination. No other signals are visible. A broad scan around the "free electron" g = 2.002 is illustrated in Figure 1. (The g is defined by $h\nu/\beta H$, where h is Planck's constant, ν is the microwave frequency, β is the Bohr magneton and H is the magnetic field at which resonance occurs. Table 1 illustrates E' concentrations for a variety of SIMOX films. In all of the oxides, a high density ($\approx 1 \times 10^{18}/cm^3$) of E' centers were generated. (A 20 hour exposure was sufficient to "saturate" the spin densities.)

Etch back experiments (shown in Figure 2) on both and argon and the nitrogen annealed samples show that the E' centers generated by VUV irradiation without bias are distributed throughout the oxide.

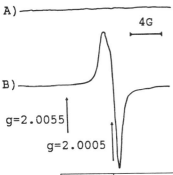

Figure 1: A broad ESR trace before (A) and after (B) exposure to VUV illumination. A strong E' signal appears at g=2.0005. No "amorphous silicon" signal appears at g=2.0055.

Sample	Anneal Ambient	Implant	Spin Density(cm^{-3})
A	Ar/0.5% O$_2$ 1325 C	Single	0.7x10^{18}
B	N$_2$/0.5% O$_2$ 1325 C	Single	1.4x10^{18}
C	Ar/0.5% O$_2$ 1315 C	Single	1.0x10^{18}
D	Ar/0.5% O$_2$ 1315 C	Double	0.35x10^{18}

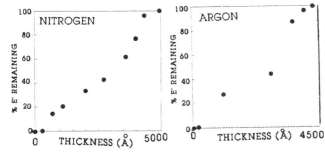

Figure 2: Etchback experiments illustrate the distribution of E' centers throughout the VUV illuminated SIMOX oxides.

Capacitance versus voltage measurements on the VUV illuminated oxides indicate virtually no buildup of net positive space charge within the oxide (~1 volt with respect to the preillumination curve). This result indicates that the E' centers are either electrically neutral or compensated by an equally high number of negatively charged centers. In order to test these possibilities, we exposed the VUV irradiated oxides to ultraviolet light (hc/$\lambda \lesssim$ 5eV) from a 100 watt mercury xenon lamp while the oxide surface was positively charged. Oxides were charged with corona ions to about 80 volts (measured with a Kelvin probe). Oxides were then briefly (~ seconds) exposed to UV light from a mercury xenon lamp. A few seconds of UV light reduced the corona potential to ~10% of the original value, indicating electron injection into the oxide. The total electron injection fluence was roughly determined from CV = Q. The process was repeated until an injected electron fluence of about $\approx 2 \times 10^{14}$ electrons/cm^2 was achieved.

Results of the electron photoinjection are shown in Figure 3. The E' density is decreased by about half (from about 13×10^{12}/cm^2 to about 7×10^{12}/cm^2). CV measurements on the devices indicate shifts of about 10 volts -- corresponding to a change in space charge of about ~10^{12}/cm^2, assuming uniform charge capture throughout the insulator. Clearly, a significant fraction (at least half) of the E' centers can capture electrons with a relatively large capture cross section.

To further test the relationship between oxide charge and E' centers, we exposed these same SIMOX oxides (previously exposed to 20 hours VUV without bias and UV electron injection) to VUV with positive bias. We applied a corona potential of about 100V, measured with a Kelvin probe electrostatic voltmeter. Then we exposed the oxide to 10.2 eV photons from a deuterium lamp. Most of the electron hole pairs are generated in the top ~100Å of the oxide. The positive bias floods the oxide with holes. The process was repeated until an injected hole fluence of about 1×10^{14} holes/cm^2 was achieved.

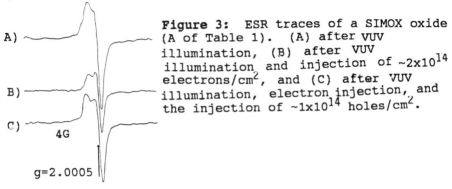

Figure 3: ESR traces of a SIMOX oxide (A of Table 1). (A) after VUV illumination, (B) after VUV illumination and injection of $\sim 2\times10^{14}$ electrons/cm^2, and (C) after VUV illumination, electron injection, and the injection of $\sim 1\times10^{14}$ holes/cm^2.

The result of this process is also shown in Figure 3. The E' density increased to almost the value before the electron injection process. Notice that the line shape of the curve has been changed slightly. This result is also qualitatively consistent with a paramagnetic E' structure corresponding to a hole trapped in an oxygen vacancy. However, coupled with the (pre-charge injection) results of high E' densities ($\sim 1 \times 10^{18}$/cm^3) and CV traces within a few volts of the origin, it suggests that the charged E' centers may be compensated by negatively charged centers.

We conclude that E' centers play an important role in SIMOX oxide trapping. E' precursors are present in high density in the SIMOX oxides explored in this study. Large changes in E' density are induced by injecting electrons or holes into VUV illuminated oxides; this shows that a high percentage of the centers are efficient traps. CV measurements show low amounts of net space charge; this suggests a compensating trap mechanism. The electronic properties of SIMOX oxides are more complex than those of thermal oxides on silicon.

REFERENCES

1. J. Stoemenos, C. Jasaud, M. Bruel, and J. Margail, J. Chryst. Growth 73, 546 (1985).
2. C. Jaussaud, J. Stoemenos, J. Margail, M. Dupuy, B. Blanchard, and M. Bruel, Appl. Phys. Lett 46, 1046 (1985).
3. H. E. Boesch, Jr., T. L. Yuglar, L. R. Hite, and W. E. Bailey, IEEE Trans NS 37, 1982 (1990).
4. R. G. Barklie, A. Hobbs, P. L. F. Hemmet, and K. Reesor, J. Phys. C. 19, 6417 (1986).
5. T. Makino and J. Takahashi, Appl. Phys. Lett 50, 267 (1987).
6. R. E. Stahlbush, W. E. Carlos, and J. M. Prokes, IEEE Trans NS 34, 1680 (1987).
7. A. Stesmans, R. A. B. Devine, A. Revesz, and H. Hughes, IEEE Trans NS 37, 2008 (1990).
8. R. A. Weeks, J. Appl. Phys. 27, 1376 (1956).
9. D. L. Griscom, Phys Rev. B 22, 4192 (1980).
10. P. M. Lenahan and P. V. Dressendorfer, IEEE Trans NS 30, 4602 (1983).
11. P. M. Lenahan and P. V. Dressendorfer, J. Appl. Phys. 55, 3495 (1984).
12. H. S. Witham and P. M. Lenahan, Appl. Phys. Lett 51, 1007 (1987).
13. H. L. Warren, P. M. Lenahan, B. Robinson, and I. H. Sathis, Appl. Phys. Lett 53, 482 (1988).
14. Z. A. Weinberg, W. C. Johnson, and M. A. Lampert, J. Appl. Phys. 47, 248 (1976).

Paper presented at INFOS '91, Liverpool, April 1991
Contributed Papers, Section 6

Oxide field dependence of bulk and interface trap generation in SiO_2 due to electron injection

A. v. Schwerin[*] and M. M. Heyns

Interuniversity Microelectronics Centre (IMEC), Kapeldreef 75, B-3001 Leuven, BELGIUM

ABSTRACT: Substrate hot electron injection into the gate oxide of CMOS n-channel MOSFETs is performed. Electron trapping and trap generation in the SiO_2 bulk and the generation of Si/SiO_2 interface states is studied. The analysis of these oxide degradation mechanisms comprises a detailed examination of their dependence on the oxide electric field, electron injection energy and oxide current density. It is found that trap generation increases about exponentially with increasing oxide field whereas the occupation of generated bulk traps with electrons decreases. Electron trap generation is found to be independent of injection energy. For the first time a dependence of electron trapping on oxide current density is reported. The energetic distribution of the generated interface states in the Si-bandgap is found to be independent of the oxide field.

1. INTRODUCTION

The problem of MOSFET gate oxide degradation due to charge carrier injection poses a severe limitation to further miniaturization of deep submicron MOS-technology. Therefore a detailed knowledge of the various dependences of oxide degradation mechanisms on the injection conditions is necessary to relieve this problem and to predict device lifetime reliably. The electron trap generation in the gate oxide layer of MOS devices has already been investigated by various workers. In particular its field dependence has been studied intensely by DiMaria (1987), DiMaria and Stasiak (1989) and Heyns *et al* (1989). To be able to draw valid conclusions about the nature of the trap generation mechanism, it is indispensable to interpret the results of trapping experiments with great care. In this paper a systematic study of the phenomena occurring during injection of electrons in the SiO_2 layer is made using homogeneous electron injection in MOS-transistors.

2. EXPERIMENTAL DETAILS

The substrate hot electron injection technique (Ning and Hu 1974) was used on CMOS samples. Electrons, drifted into the p-well of the n-MOSFET by slightly forward biasing the n-substrate/p-well junction, are accelerated towards the Si/SiO_2 interface by a reverse well bias (V_{pwell}) while the source and drain of the transistor are grounded. Electrons which gain enough energy to overcome the barrier between Si and SiO_2 are injected into the gate oxide.

The 20nm thick oxide of the n^+-poly-Si gated transistors was grown at 900°C in dry O_2 under addition of HCl. The p-type channel doping of the devices was about $5 \cdot 10^{16} cm^{-3}$. MOSFETs with a large gate area of 40x40µm were used. Thus, the region of nonuniform electric field due to the space charge regions around source and drain, is small (less than 0.5µm) and can be neglected. It was carefully checked using 2-D simulations, that the current, which is driven from the substrate through the channel to source and drain during injection, was in all cases low enough not to disturb (due to the limited

---

[*] now with: Siemens, Corporate Research and Development, Otto-Hahn-Ring 6, D-8000 München 83, Germany

© 1991 IOP Publishing Ltd

channel conductivity) the uniformity of the channel potential by more than 0.1 V. The injected charge density is determined from the gate current, which is measured during injection using an integrating electrometer. In between two subsequent injection steps the oxide damage is characterized. The trapped oxide charge is obtained from the gate voltage shift in deep subthreshold at a fixed drain current and drain voltage. The integral interface state density (N_{it}) and its energetic distribution (D_{it}) is measured using the charge pumping technique (Groeseneken et al 1985).

3. RESULTS

Oxide bulk traps were generated by the injection of a large amount of electrons (e.g. $2...5 \cdot 10^{19}$ e/cm^2) at fixed oxide field. It was found that a subsequent injection of small amounts of electrons at varying oxide electric field (E_{ox}), changes the apparent amount of trapped electrons (as detected in an electrical measurement) depending on E_{ox}, respectively. With other words, the fraction of present traps which is actually charged, due to trapped electrons, is in equilibrium with the field which is applied during electron injection. The respective equilibrium level of trap occupation is reached after switching to the respective field value upon injection of $2...4 \cdot 10^{16}$ e/cm^2. Switching of the electric field without simultaneous injection of electrons does not change the trap occupation considerably. The effect is reversible. The occupation of traps with electrons is a decreasing function of E_{ox}, as shown in figure 1. As a consequence, evidence for the field enhancement of the trap generation process at $E_{ox} > 4$MV/cm is found only when a short low-field injection is carried out after the stressing. This is necessary in order to fill eventually generated, but empty, traps with electrons, and thus enabling the detection of these traps in an electrical measurement. Only in this way a proper comparison of electron trap *generation* for different fields is possible. The result (fig. 2) differs significantly from previously reported results where the trap occupation effect was not taken into account.

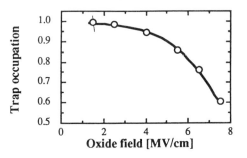

Fig.1: *Normalized equilibrium trap occupation level as a function of the oxide field*

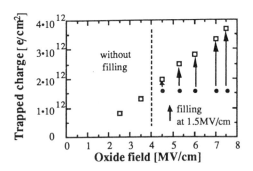

Fig.2: *Trapped electron density as a function of the oxide field after injection of about $4 \cdot 10^{19}$ electrons/cm^2. At fields above 4 MV/cm traps were 'filled' after the stress by injecting $1 \cdot 10^{18}$ electrons/cm^2 at 1.5 MV/cm.*

A second important result is shown in figure 3. If the current flow through the oxide is kept fixed by adjusting the carrier supply from the n-substrate, the p-well bias (V_{pwell}) does not affect electron trapping. Thus, the energy of the electrons at the moment of injection (which is controlled by V_{pwell}) has apparently no effect on electron trap generation. This is most likely due to a fast thermalization of the injected electrons in the SiO$_2$. This finding is in contrast to results on the substrate bias dependence of electron trapping measured by DiMaria and Stasiak (1989). However, without special precautions, an increase in substrate voltage will also cause an increase in oxide current density (with other words, the injection of the same amount of electrons takes less time for higher V_{Sub}).

The effect of this is demonstrated in figure 4, which shows the result of the complementary experiment, where V_{pwell} was kept fixed but the oxide current density (j_g) was set to three different values by adjusting the substrate/p-well forward bias respectively (by this actually changing the carrier supply to the accelerating space charge region). It is found that for medium range oxide fields between 1.5 and 4MV/cm, electron trapping indeed increases with increasing oxide current density. With other words, the less time it takes to inject a certain amount of electrons, the more electrons are trapped. This effect disappears for E_{ox} well above 4MV/cm, both before and after trap filling. The j_g-dependence was found likewise in experiments, where optically stimulated substrate hot electron injection on NMOS samples was used, when the light intensity was changed, in order to change the carrier supply to the accelerating space charge region in the Si, thus changing j_g. The knowledge of the current density dependence of electron trapping is important because j_g depends on V_{pwell} as well as on E_{ox} in the injection experiment, if no special precautions are taken. Therefore, special care is needed not to misinterpret the j_g-dependence at low oxide fields as a dependence on E_{ox} or on V_{pwell}. The j_g-dependence can *not* be explained by a simple dynamic trapping-detrapping argument. This was verified in the following experiment. Between every two subsequent characterization measurements, electrons were injected with a high current density for part of the time, whereas the electric field was applied,

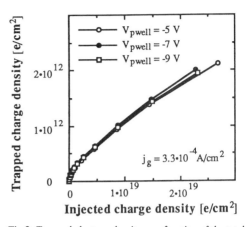

Fig.3: *Trapped electron density as a function of the total density of electrons injected at an oxide field of 3.5 MV/cm, a gate current density of $3.3 \cdot 10^{-4}$ A/cm^2 and with three different p-well biases (V_{pwell}).*

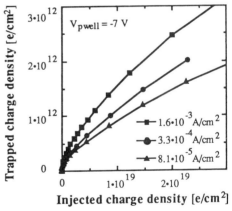

Fig.4: *Trapped electron density as a function of the density of electrons injected at an oxide field of 3.5 MV/cm a p-well bias of -7V and different gate current densities (j_g).*

respectively, as long as in a corresponding low j_g experiment; i.e. for part of the time between two subsequent characterization steps, the electric field was applied without electron injection. In this case no considerable difference in the amount of trapped electrons was found compared to the standard high j_g measurement.

A conclusive model for the origin of the current density dependence of electron trapping is still missing. We suggest as a possible explanation that a shallow trap is transformed into a deep one when an electron is trapped and relaxes into a shallow one after detrapping of the electron. A key parameter in this model would be the time constant of the latter relaxation process, which is in competition with the mean capture time (defined by the capture cross section times the current density).

In contrast to the rather complicated evaluation of bulk trap generation experiments (due to trap occupation effects), the analysis of interface state (D_{it}) generation and its dependence on oxide field, substrate bias and current density is rather straight forward. No evidence is found for a current density dependence of D_{it}-generation. Also, no increase in interface state generation with increasing energy of the incoming electrons (as varied by V_{pwell}) was found. This is in contradiction to results previously reported by Zekeryia and Ma (1983). However, in their photo injection experiments not only electrons are involved but also photons with energies well above 4 eV, which might have a direct impact on the D_{it}-generation.

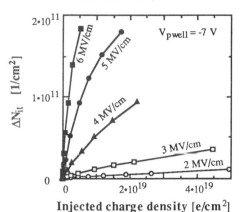

Fig.5: *Increase in Si/SiO$_2$ interface states as a function of the total density of electrons injected at different constant oxide fields between 2 and 6 MV/cm*

The D_{it} generation rate strongly increases with increasing oxide field as shown in figure 5. Extracting the generation *rate* from the starting slope of the curves in figure 5 shows that D_{it}-generation is about exponentially dependent on the magnitude of E_{ox}. Figure 6 shows the distribution of the interface state density as a function of energy in the Si-bandgap. It was measured using charge pumping where rise and fall time of a pulse with fixed amplitude were varied (Groeseneken et al 1985). The shape of the interface state density distribution is independent of the oxide field and shows a larger density in the upper half of the bandgap similar to results found in irradiation and high field Fowler-Nordheim injection experiments.

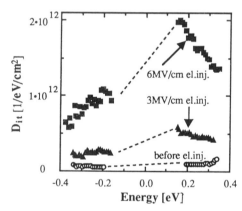

Fig.6: *Distribution of interface states (D_{it}) as a function of energy in the Si-bandgap before electron injection and after electron injection at oxide fields of 3MV/cm and 6MV/cm (new sample). In both cases around $3 \cdot 10^{19}$ electrons/cm^2 were injected.*

REFERENCES

DiMaria D J 1987 Appl. Phys. Lett. **51** 655
DiMaria D J and Stasiak J W 1989 J. Appl. Phys. **65** 2342
Groeseneken G, Maes H E, Beltran N and De Keersmaecker R F 1985 IEEE Trans. Electron Devices **ED-32** 375
Heyns M M, Krishna Rao D and DeKeersmaecker R F 1989 Appl. Surf. Sci. **39** 327
Ning T H and Hu H N 1974 J. Appl. Phys. **45** 5373
Zekeryia V and Ma T P 1983 Appl. Phys. Lett. **43** 95

The role of the two-step process in hot-carrier degradation

Martin Brox and Werner Weber
Siemens AG, Corporate Research and Development, ZFE ME MS34,
Otto Hahn Ring 6, 8000 München 83, Germany

Abstract. Interface trap generation by hot holes during hot-carrier stressing of Si-MOS transistors is studied with the aim of differentiating effects of the hole injection from effects of trapping and electron-trapped hole recombination. Using a dynamic stressing sequence which allows hole trapping and injection to be controlled independently, we find interface trap generation to correlate well with the injection, but poorly with the trapping. Thus hole trapping and electron-hole recombination is ruled out as the rate-limiting step in interface trap generation by hole injection.

Interface trap generation by hole injection is of major interest for the degradation of SiO_2 films on silicon. One of the models presently under consideration is known as the two-step process (Lai 1981), where interface traps are believed to be generated when holes are first trapped in the vicinity of the Si/SiO_2 interface and subsequently recombine with injected electrons. In the static hot-carrier degradation of MOS-transistors the two-step process can hardly be distinguished from interface trap generation by hole injection without trapping at deep trapping sites (e.g. Heremans 1988), as the capture of holes cannot be controlled independently of the injection and a sufficient number of hot electrons is usually available to complete the recombination step. In this work we introduce and analyze a carrier injection method for n-MOSFETs which allows for an independent tuning of hole injection and trapping, thus making it possible to clarify the relative role of both components.

1. The injection sequence

In the injection sequence, we utilize the low mobility and, according to the Einstein relation, small diffusion constant for holes in the oxide after small polaron formation.

Figure 1 outlines the time dependence of the stressing pulses, the injection current and the oxide electric field above the junction edge of the drain. At the beginning of each cycle the gate voltage is kept around threshold, while the drain voltage is ramped up to a large value so that holes are injected at the drain edge. In less than a picosecond holes in the oxide form small polarons with a room-temperature mobility μ of about $2 \cdot 10^{-5} cm^2/Vs$ (Hughes 1978) and hence a drift velocity of only $0.2 nm/ns$ in a field of $1 MV/cm$. If the injection time is chosen in the low nanosecond range, therefore, injected holes are unable to traverse the oxide during the injection pulse, but stay in the vicinity of the interface. When the drain voltage is ramped down, further hole injection is stopped and the drift velocity of the holes in the oxide drops to nearly zero due to the low oxide electric field. In this phase, holes in the oxide are mainly moving by diffusion, eventually being trapped if they get into the vicinity of a deep trapping site. As holes in the oxide perform a random walk (Hughes 1978) the time constant of deep trapping can be estimated

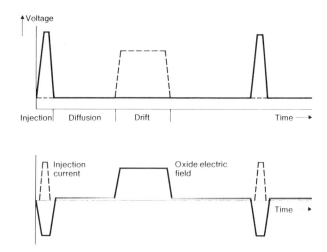

Figure 1: Outline of the pulse sequence as used in the experiments and the corresponding time dependence of injection currrent and oxide electric field.

to be in the range of $10 \cdots 100 ns$ using typical values for the hole trap density, capture cross section and diffusion constant. As the last step in the sequence, the gate voltage is ramped up so that holes which are not yet trapped are driven out of the oxide by the positive oxide electric field.

Using the injection sequence as shown in Figure 1, it is thus possible to inject holes and independently control the trapping by varying parameters of the diffusion or the drift phase. This obviously implies the possibility to tune the number of electron-hole recombination events which is of basic interest in the two-step process. As we cannot determine this number experimentally, electrical characterization allows to access the number of *permanently* trapped holes only, a theoretical analysis of hole trapping and recombination during the dynamic stressing was performed to confirm this implication.

2. Analysis and experimental test of the injection sequence

To be able to analyze the dependence of the hole trapping on the various parameters of the pulse sequence, several physical effects have to be taken into account.

To estimate the injection currents j_h and j_e of holes and electrons, respectively, the lucky electron model is applied, using standard values from the literature for the mean free paths of the carriers. As the electric field is repelling for hot electrons during the injection phase, we assumed them to require an additional energy of $qE_{ox}z$ in excess of the barrier to penetrate by z into the oxide so that j_e decreases exponentially with the distance to the interface. Because of the opposing field, we use a low-field capture cross section of $\sigma_{rec} = 5 \cdot 10^{-13} cm^2$ for the electron-hole recombination. Use of the high field cross section yielded a significant overestimation of the permanent hole trapping as observed experimentally.

For the small polaron formation of the injected holes, in this first approach we use an exponential law with a mean free path $\lambda_{polaron}$. After polaron formation, the holes are assumed to follow a drift-diffusion equation where deep trapping was accounted for by the introduction of a trapping time constant τ_{trap}. Additionally, both mobile and trapped holes can recombine with electrons tunneling from the silicon valence band. For mobile holes the respective time constant $\tau_{p,tunnel}$ was calculated using a trap depth of $0.2 eV$ for the polarons in the theory of Lundström

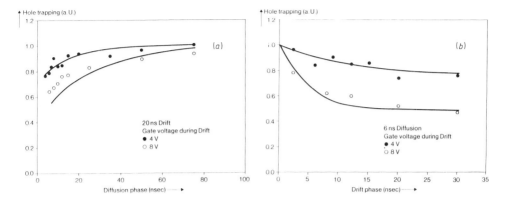

Figure 2: Dependence of permanent hole trapping on the length of the diffusion (a) and the drift phase (b). The trapped positive charge was characterized by the DC substrate current decrease at low drain voltage. The local flatband voltage shift was estimated to be less than 100mV at the drain edge. Solid lines are calculated using the theory as discussed in the text. Total stressing time was 100sec at a frequency of 5MHz and an injection phase of 1.5ns at 8V drain voltage and gate voltage equal to threshold.

(1972), whereas for trapped holes $\tau_{h,tunnel}$ was inferred by comparison with static hole detrapping experiments (Brox 1990).

Considering that in the experiments the trapped hole density is small compared with the hole trap density, these assumption yield the following for the two properties of interest, the density of mobile holes $p(z,t)$ and of trapped holes $h(z,t)$:

$$\frac{\partial p}{\partial t} = D\frac{\partial^2 p}{\partial z^2} - \mu E_{ox}\frac{\partial p}{\partial z} - \frac{p}{\tau_{trap}} - \frac{p}{\tau_{p,tunnel}} + \frac{j_h}{\lambda_{polaron}}\exp\left(-\frac{z}{\lambda_{polaron}}\right) \quad (1)$$

$$\frac{dh}{dt} = \frac{p}{\tau_{trap}} - \frac{h}{\tau_{h,tunnel}} - j_e\sigma_{rec}h \quad (2)$$

Due to the greatly differing time scales, ranging from nanoseconds as typical cycle- or injection times to seconds as typical stressing times, a complete numerical treatment of (1) and (2) is not reasonable. However, assuming that the mobile hole density in the oxide is periodic with the cycle time T_{cycle} and T_{cycle} is small compared with the stressing time T_{stress} allows the trapped hole density to be expressed as a sum over the 4 phases of the injection sequence:

$$h(t,z) = \sum_{4\,phases} \frac{T_{phase}}{T_{cycle}}\left(1 - \exp\left(-t\left(j_e\sigma_{rec} + \frac{1}{\tau_{h,tunnel}}\right)_{phase}\right)\right)\int_{phase}dt'\frac{p(t')}{\tau_{trap}} \quad (3)$$

so that a numerical integration of (1) over one cycle suffices.

As an experimental test, hole-trapping experiments using the injection sequence as given in Fig. 1 were performed on conventional n-MOSFETs with a gate oxide thickness of $42nm$ and an effective channel length of $1.7\mu m$. Results for the dependence of the permanent hole trapping on the length of the diffusion and the drift phase are given in Figure 2, where the model calculations have been included as solid lines. As fitting parameter we used $\lambda_{polaron} = 2nm$ and $\tau_{trap} = 20ns$, so that the trapping time constant is well within the expected region.

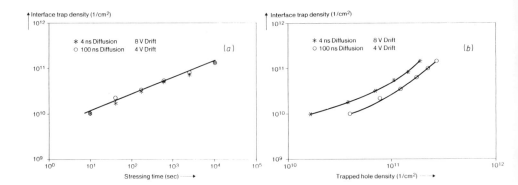

Figure 3: Dependence of the density of interface traps on stressing time (a) and number of permanently trapped holes (b) for two samples stressed with identical injection but different diffusion and drift. The length of each injection phase was 3ns at a drain voltage of 9V and gate voltage equal to threshold. Cycle time was 150ns.

3. Application to the two step process

Using the theory as discussed above, it was confirmed that for a given number of injection pulses the non-measurable number of electron-hole recombination events is approximately proportional to the measurable number of permanently trapped holes. Thus, we are able to apply our method to the question of the two-step process in hot-carrier degradation.

Two devices were stressed using identical injection conditions but pulse sequences which differ in diffusion and drift phase, so that a different number of holes is trapped and hence a *different* number of electron-hole recombination events occur for an *identical* number of injected holes. Although the number of recombination events is thus different, Figure 3a shows a unique power law between interface trap generation and number of injected holes being proportional to the stressing time. On the other hand, we do not observe a unique relation between interface trap generation and hole trapping (Fig. 3b).

The fact that only the two measurements in Fig. 3a correlate well allows us to conclude that interface trap generation by hole injection is triggered by the injection process alone. The two-step process involving trapping at deep trapping sites cannot impose a rate-limiting step. A possible explanation for the much higher interface trap generation efficiency of hot holes as compared to hot electrons might be the stronger coupling to the lattice as evidenced by the small polaron formation.

References

M. Brox and W. Weber, Proc. 20th ESSDERC, Nottingham, 295(1990)
P. Heremans et al., IEEE Trans. Elec. Dev. ED-35, 2194(1988)
R. C. Hughes, Phys. Rev. B15, 251(1978)
S. K. Lai, Appl. Phys. Lett. 38, 58(1981)
I. Lundström and C. Svensson, Journ. Appl. Phys. 43, 5045(1972)

Paper presented at INFOS '91, Liverpool, April 1991
Contributed Papers, Section 7

AC hot carrier degradation behaviour in n- and p-channel MOSFETs and in CMOS invertors

R Bellens, P Heremans, G Groeseneken, HE Maes
Imec, Kapeldreef 75, B-3001 Leuven

ABSTRACT: In this paper the AC hot carrier degradation behaviour of MOSFET's will be discussed. NMOSFET's in general proved to behave quasi-statically. In pMOSFET's however, detrapping of trapped electrons can result in a frequency dependent interface trap generation. The combination of lower net electron trapping and higher interface trap generation can lead to a different degradation behaviour in p-channel devices under certain AC operation conditions compared to static degradation.

1. INTRODUCTION

Since many years, hot carrier degradation in MOSFET devices has been studied extensively. While for degradation under DC conditions the experimental observations of the different research groups agree and the physical explanation for the behaviour can be classified into a few categories, this is certainly not the case for the degradation under AC stress conditions. It has recently been shown by Bellens *et al.* (1990) and confirmed by Izawa *et al.* (1990) that many of the reported enhanced degradation effects are caused by shortcomings of the measurement set-up. In this paper AC degradation behaviour of n- and p-channel transistors, stressed separately as well as in invertor circuits, whereby all necessary precautions were taken, will be discussed. The charge pumping technique, as described by Heremans *et al* (1989), was used as a tool for analysis. It will be demonstrated that the AC degradation of n-channel transistors in general behaves quasi-statically, but pMOSFET AC degradation is different from that induced during DC stressing. This deviating behaviour will show to be important in invertor operation.

2. AC DEGRADATION BEHAVIOUR OF NMOS TRANSISTORS

On n-channel transistors, experiments with pulsed gate and constant or pulsed drain voltage are carried out. Fig. 1 shows the increase of the maximum charge pumping current (ΔI_{cp}), which is proportional to the number of fast interface traps generated during stress, as a function of frequency for a stress with constant drain voltage and with the gate pulsed between 0 and V_d. Rise and fall times of the gate pulse are each 10% of the period and the duty cycle is 50%. This means that pulses with a constant shape are applied such that, after a fixed stress time, the total time that the gate is biased at any voltage between 0 and V_d is equal for all frequencies. The rise and fall times become smaller with increasing frequency. The total stress time for each stress condition is 10000s. Since on Fig. 1 ΔI_{cp} is found to be independent on frequency, any effects due to the edges of the gate pulse can therefore be excluded.

On Fig.2, results of n-channel degradations for both DC stress and AC stress using constant pulse shapes or in invertor operation are compared using the duty cycle calculation procedure, suggested by Weber (1988), but using ΔI_{cp} as the degradation monitor. It can

© 1991 IOP Publishing Ltd

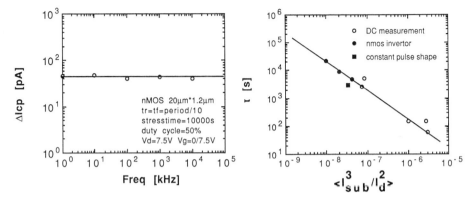

Fig. 1: ΔIcp - Freq for nMOS transistors stressed with constant pulse shape

Fig. 2: Lifetime (ΔI$_{cp}$=20pA) as a function of $<I_{sub}^3/I_d^2>$.

be concluded that for these AC stress conditions, the degradation can be predicted from DC measurements.
Fig. 3 and 4 illustrate the case where both gate and drain voltages are pulsed. Fig. 3 compares the ΔI$_{cp}$ increase during two degradation measurements: the curve labeled 'gate delay'

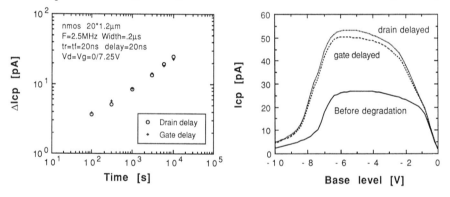

Fig. 3: ΔIcp - time for nMOS transistors with gate and drain pulsed

Fig. 4: Charge pumping curves for nMOS transistors with gate and drain pulsed

shows the case when the gate pulse is delayed 20ns with respect to the drain pulse, the curve 'drain delay' illustrates the case with the drain pulse delayed to the gate pulse. No difference in interface trap generation can be observed between both conditions. Fig. 4 shows that there is also no difference in trapped charge between both conditions. Therefore both stress conditions result in the same degradation, which confirms the quasi-static behaviour of the degradation.

3. AC DEGRADATION BEHAVIOUR OF PMOS TRANSISTORS

For p-channel devices "constant pulse shape" experiments, similar to the case of n-channel transistors, show a different frequency dependence, as illustrated on Fig. 5. ΔI$_{cp}$ decreases with increasing frequency. This behaviour can be explained by detrapping of the negative charge, trapped during the rising and falling edges of the gate pulses, an effect which has been confirmed by Weber et al (1991). The detrapping occurs during the phases of the

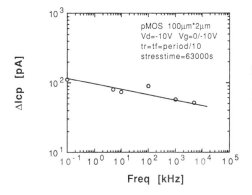
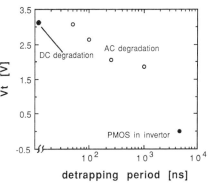

Fig. 5: ΔIcp - Freq for pMOS transistors stressed with constant pulse shape

Fig. 6a: Threshold voltage of degraded region, obtained from charge pumping measurements, for different detrapping periods

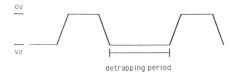

Fig. 6b: Gate voltage waveform with detrapping period

pulse where $V_g=0$ and $V_g=V_d$. In order to demonstrate this effect, the following experiment was set up: Gate pulses, as shown on Fig. 6, with $t_r=t_f=10$ns, $t_{(Vg=0V)}=50$ns and different detrapping periods at $V_g=V_d$ are applied under constant drain voltage. The total stress time in this case is adjusted to obtain equal ΔI_{cp}. Fig. 6 shows the resulting local threshold voltage of the degraded region, as extracted from charge pumping curves, see Heremans et al (1989). With increasing time of $V_d=V_g$, the local threshold voltage decreases, indicating more detrapping. For comparison, the local V_t after a DC stress for equal ΔI_{cp} is shown too, and is in the line with what is expected for zero detrapping time.

In the case of Fig. 5, a larger electron detrapping time (i.e. lower frequencies) results in a lateral field that is larger on the average and thus leads to a larger ΔI_{cp} after a fixed stress time. This detrapping also results in a time behaviour different from that during DC degradation. On Fig. 7, the curve labeled 'DC', results from a DC stress at maximum degradation. The slope of the curve on a log-log scale is 0.2 and a saturation effect is observed. The curve labeled 'AC' is measured for an AC stress condition with a wait time that is sufficient to detrap the negative charge. For AC stress, the slope is 0.45 and no saturation is observed. It is clear from Fig. 7 that because of the electron detrapping, extrapolation of DC degradation measurements towards AC stress conditions is not allowed.

4. DEGRADATION BEHAVIOUR OF INVERTOR CIRCUITS

This last section deals with the degradation of n- and p-channel transistors in invertor circuits. Fig. 8 shows the ΔI_{cp} degradation of the n- and p-channel transistor in an invertor as a function of stress time. The n-channel transistor behaves quasi-statically and its degradation can be extrapolated based on DC measurements, in agreement with the results of section 2. For p-channel transistors in the invertor, the detrapping time (now $V_g=0V$) is very long as compared to the effective stress time which is restricted to the time during switching. The local V_t of the degraded region for a pMOS used in invertor operation has

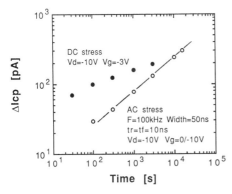
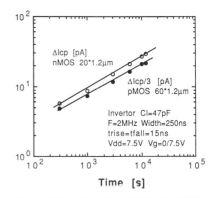

Fig. 7: ΔIcp - time for pMOS transistors stressed under DC and AC stress conditions

Fig. 8: ΔIcp - time for nMOS and pMOS transistors stressed in invertor operation

been added to Fig. 6. Almost no electron trapping is observed. Consequently, if we compare the time dependence of the generation of interface traps, the slope of the curve for the p-channel transistors on Fig. 8 is equal to the slope of the curve for the AC case on Fig. 7 and is equal to the slope of the curve for the n-channel transistor on Fig. 8.

5. CONCLUSIONS

In conclusion we can state that the AC degradation of n-channel transistors behaves quasi-statically and can be predicted based on DC measurements for the described pulses and for invertor operation. For the AC degradation of the p-channel transistors the detrapping phenomenon is very important and DC based extrapolations lead to erroneous results. The interface trap generation in p-channel transistors becomes equally important as that in n-channel transistors and the time dependence of the interface trap generation strongly depends on the detrapping of electrons.

ACKNOWLEDGEMENT

Many of the results presented in this work were obtained on devices which we obtained from Alcatel-Bell Telephone within project IWONL 89.021(I)-I352/22. We acknowledge Alcatel-Bell for receiving these devices and more in particular ir. J. Vandenbroeck for the fruitful collaboration in this field.

REFERENCES

Bellens R, Heremans P, Groeseneken G, Maes H E 1990 IEEE TED **37** 1 pp 310
Heremans P, Witters J, Groeseneken G, Maes H E 1989 IEEE TED **36** 1 pp 1318
Izawa R, Umeda K, Takeda E 1990 Tech. Dig. IEDM pp 573
Weber W 1988 IEEE TED **35** 6 pp 1476
Weber W, Brox M, Bellens R, Heremans P, Groeseneken G, v Schwerin A, Maes H E
    1991 to be published by Van Nostrand, ed CT Wang

Paper presented at INFOS '91, Liverpool, April 1991
Contributed Papers, Section 7

Correlation of generated electron traps in degraded silicon dioxide

J.F.Zhang, S.Taylor and W.Eccleston
Department of Electrical Engineering & Electronics, University of Liverpool, Brownlow Hill, P.O.Box 147, Liverpool, L69 3BX, U.K.

Abstract It has been found that the behaviour of generated electron traps under irradiation, avalanche hole injection(AHI) and Fowler-Nordheim(FN) stress is similar. The interaction of free holes with the oxide is responsible for the acceptor-like trap generation, while hole trapping produces donor-like traps in the oxide under these three different stress conditions. The nature of the generated traps is not sensitive to the electrical field strength during stressing.

1. Introduction

New electron trapping sites can be generated in SiO_2 under irradiation (Aitken and Young 1976, Sah et.al 1983), AHI (Lai 1983, Ogawa et.al 1990) and FN (Uchida and Ajioka 1987) stress. The electrical field strength is normally kept relatively low (e.g., 1.8MV/cm, Aitken and Young 1976) during irradiation. Under AHI, the average electrical field is less than 6MV/cm. For FN stress, the electrical field is usually over 6MV/cm. Although there is such a large difference in the electrical field under different stress conditions, it is found that the behaviour of the generated traps is similar. The objective of this paper is to find the essential conditions for trap generation and to suggest the generation mechanism by comparing the traps generated under irradiation, AHI and FN stress. The data of trap generation under FN stress are obtained from our own experiment described below , while those under irradiation and AHI are taken from literature.

2. Trap generation under FN stress

2.1 Experimental

Thermally grown 48nm oxide at 960^0C on boron doped silicon $(0.17\sim 0.23\Omega cm)$ was made into capacitors with filament evaporated aluminium dots. FN injection was performed for variable times and the traps were then filled with electrons by avalanching the silicon surface. The midgap voltage, V_{mg}, was monitored to assess trapped charges. V_{mg} on unstressed capacitors (curve B, fig.1) were subtracted from those of stressed capacitors (curve A, fig.1) to yield the effect of the new traps (curve C, fig.1).

Both FN and avalanche electron injection(AEI) were carried out at room temperature. The effects of the anomalous positive charge, fast interface state generation and the partial filling of the newly created traps during FN stress on the mid-gap voltage shift, ΔV_{mg}, are controlled to be within 27%, which is considered as acceptable for present objective.

2.2 Results

Using the analytical procedure developed by Ning (1978) and Young(1980), we can extract the effective density and the capture cross section of the generated traps under FN stress from curve C in fig.1. Care has been exercised to ensure the extracted parameters are unique and justification for the used curve fitting procedure is given elsewhere (Zhang et.al, 1991) and will not be repeated here.

Four electron capture cross sections were unambiguously detected for the generated traps. One of these is greater than $10^{-14} cm^2$, whose precise value is not known. The other three are

© 1991 IOP Publishing Ltd

5×10^{-16}, 8×10^{-17} and $1.5 \times 10^{-18} cm^2$, which we label N16, N17, N18 respectively, with the unresolved N14. The effective densities corresponding to each group of generated traps are shown in Table.1.

Table 1 The effect of annealing at room temperature on the generated traps by FN stress at 8.61MV/cm for $0.1C/cm^2$.

q/cm^2	N_p	N14	N16	N17	N18	N_n
A	6.32×10^{12}	3.84×10^{12}	1.5×10^{12}	9.7×10^{11}	1.1×10^{12}	8.1×10^{11}
B	4.37×10^{12}	2.43×10^{12}	1.16×10^{12}	Negligible	1.1×10^{12}	Negligible
C	69%	63%	77%	Negligible	100%	Negligible

A = Without annealing; B = Electron traps annealed out;
C = Percentage of traps annealed out.

The capture cross sections of the generated traps have been found to be insensitive to either the stressing electrical field strength or the electronic charge density injected under FN stress, Q_{FN}. As Q_{FN} increases, only the effective density of N17 increases continuously, while those of N14, N16 and N18 saturate at higher Q_{FN}.

To determine the electrical status of the generated traps, we examine the annealing behaviour of each group of traps at room temperature. One capacitor was stressed and left for annealing with gate floating. The annealing of the positive charges in the oxide can be observed from the positive shift of the capacitance verse voltage characteristics with annealing time, as shown in fig.2. It is believed that electrons from silicon substrate reach the positive charge in the oxide by tunnelling (Lakshmanna and Vengurlekar 1988). The characteristic time for such a tunnelling process increases exponentially with the tunnelling distance, which explains the large time dispersion observed in fig.2.

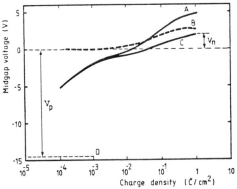

Fig.1 The results of avalanche electron injection for a MOSC after FN stress (8.61MV/cm, $0.1C/cm^2$). Level D is V_{mg} after FN and before AEI stress.

After 988 hours annealing at room temperature, the stressed capacitor was then subjected to AEI. Compared with the capacitor without such an annealing, the differences in the mid-gap voltage shift is shown in fig.3, which corresponds to the traps annealed out. By analysing fig.3, the annealing of each group of traps is summarised in Table.1. It was found that the generated N18 disappeared, N14 and N16 reduced, accompanied by a reduction in the positive charge in the oxide, N_p (corresponding to V_p in fig.1). However, neither N17 nor the net negative charge, N_n (corresponding to V_n in fig.1) was changed by this prolonged annealing. Furthermore, the density of N17 is approximately the same as that of N_n. Thus, we conclude that N17 is the only generated acceptor-like (neutral) electron trap and N14, N16 and N18 are the donor-like (positive charge related) traps.

3. Correlation of the generated traps under different stressing conditions

3.1 The acceptor-like traps

The creation of acceptor-like traps in the oxide has been found not only during FN stress, but also in other degradation test. Ogawa et.al(1990) reported that acceptor-like traps with capture cross section, σ, in the order of $10^{-17} \sim 10^{-16} cm^2$ were generated during AHI stress. Creation of acceptor-like trap with similar σ was also observed by Sah et.al(1983) when the oxide was exposed to 8KeV electron beam. These values of σ agree well with our observation under FN stress. Furthermore, Ogawa et.al(1990) observed that as the stressing time increased, the positive charge saturated, while the generation of the acceptor-like trap continued. This phenomenon was also observed by us under FN stress. These similarities in the behaviour of the acceptor-like traps indicate that the generation mechanism for the acceptor-like trap could be the same under different stressing conditions.

The difference in the electrical field strength, F, under irradiation, AHI and FN stress can be as large as 8MV/cm. The fact that similar acceptor-like traps can be generated over such a large range of F indicates that high electrical field is not a necessary condition for the generation of the acceptor-like traps.

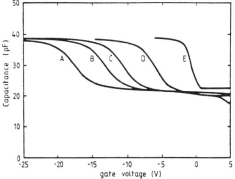

Fig.2 The capacitance verse voltage at 1MHz. A--Immediately after FN stress (8.61MV/cm, 0.1C/cm^2); B--12 hours after A; C--105 hours after A; D--988 hours after A; E--Before FN stress.

The common feature for AHI, irradiation and FN stress is the presence of holes in the oxide. Under AHI and irradiation stress, holes are injected into and generated within the oxide, respectively. For FN stress, the energetic electrons entering the anode create surface plasmons. These surface plasmons rapidly decay into electron-hole pairs and some of the holes are then injected into the oxide (Fischetti et.al 1985). Heyns et.al (1989) observed that there was a threshold in the electrical field strength around 4MV/cm for generating electron traps, which was attributed to the hole injection from the anode. Thus, the presence of holes in the oxide could be responsible for the generation of the acceptor-like traps.

After holes were injected into the oxide, it was suggested that hole trapping and subsequent recombination with the electrons caused damage to the oxide (Chen et.al 1987). However, as mentioned earlier, under both AHI (Ogawa et.al 1990)

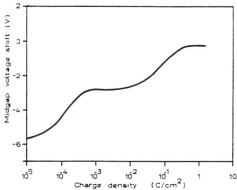

Fig.3 ΔV_{mg} during AEI due to the annealing at room temperature for 988 hours.

and FN stress conditions, it has been found that the saturation of the positive charge does not prevent further generation of the acceptor-like traps. Thus, it is concluded that the generation of acceptor-like traps under different stressing conditions is caused by the interaction of free holes with the oxide, rather than by hole trapping and annihilation.

3.2 The donor-like traps

It is well known that positive charges formed in the oxide can act as coulombic attractive electron trapping centres with $\sigma > 10^{-14}$cm^2 (Aitken and Young 1976, Ning 1978). However, it was reported for both AHI (Lai 1983) and irradiation (Aitken and Young 1976) stress only part of the positive charge could be neutralised by the injected electrons with $\sigma > 10^{-14}$cm^2. Aitken and Young (1976) observed that the σ of the generated traps by irradiation can spread from 10^{-18} to greater than 10^{-14}cm^2. We found that σ of the generated donor-like traps under FN stress covered the same range as that for irradiation. Furthermore, the saturation of positive charges has been observed under both AHI and FN stressing. It has also been shown that the saturation value of the positive charge in the oxide is independent of the electrical field strength (Heyns and DeKeersmaecker 1988). Thus, the saturation density of the positive charge in the oxide is fixed by the fabrication process of the sample. We therefore conclude that the donor-like traps generated under different stress conditions have the same origin.

A multiplicity of mechanisms have been proposed for the formation of the positive charges in the oxide. These include hole injection from the anode (Fischetti et.al 1985), hydrogen migration (DiMaria and Stasiak 1989), impact ionisation at (Nissan-Cohen et.al 1985), or tunnelling of electron from (Breed 1974) defects in the oxide. Tunnelling of electron from the trapping sites can be ruled out since a high electrical field alone can not generate positive charges in the oxide (Fischetti et.al 1985). Impact ionisation is unlikely since positive charge can be formed near SiO_2/Si interface even silicon substrate is cathode in FN stress. Hydrogen migration under electrical stress has been observed experimentally (Gale et.al 1983) and is generally believed to be responsible for the interface state generation after irradiation (Saks and Brown 1989). However, it is not clear how hydrogen-species react with SiO_2 to form positive charge. Although the hydrogen migration model cannot be ruled out yet, we believe that hole trapping is a more likely process for the positive charge formation. The vast differences in σ for the generated donor like traps indicate that there certainly exist more than one type of hole trapping sites in the oxide. The oxygen vacancy, strained Si-O bond and Si-H bond have been suggested to be the candidates for hole trapping sites.

4. Conclusion

It has been shown that the electrical status of the generated electron traps under FN stress can be both positive and neutral. The generated donor-like traps can be annealed out at room temperature, but the generated acceptor-like traps are stable. The capture cross section of the generated acceptor-like trap is around $8 \times 10^{-17} cm^2$, while they spread from 10^{-18} to greater than $10^{-14} cm^2$ for donor-like traps.

A comparison of the generated traps under irradiation, AHI and FN stress indicate that their behaviour is similar. The nature of the generated traps is insensitive to the electrical field strength during stressing. The saturation density of the donor-like traps is fixed by the sample fabrication process, but no up-limit is observed for the acceptor-like trap generation. The trap generation mechanism under irradiation, AHI and FN stress is believed to be the same: the interaction of free holes with the oxide for acceptor-like trap generation and hole trapping for the donor-like trap generation.

References

Aitken J M and Young D R 1976 J.Appl.Phys. 47 1196
Breed D J 1974 Solid-State Electronics 17 1229
Chen I C, Holland S, and Hu C 1987 J.Appl.Phys. 61 4544
DiMaria D J and Stasiak J W 1989 J.Appl.Phys. 65 2342
Fischetti M V, Weinberg Z A, and Calise J A 1985 J.Appl.Phys. 57 418
Gale R, Feigl F J, Magee C W, and Young D R 1983 J.Appl.Phys. 54 6938
Heyns M M and DeKeersmaecker R F 1988 Mater.Res.Soc.Symp.Proc. 105 205
Heyns M M, Krishna Rao D, and DeKeersmaecker R F 1989 Appl.Surface Sci. 39 327
Lai S K 1983 J.Appl.Phys. 54 2540
Lakshmanna V and Vengurlekar A S 1988 J.Appl.Phys. 63 4548
Ning T H 1978 J.Appl.Phys. 49 4077
Nissan-Cohen Y, Shappir J, and Frohman-Bentchkowsky D 1985 J.Appl.Phys. 58 2252
Ogawa S, Shiono N, and Shimaya M 1990 Appl.Phys.Lett. 56 1329
Sah C T, Sun J Y C, and Tzcu J J T 1983 J.Appl.Phys. 54 4378
Saks N S and Brown D B 1989 IEEE Trans. Nuclear Sci. 36 1848
Uchida H and Ajioka T 1987 Appl.Phys.Lett. 51 433
Young D R 1980 Inst.Phys.Conf.Ser. 50 INFOS-80 28
Zhang J F, Taylor S and Eccleston W 1991 to be published.

Paper presented at INFOS '91, Liverpool, April 1991
Contributed Papers, Section 7

Defects at oxidised <100> silicon surfaces induced by high electric field stress from low (100 K) to high (450 K) temperatures and relation with the trivalent silicon defect

D. VUILLAUME[a], R. BOUCHAKOUR, M. JOURDAIN[b], G. SALACE, A. EL-HDIY[c]

(a) Laboratoire d'Etude des Surfaces et Interfaces, URA 253 CNRS,
    ISEN, 41 Bd. Vauban, 59046 Lille cedex, France.
(b) LAM, U. de Reims, BP 347, 51062 REIMS cedex, France.
(c) LASSI, U. de Reims, BP 347, 51062 Reims cedex, France.
    (R. Bouchakour is now with ENST, Paris, France).

ABSTRACT : The interface states produced at the Si-SiO$_2$ interface by high electric field electron injection in the SiO$_2$ are studied for stress temperatures ranging from 100 K to 450 K. At stress temperatures above 180 K, our results are consistent with the hydrogen-related diffusion model, while below 180 K, they seem consistent with the trapped-hole model. We have determined a diffusion coefficient of the hydrogen-related species in the temperature range 200-250 K. A carefull analysis of the electronic properties of these defects is performed and compared with those of the trivalent silicon defect at the Si-SiO$_2$ interface (P$_b$ center).

1. INTRODUCTION

The electronically active defects generated in MOS devices are known to reduce the lifetime of VLSI circuits. As the channel lengths, gate oxide thicknesses and voltages are not reduced accordingly, the electronic properties of the generated defects have gained a great interest these last years. However, the mechanism for the creation of interface states has not been fully understood. Two main mechanisms involve the role of interfacial trapped-holes (Hu et al 1980, Wang et al 1988) or the diffusion of hydrogen-related species (Sah et al 1983, 1984, DiMaria 1987, DiMaria et al 1989) as the precursors for the interface state creation. Also the exact atomic nature of the generated defects is not fully determined. Several work have pointed-out the possible relation with the trivalent silicon defect (P$_b$ center) at the Si-SiO$_2$ interface (Warren at al 1986, Miki et al 1988, Gerardi et al 1990, Vuillaume et al 1991) but no definitive picture emerges at the moment.

This paper presents data on the temperature dependence of the generation of fast interface states by high electric field stress (HEFS) and discuss them as regards the two main models cited above. Also, the possible nature of these interface defects will be discussed in the light of the P$_b$ center properties.

2. EXPERIMENTS

The n-type MOS capacitors (MOSC) were fabricated on <100> oriented Si surfaces by dry oxidation at 1050°C resulting in an oxide thickness of

© 1991 IOP Publishing Ltd

750 Å. Oxidation was followed by an anneal in N_2+H_2 at 400 °C during 1 h. Al-gate was evaporated and followed by an anneal in forming gas. The HEFS were performed by Fowler-Nordheim tunneling (FNT) injection of electrons from the Si substrate (positive gate bias). The FNT stresses were carried out in the constant voltage mode with the average oxide fields of 8.3-8.6 MV/cm. The interface state densities are measured by Deep Level Transient Spectroscopy (DLTS), high-low frequency capacitance-voltage method (HLFCV) and admittance spectroscopy (Gω) after the samples have been warmed-up at room temperature, or by the low-temperature (100 K) high-frequency CV method of Jenq (1977) to avoid the extra-generation of interface states during the warm-up (Hu et al 1980).

3. RESULTS AND DISCUSSION

Figure 1 shows the Arrhenius plot of the increase of the fast interface states ΔD_{it} versus the inverse of the stress temperature (T_s) measured by DLTS at several energies below the conduction band. The DLTS measurements were performed after the samples have been warmed-up and stored at room temperature during few months in order to have obtained a stabilized fast state density resulting from the extra-creation during the warm-up (Hu et al 1980) The DLTS measurements were reperformed after one year storage at room temperature and have given the same results. ΔD_{it} was also measured at low temperature by the Jenq method immediately after the

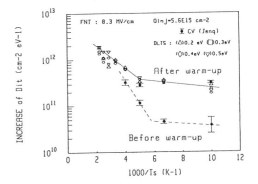

Fig. 1. Arrhenius plot of the creation of fast interface states with and without the warm-up effect

HEFS and before any warm-up. For the ΔD_{it} measured after stress and warm-up, we clearly observed two regimes. At $T_s>250$ K, the generation is thermally activated with an activation energy E_a=40-60 meV, while the generation mechanism is not thermally activated below T_s=250 K (E_a is lower than kT/q). This activation energy is independent on the energy level of the fast states in the band-gap and independent on the electron injected fluences. Similar E_a have been measured from HLFCV and Gω results. When the warm-up effect is avoided, we have observed a weak creation below T_s=180 K , D_{it} is increased by almost a factor 2. A new feature appears for the HEFS performed at medium temperature range, 180 K<T_s<300 K, where the interface state creation seems to follow a thermally activated law with $E_a \approx 0.1$ eV. These results extends the earlier results only obtained at low temperatures (T_s<150 K) and room temperature.

This warm-up effect has been ascribed to a transformation process from interfacial positive charges (supposed to be trapped holes) which turn into interface states during the warm-up (Hu et al 1980, Wang at al 1988). Also, hydrogen-related model has been proposed (Sah et al 1983,1984, DiMaria 1987, DiMaria et al 1989) in which mobil hydrogen-related species released by hot-electrons pill-up at the $Si-SiO_2$ interface and create interface states. In the present case, we have failed to detect any significant interfacial positive charges after the stress from the Jenq CV measurements. Only a weak positive charge has been generated $\Delta N_+ = 1.3-4.8 \times 10^{10}$ cm^{-2} in our n-type MOSC stressed from 100 to 450 K. This is one decade lower than ΔD_{it} resulting from the warm-up (fig. 1) and no correlation was found between ΔN_+ and ΔD_{it}. We conclude that the trapped holes model does not satisfactorily explain our results. We cannot infer that such a mechanism has not occured during our HEFS experiments but it is not the dominant mechanism.

Our results seems consistent with the hydrogen-related species diffusion model. From experimental results reported in fig. 1, we are able to determine the value of the diffusion coefficient of the hydrogen-related species responsible for the interface state creation. At $T_s > 300$ K, the diffusion proceeds very fast and the creation of interface states is limited by the rate at which hydrogen-species are released by the hot-electrons at the anode and the rate at which defects are created by these species pilling up at the Si-SiO$_2$ interface. This gives the exponential law $k_0 \exp(-qE_a/kT)$ with the determined E_a value of ≈ 50 meV. At intermediate temperature (180 K$< T_s <$300 K), the diffusion limits the number of hydrogen-related species pilling-up at the interface and ΔD_{it} deviates from the high-temperature law by a factor $k_1 = \mathrm{erfc}\left(t_{ox}/2\sqrt{Dt_s}\right)$ where t_{ox} is the oxide thickness, t_s the stress time and D the diffusion coefficient. From the results in fig. 1, we estimate the k_1 values from the ratio of ΔD_{it} measured in the range 180-300 K and the extrapolated ΔD_{it} from the exponential high-temperature law which correspond to ΔD_{it} measured after the warm-up of the samples. We have obtained $D = (9.3 \pm 0.5) \times 10^{-15}$ cm^{2}s^{-1} and $D = (2.4 \pm 0.2) \times 10^{-14}$ cm^{2}s^{-1} at 200 and 250 K, respectively. These values are reported in figure 2 and compared with typical diffusion coefficient for hydrogen in SiO$_2$ (Rigo 1986). We observe a good agreement with the extrapolated data below room temperature. This supports the hydrogen-related species diffusion model to explain the generation interface states in our MOSC submitted to HEFS experiments.

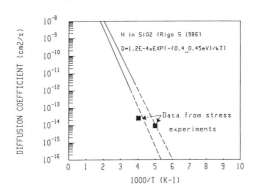

At low temperature ($T_s <$180 K), the creation of interface states is not thermally activated and a tunneling mechanism should be responsible for the weak creation before any warm-up of the samples. The energy integrated number of interface states measured by the Jenq method immediately after the

Fig. 2. Diffusion coefficient data of hydrogen-related species and hydrogen in SiO$_2$

stress is of the same order of magnitude than the number of interfacial positive charges observed simultaneously. We can speculate that capture of electrons accumulated at the Si-SiO$_2$ interface by interfacial trapped holes located at a tunneling distance (≈ 20 Å) from the interface (Chang et al 1986) should be the mechanism responsible for this interface state generation as suggested by the two steps model (Lai 1981).

We have reported elsewhere a carefull comparison of the electronic properties (energetical distribution of levels in the band gap, capture cross-sections and annealing properties) of the HEFS-induced defects at room temperature and up-dated data for the P_{b1} and P_{b0} center at the <100>Si-SiO$_2$ interface (Vuillaume et al 1990, 1991). We have suggested that HEFS-induced defects should be a nonisolated P_h center, i.e. a P_b-like defect [P_b-X] with an unknown species X weakly interacting with the Si dangling bond in the case of our dry oxide MOSC stressed under positive gate biases. Preliminary results on the capture cross-sections of the interface states created at different temperatures will be now given (figure 3). Capture cross-sections have been measured by Gω for states located at 0.27 eV below the conduction band. A constant value of about $\sigma_n = (2\pm 1) \times 10^{-16}$ cm^2 was found in agreement with our earlier results measured by Energy-Resolved DLTS (Vuillaume et al 1989, 1990, Goguenheim et al 1990). This should suggest that defects generated at

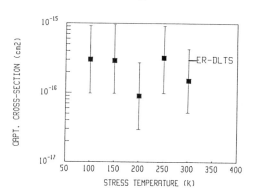

Fig. 3. Capture cross-sections of interface states created by HEFS at different temperatures

different temperatures by HEFS probably have the same microscopic nature as we have suggested above. More strengthen results are in progress.

Similar measurements have also been performed on p-type MOSC submitted to HEFS by FNT electrons injection from the gate and we have basically obtained similar conclusions (Vuillaume et al unpublished).

ACKNOWLEDGEMENTS : This work was supported by the French "GCIS".

Chang S T and Lyon S A 1986 *Appl. Phys. Lett.* 48 136
DiMaria D J 1987 *Appl. Phys. Lett.* 51 1431
DiMaria D J and Stasiak J W 1989 *J. Appl. Phys.* 65 2342
Gerardi G J, Poindexter E H, Caplan P J, Harmartz M, Buchwald W R and Johnson N M 1990 *J. Electrochem. Soc.* 136 2609
Goguenheim D, Vuillaume D, Vincent G and Johnson N M 1990 *J. Appl. Phys.* 68 1104
Hu G and Johnson W C 1980 *Appl. Phys. Lett.* 36 590
Jenq C S 1977 PhD Thesis, Princeton University, unpublished.
Lai S K 1981 *Appl. Phys. Lett.* 39 58
Miki H, Noguchi M, Yogokawa K, Kim B, Asada K and Sugano T 1988 *IEEE Trans. Electron Devices* 35 2245
Rigo S 1986 "Instabilities in silicon devices" ed G Barbottin and A Vapaille (Amsterdam: North-Holland) pp 31-2
Sah C T, Sun J Y C, Tsou J J 1983 *J. Appl. Phys.* 54 5864
Sah C T, Sun J Y C, Tsou J J 1984 *J. Appl. Phys.* 55 1525
Vuillaume D, Bouchakour R, Jourdain M and Bourgoin J C 1989 *Appl. Phys. Lett.* 52 153
Vuillaume D, Goguenheim D and Vincent G 1990 *Appl. Phys. lett.* 57 1206
Vuillaume D, Goguenheim D and Bourgoin J C 1991 *Appl. Phys. Lett.* 58 490
Vuillaume D, Bouchakour R, Jourdain M, Salace G and El-Hdiy A submitted to *J. Appl. Phys.*
Wang S J, Sung J M and Lyon S A 1988 *Appl. Phys. Lett.* 52 1431
Warren W L and Lenahan P M 1986 *Appl. Phys. Lett.* 49 1296

Paper presented at INFOS '91, Liverpool, April 1991
Contributed Papers, Section 7

Homogeneous hole injection into gate oxide layers of MOSFETs: injection efficiency, hole trapping and Si/SiO$_2$ interface state generation

A. v. Schwerin[*] and M. M. Heyns

Interuniversity Microelectronics Centre (IMEC), Kapeldreef 75, B-3001 Leuven, BELGIUM

ABSTRACT: Homogeneous injection of holes into the gate oxide layer of p-channel MOSFETs is performed using an experimental technique in analogy to the well known substrate hot electron injection. A hole gate current is measured during injection which is in good agreement with a proposed simple lucky hole model. This injection technique is used to study the oxide field dependence of hole trapping in the SiO$_2$ bulk and of Si/SiO$_2$ interface state generation. No evidence is found for hole trap generation by hole injection. Hole trapping is found to be slightly decreasing with increasing oxide field (E_{ox}). An effective capture cross section of about $3 \cdot 10^{-14}$ cm^2 is extracted. The efficiency for direct interface state generation is about 1 state / 100 injected holes at $E_{ox}=2$MV/cm and decreases with increasing oxide field approximately like $1/(E_{ox})^2$. After hole injection a delayed build-up of interface states, without saturation for up to 10^6s, is observed.

1. INTRODUCTION

Holes, injected into the gate oxide layer of MOS devices, are known to cause severe reliability problems, as they degrade gate oxide properties. Holes are trapped efficiently in the oxide and, in addition, cause the build-up of Si/SiO$_2$-interface states. None of the methods, which are commonly used to intentionally introduce holes into SiO$_2$ layers, permits to inject holes at well controlled dc oxide fields (E_{ox}) without any simultaneous injection of electrons. Therefore, either it is not clear wether the recombination of holes with electrons plays an important role in the oxide degradation process, like in irradiation experiments, or the dependence of the degradation processes on E_{ox} is experimentally not accessible, like in avalanche injection experiments. In this work an experimental technique is presented where holes are injected from the silicon into the gate oxide layer, uniformly over the channel area of a MOSFET, under well controlled oxide field conditions. The simultaneous presence of electrons in the oxide during hole injection is completely excluded. Thus, this technique is very well suited for a systematic investigation of gate oxide degradation due to holes.

2. EXPERIMENTAL DETAILS

Holes are injected from the silicon substrate uniformly over the gate area into the gate oxide layer of MOS-transistors by using the analogon of the well known substrate hot electron injection technique (Ning and Hu 1974). Holes, generated by illumination in the n-type Si-substrate of p-channel MOSFETs, are accelerated towards the Si/SiO$_2$ interface by applying a reverse bias to the substrate, while source and drain are grounded. As long as the channel is kept in inversion the oxide field is well

---

[*]now with: Siemens, Corporate Research and Development, Otto-Hahn-Ring 6, D-8000 München 83, Germany

controlled by the gate voltage. This technique permits to introduce holes into the SiO$_2$ layer without any simultaneous injection of electrons under perfect control of E$_{ox}$.

The gate oxide of the samples was grown at 900°C in dry O$_2$ under addition of HCl to a thickness of 20nm. The n-type channel doping of the n$^+$-polySi gated transistors was about $3 \cdot 10^{17}$ cm^{-3}, the p-type compensation doping, usually necessary for the threshold voltage adjust, was omitted to provide for a high transverse field next to the oxide interface.

In between short hole injection steps, where the gate current is measured using an integrating electrometer, the oxide degradation is characterized. The trapped oxide charge is obtained from the gate voltage shift in deep subthreshold at a fixed drain current and low drain voltage. The interface state density is measured using the charge pumping technique (Groeseneken et al 1985). The injection probability is determined by dividing the gate current through the sum of source and drain current, which are measured simultaneously during injection. The resulting value is an upper limit to the actual injection probability.

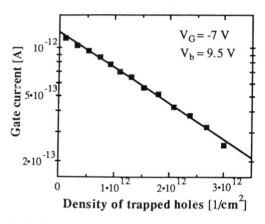

Fig.1: *Gate current, measured during hole injection, as a function of the density of trapped holes, as measured in between subsequent injection steps.*

3. HOLE INJECTION EFFICIENCY

Figure 1 shows the gate current which is measured during hole injection, plotted as a function of the trapped positive charge, as extracted from I$_D$/V$_G$-measurements in between two hole injection steps, respectively. The measurement was done at a gate voltage (V$_G$) of -7V and a substrate bias (V$_b$) of 9.5V using a MOSFET with a gate area of 40x40µm^2. Because of very efficient trapping of holes in SiO$_2$ the gate current change due to the trapped positive charge cannot be neglected after a few seconds of hole injection. Therefore the measurements shown in figure 2 and 3 of the gate and substrate voltage dependence of the hole injection probability, where done on a new device for every voltage condition, respectively, and plots like the one in figure 1 were used to extrapolate the measured gate current to 0 trapped charge.

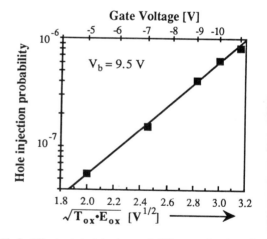

Fig.2: *Effective hole injection probability as a function of the gate voltage.*

In the following it is demonstrated that the results presented in figures 1 to 3 are consistent with a simple lucky hole model, that we suggest here in analogy to the lucky electron model of Ning et al (1977). If tunneling of holes through the barrier edge is neglected, the injection probability is given by P$_{inj}$=A·exp(-d/λ), where A is a constant of the order of 1 and λ is the inelastic mean free path for holes in Si. The injection length d is the distance from the Si/SiO$_2$ interface at which the electrostatic potential Ψ in the Si is equal to Ψ at the interface plus the effective barrier height Φ_b: $\Psi(d)-\Psi(0)=\Phi_b/q$. Φ_b is given by the valence band offset between Si and SiO$_2$ minus the Schottky barrier lowering $\beta(E_{ox})^{1/2}$. Where β is equal to $2.59 \cdot 10^{-4}$ e(Vcm)$^{1/2}$ (Ning et al 1977). Therefore, plotting P_{inj} on a logarithmic scale as a function of $(E_{ox})^{1/2}$ should result in a straight line (compare figure 2). The trapped positive charge (areal density Q_{ot}) decreases the oxide field at the injecting interface by Q_{ot}/ε_{ox}, therefore $(E_{ox})^{1/2}$ at the interface is in good approximation given by $(E_{ox})^{1/2} \approx (E_{av})^{1/2} - Q_{ot}/(2\varepsilon_{ox}E_{av})$, where E_{av} is the average oxide field. Therefore, the gate current plotted on a log scale as a function of the trapped charge as in figure 1

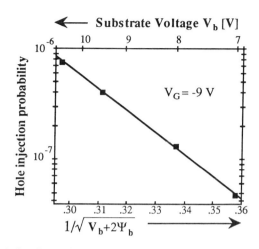

Fig.3: *Effective hole injection probability as a function of the substrate voltage.*

shows a straight line. Finally, the electric field in the Si near the interface is defined by the substrate bias. It can be shown that as long as the substrate voltage V_b is considerably larger than Φ_b/q, the injection length d will vary in first approximation linearly with $1/(V_b+2\Psi_b)^{1/2}$, where $(V_b+2\Psi_b)$ is the band bending from the Si bulk to the oxide interface. Accordingly, $\log(P_{inj})$ plotted as in figure 3 is expected to show a straight line. Assuming a homogeneous channel doping of $3 \cdot 10^{17}$cm^{-3} and a zero field barrier height of 4.8eV, an inelastic mean free path for holes in Si of about 6.5nm can be extracted from figure 3.

As full consistency is found between measurement results and the predictions of our simple model, we are confident that the current measured in our experiment is due to hole injection from the Si substrate without any parasitic electron contributions. This is verified independently by the experimental fact that no gate current at all is measured when either the light is switched off or when the substrate bias is set to zero. Therefore, this technique is ideally suited for the homogeneous injection of holes into gate oxide layers and thus for the investigation of the oxide degradation mechanisms due to injected holes.

4. HOLE TRAPPING AND INTERFACE STATE GENERATION

This injection technique was used to study the oxide field dependence of hole trapping and of Si/SiO$_2$ interface state generation du-

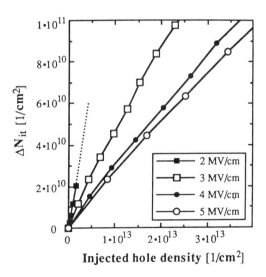

Fig.4: *Increase in Si/SiO$_2$ interface states (ΔN_{it}) as a function of the effective density of injected holes for different oxide fields between 2 MV/cm and 5 MV/cm*

ring hole injection. As shown in a recent work (Schwerin et al 1990), the trapping efficiency is found to decrease slightly with increasing oxide field, whereas the same trapping saturation level is reached, independently of the oxide field. An effective capture cross section of the dominant hole trap could be extracted. A value between $4 \cdot 10^{-14}$ cm^2 (at about 2MV/cm) and $2.7 \cdot 10^{-14}$ cm^2 (at 5MV/cm) was found.

Figure 4 presents the result for the Si/SiO$_2$ interface state (N_{it}) generation due to hole injection for E_{ox} between 2 and 5MV/cm. Here, ΔN_{it} was measured immediately after the hole injection steps, respectively. From the starting slope of the curves in figure 4 the direct interface state generation rate can be extracted. It is found to be decreasing with increasing oxide field approximately like $1/(E_{ox})^2$, with a generation efficiency of about 1 interface state per 100 injected holes at E_{ox}=2MV/cm.

After the end of hole injection, an additional delayed build-up of interface states is observed, together with the simultaneous detrapping of positive charge. However, as illustrated in figure 5, no one-to-one-correlation of hole detrapping and interface state build-up is found. The time dependence of the trapped charge decrease after hole injection is consistent with the assumption of direct tunneling of charge carriers between Si and SiO$_2$. Consistently, the hole detrapping depends on *magnitude* and *polarity* of the applied gate voltage, whereas the simultaneous build-up of interface states is apparently only dependent on field *polarity*. The missing one-to-one correlation is in contrast to predictions of the trapped-hole-to-N_{it} transformation model of Wang et al (1988). The missing dependence of the N_{it} build-up on the field magnitude is in contrast to the H$^+$ model of McLean (1980), which is quite successful in predicting the result of the delayed interface state build-up after irradiation of MOS-devices.

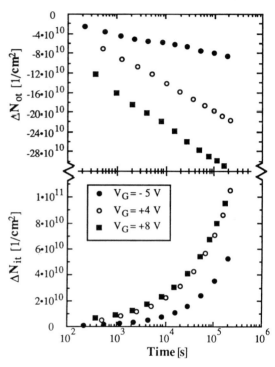

Fig.5: *Change in trapped holes desity (ΔN_{ot}) and interface state density (ΔN_{it}) as a function of time after hole injection. A constant voltage was applied to the gate after hole injection. Please, note the different scales for detrapping and interface state build-up.*

REFERENCES:

Groeseneken G, Maes H E, Beltran N and De Keersmaecker R F 1985 IEEE Trans. Electron Devices **ED-32** 375
McLean F B 1980 IEEE Trans. Nucl. Sci. **NS-27** 1651
Ning T H and Hu H N 1974 J. Appl. Phys. **45** 5373
Ning T H, Osburn C M and Yu H N 1977 J. Appl. Phys. **48** 285
Schwerin A v, Heyns M M and Weber W 1990 J. Appl. Phys **67** 7595
Wang S J, Sung J M and Lyon S A 1988 Appl. Phys. Lett. **52** 1431

Paper presented at INFOS '91, Liverpool, April 1991
Contributed Papers, Section 7

Fast and slow interface state changes in Fowler–Nordheim stressed capacitors

M J Uren

Royal Signals and Radar Establishment, St Andrew's Road, Great Malvern, Worcs. WR14 3PS, UK.

ABSTRACT: Fast and slow Si/SiO$_2$ interface state distributions have been measured before and after Fowler-Nordheim stress to a thermal gate oxide. The slow state distribution is changed from a rise towards the conduction band edge to show a peak above midgap. After stress, a part of the defect density could be reversibly created or destroyed using a small applied bias.

1. INTRODUCTION

Fowler-Nordheim (FN) conduction of electrons through oxides occurs at high electric fields and is well known to generate large densities of Si/SiO$_2$ interface states as well as trapping and de-trapping charges from the bulk of the oxide in a complex way. This paper will concentrate on the electrical characteristics of the interface states generated by the stress. Most workers have used the high/low frequency capacitance voltage technique (LFHF CV) to measure the interface state density, but this has the disadvantage of being prone to artefacts near the band edges (Miller et al 1990). Since the method does not yield capture cross section information, there is no way to separate the fast states (located at the Si/SiO$_2$ interface) from the slow states (normally assumed to be located in the bulk of the SiO$_2$). Vuillaume et al (1986, 1988) used DLTS to separate the two classes of defect following stress, measured the creation kinetics of the slow states and found different saturation behaviour from the fast states. Here I have used the conductance technique (Nicollian and Brews 1982) to perform the separation (Uren et al 1989). After FN stress, the slow state distribution was found to have changed dramatically and to be different from that of the fast states. It was also found that *reversible* changes to both the fast and slow state distributions were caused by small applied biases.

2. SAMPLES

The experiments reported here were carried out using n-type (100) silicon with a dry thermal oxide grown at 900°C to a thickness of 28nm, followed by a post oxidation anneal in nitrogen. Aluminium gates with guard rings of area 6.97x10^{-3}cm^2 were defined photolithographically. Before the post metallisation anneal in forming gas, the capacitors showed a negative bias stress instability suggesting that some traces of water were present in the growth ambient (Blat et al 1991). However after PMA, the capacitors were quite stable with a respectable interface state density of 2x10^{10}cm^{-2}eV^{-1} at midgap.

The sample discussed in detail here was Fowler-Nordheim stressed with a positive-gate constant-current of 100nA for a total flux of 10^{-2}Ccm^{-2}. It was then left for a month to allow the fast element of the annealing process to occur before detailed measurements were made. Over a period of 6 months, the annealing did not show any strong 'transformation' (Ma 1989) but rather a gradual proportionate reduction in the density of states throughout the bandgap of about 20%.

3. CAPACITANCE VOLTAGE MEASUREMENTS

LFHF CV measurements were carried out using a sequential quasi-static and then 1MHz sweep from inversion to accumulation. The density of interface states was then extracted from the LF data rather than the more normally used LFHF approach which underestimates the density near the bandedge (Miller et al 1989). The band bending was also extracted from the LF CV data, but using the HF CV data at midgap to

© Controller, Her Majesty's Stationery Office, 1991

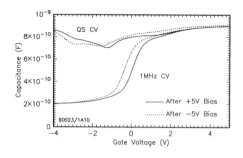
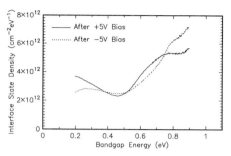

Figure 1.

Figure 2.

fix the integration constant. The use of the midgap point rather than the flatband point ensured that the large number of fast states were all responding slower than the 1MHz frequency used.

A new feature that made this sample interesting is illustrated in figures 1 and 2. The LFHF CV curves could be reversibly cycled between two states by the application of a small positive or negative bias for a period of a day. The negative bias had the effect of returning the curves to approximately the state before any bias was applied. After the removal of the positive bias, the curve relaxed back to correspond roughly to the -5V biased curve over a period of a few days. Figure 1 shows that one effect of the bias switching was to produce a small positive translation of the HF curve relative to the unbiased case indicating about $3.5 \times 10^{11} cm^{-2}$ of electron trapping. Accompanying this shift was a reversible change in the distribution of interface states through the gap. After positive bias, figure 2 shows that this was rather complex with a fall in interface state density near midgap and the appearance of a shoulder (peak?) in the upper and lower halves of the gap after positive bias. 6 months after the FN stress, the reversible component had annealled, representing only about 1/4 of that apparent in figures 1 and 2.

4. CONDUCTANCE MEASUREMENTS

A problem with the interpretation of results such as those shown in figure 2 is that the CV technique simply measures the total number of states responding faster than the equivalent frequency of the quasi-static ramp. This lumps together fast and slow states which can only be separated if some additional spectroscopic information is available. This is important for damaged samples where the relative number of slow states is normally higher following the damage process. Here the two classes of defect were separated using the conductance technique (Nicollian and Brews 1982).

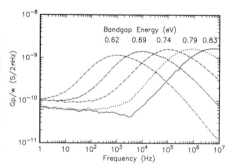

Figure 3.

Figure 3 shows the equivalent parallel conductance of the silicon part of the capacitor, G_p/ω, measured following the positive bias for a range of band bendings in depletion. In depletion, these curves are relatively easy to interpret with the area under the peak giving the density of fast states and the height of the plateau on the low frequency side of the peak giving a measure of the slow state density (Uren et al 1989). Since the slow states have a wide range of capture cross section, their number is measured on the assumption of a constant plateau level (which is only roughly true in practice), so the number of states is quoted as the number per unit area measured in a factor of 10 change in frequency (cross section).

In figure 4, the densities of fast and slow states are shown prior to the FN stress. For both types, the measured density rose towards the band edge in accordance with the standard 'U' shaped distribution for the fast states. Following the stress, large numbers of both types of states were found. Figures 5 and 6 give the separated fast and slow state distributions in the upper half of the gap. The fast states still showed a rise towards the band edge, but in contrast to this the slow states showed a peak near midgap. The fast state capture cross section was also measured and was found to be independent of bandgap energy or bias with a value of about $2 \times 10^{-15} \text{cm}^2$. This is in marked contrast to the results of Vishnubhotla et al (1990) who found a large change for radiation damaged samples. The effect of the bias switching was complex, positive bias increased the number of fast states near midgap, lowered it near the conduction band and lowered the number of slow states.

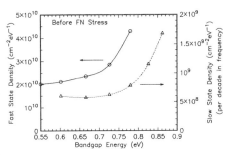

Figure 4.

5. DISCUSSION

Slow states are normally assumed to be distributed into the SiO_2 and to communicate with the bulk silicon by tunnelling. Vuillaume et al (1986) used this model with the assumption of a constant capture cross-section to determine the distribution of slow states as a function of distance into the oxide. However, the evidence from telegraph noise measurements (Kirton and Uren 1989, Cobden et al 1990) showed that the slow states have a wide range of cross-sections due to lattice relaxation. Hence, no attempt was made to use the G_p/ω information to deduce the distance of the defects into the oxide.

Let us obtain a rough estimate of the number of active slow states. If we assume that defects are accessible over a range of perhaps 10 decades in frequency at midgap in the period of a day, then the number would be around $2 \times 10^{12} \text{cm}^{-2} \text{eV}^{-1}$ which is less than, but comparable to the fast state density. Further assuming that the fast states are P_b centres and hence in their neutral state at midgap (Poindexter 1989), then, since the midgap voltage was relatively little changed by the FN stress, the slow states must be mostly in their neutral state. This implies that the slow states generated by the FN stress are acceptor like (0/-).

Turning to the effect of the bias switching, it is apparent that the changes in defect distributions could not have been due to changes in occupation of the slow states. This is because any change in the internal oxide potential would not affect the fast states which are located at the Si/SiO_2 interface. The only way this could happen would be if the fast states were physically collocated with the slow states. It seems more likely that a reversible electrochemical reaction was taking place perhaps in a similar way to that suggested by Blat et al (1991) to explain the negative bias stress instability. For the P_b centre, they picture a reaction of the form:

$$\equiv Si-H + A + p^+ \rightarrow \equiv Si\cdot + B^+ \qquad (1)$$

here p^+ is a hole and A and B are water related species. A similar reaction might be consistent with the results for the slow states where $\equiv Si\cdot$ would correspond to an E' centre. Negative bias would make holes available so that E' centres could form but would have to have their 0/- level at a position that lined up with

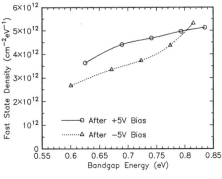

Figure 5.

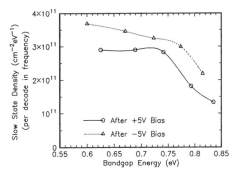

Figure 6.

the silicon midgap energy. This is in contrast to the calculation of Chu and Fowler (1990) which placed the +/0 level in this position. The picture for the fast states is even more complicated and more work would be needed before any speculative model could be attempted.

6. CONCLUSION

It has been shown that the study of the slow states is important for any understanding of the mechanisms of FN damage to the Si/SiO_2 interface. The slow state distribution as a function of silicon bandgap energy was found to be quite different after stress, showing a peak above midgap. Following the stress, an instability was found that produced significant changes to both the fast and slow state distributions following small potential biases.

I would like to thank A M Hodge and the RSRE silicon processing laboratory for the samples used here and M J Kirton for helpful discussions.

7. REFERENCES

C Blat, E H Nicollian and E H Poindexter, J. Appl. Phys. **69** 1712 (1991).
A X Chu and W Beall Fowler, Phys. Rev. **B41** 5061 (1990).
D H Cobden, M J Uren and M J Kirton, Appl. Phys. Lett. **56** 1245 (1990).
M J Kirton and M J Uren, Adv. in Phys. **38** 367 (1989).
T P Ma, Semicond. Sci. Technol. **4** 1061 (1989).
J A Miller, C Blat and E H Nicollian, J. Appl. Phys **66** 716 (1989).
E H Nicollian and J R Brews, MOS Physics and Technology, Wiley 1982.
E H Poindexter, Semicond. Sci. Technol. **4** 961 (1989).
M J Uren, S Collins and M J Kirton, Appl. Phys. Lett. **54** 1448 (1989).
L Vishnubhotla et al, Appl. Phys. Lett. **57** 1778 (1990).
D Vuillaume, J C Bourgoin and M Lannoo, Phys. Rev. **B34** 1171 (1986).
D Vuillaume et al, Mat. Res. Soc. Symp. Proc. **105** 235 (1988).

© British Crown Copyright 1991/MOD, Published with the permission of the Controller of Her Brittanic Majesty's Stationery Office.

Paper presented at INFOS '91, Liverpool, April 1991
Contributed Papers, Section 7

Detrapping of trapped electrons in SiO_2 under Fowler–Nordheim stress

J.F. Zhang, S. Taylor and W. Eccleston.

Department of Electrical Engineering and Electronics,
University of Liverpool, P.O. Box 147, Liverpool, L69 3BX.

Abstract: Detrapping of electrons following avalanche injection is reported. The results elucidate the significant role of detrapping as an instability/degradation mechanism in SiO_2.

1. Introduction

Under FN high field stress, there are at least three important physical processes occurring in the layer: trapping of free charge carriers, trapping site generation and release of carriers from traps i.e. detrapping. The net trapping of electrons (Young et al, 1979) and the trapping site generation (Hsu and Sah, 1988) under FN stress have been intensively investigated. However information on detrapping of trapped carriers is relatively scarce. The objective of this paper is to investigate electron detrapping from the oxide bulk under conditions of constant voltage high field FN stress. Prior to the FN stress schedule the as-grown traps in the oxide were filled using conventional Avalanche Electron Injection (AEI). AEI prior to FN stress greatly enhances the observed detrapping effects during the subsequent stress and facilitates the distinction between trapping and detrapping. A schematic diagram of the experimental method is shown in fig. 1.

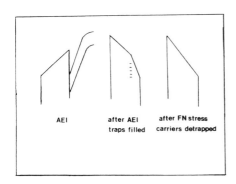

Fig.1 Experimental method for detrapping studies.

2. Experimental

The silicon wafers used in this study were of (100) orientation, boron doped and 0.17-0.23 Ω cm resistivity. The oxide was thermally grown to 46nm in dry O_2 at 900°C. Post oxidation annealing was not performed. MOS capacitors (MOSC) were fabricated by evaporating aluminium through a shadow mask to form 250μm diameter gate electrodes. The sample was then post metal annealed (PMA) at 400°C for 30 minutes in forming gas. The oxide on the back of the wafer was etched off using HF and aluminium was deposited by evaporation for the back-side electrical contact. AEI was performed using a 50 kHz rectangular waveform with the amplitude automatically adjusted to maintain a constant average current density. The midgap voltage shift versus charge injected was sensed and plotted. The effect of anomalous positive charge on the midgap voltage shift, ΔVmg, was kept to a negligible level by performing AEI with the magnitude of the negative going voltage pulse kept at a low level compared with the magnitude of the positive voltage pulse. I-V measurements were performed using an electrometer and voltage source.

To determine the amount of detrapping the variation of FN tunnelling current density, J, with the cumulative injected charge during stress, Qfn, was recorded and a typical plot is shown in fig. 2. Curve A shows J versus Qfn for a MOSC not subjected to AEI prior to stress and curve B shows the curve for a MOS pre-charged by AEI. The difference in the curves clearly shows the detrapping effect. The increase in J with Qfn in curve B is due to enhancement of the electric field near the cathode, ΔF, which occurs as electrons are detrapped. This field enhancement near the cathode may be computed from the measured increase in current density using the equation for FN tunnelling (Lenzlinger and Snow, 1969). As Qfn increases, the enhancement of the electrical field strength near the cathode ΔF (Qfn) relative to its initial value, F(0), is given by:

$$\Delta F(Q_{fn}) = F(Q_{fn}) - F(0) \qquad (1)$$

This is a direct measure of the amount of the electron detrapping through the relation:

$$\Delta F(Q_{fn}) = (q/d\varepsilon_o\varepsilon_r)(N(0)(d-x_c(0)) - N(Q_{fn})(d-x_c(Q_{fn}))) \qquad (2)$$

where N is the instantaneous areal density of trapped electrons in the oxide, N(0) is the value of N before the detrapping process starts, xc the centroid of the trapped charge as measured from the cathode and d is the oxide thickness. The computed ΔF versus Qfn curve is shown plotted in fig. 3. As a check on whether ΔF is proportional to the amount of

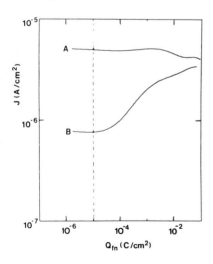 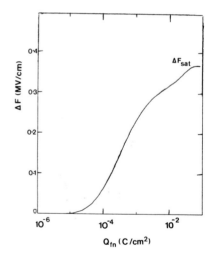

Fig.2 Current density transients during FN stress with an average electric field of 8.44 MV/cm.
Curve A: MOSC not pre-charged
Curve B: MOSC pre-charged by AEI with $Q_{aei} = 0.18$ C/cm^2

Fig.3 Increase in electrical field strength near the cathode during FN stress

Degradation and Trapping in SiO_2

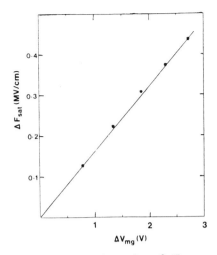

Fig.4 The saturation value of ΔF near the cathode during FN stress at an average electric field of 9.1 MV/cm. ΔVmg is the midgap voltage shift at the end of the AEI step.

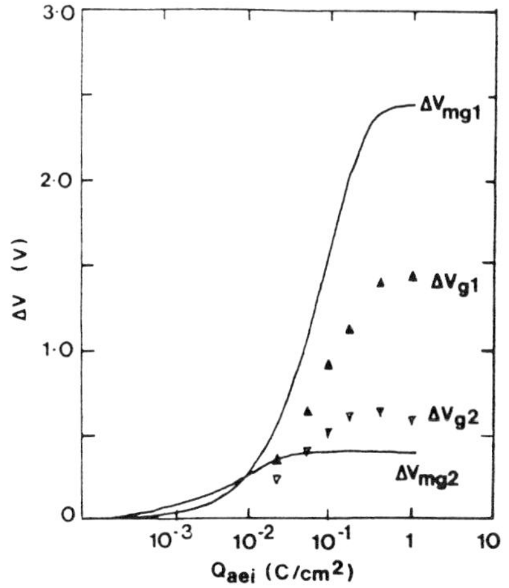

Fig.5 Voltage shift versus injected charge during AEI. $\Delta Vmg1$ and $\Delta Vmg2$ correspond to trapping capture cross sections, $\Delta Vg1$ and $\Delta Vg2$ correspond to detrapping mechanisms.

electrons detrapped, the saturation value of ΔF, is shown plotted against the measured shift in midgap voltage ΔVmg caused by AEI. Fig. 4 clearly shows such a proportionality and we conclude that ΔF can therefore be used to determine the amount of electron detrapping.

After obtaining the ΔF versus Q_{fn} curve, we can evaluate the detrapping efficiency, σ_D (or the characteristic Q_{fn}) for a detrapping mechanism using the analytical procedure similar to that used for the extraction of the capture cross sections (Ning, 1978). The ΔF versus Q_{fn} curve may be fitted to the expression:

$$\Delta F = \sum_{i=1}^{M} \Delta F_i \left(1 - \exp\left(-Q_{fn}/Q_{fni}\right)\right) \qquad (3)$$

where M is the total number of detrapping mechanisms, and $\sigma_{Di} = q/Q_{fni}$.

It is well known that positive charge will be formed both in the bulk of the oxide and near the SiO_2/Si interface under FN stress. Although the positive charge at the SiO_2/Si interface does not affect the electrical field at the Al/SiO_2 interface, an increase of the positive charge in the bulk of oxide can also enhance the electrical conduction current. This is also seen in fig. 2 (curve A), which shows that J increases with Q_{fn} and reaches a maximum around 2×10^{-3} C/cm². The corresponding shift in the electrical field near the cathode is 0.061 MV/cm, which is caused by positive charge formation in the oxide bulk.

For MOSC which were pre-charged with $Q_{aei} > 0.178$ mC/cm² the ΔF corresponding to the detrapping of captured electrons was greater than 0.4 MV/cm. Thus, the contribution to ΔF by positive charge formation is less than 15% of the total ΔF. This contribution can be corrected for by subtracting the shift in the electrical field strength near the cathode due to the positive charge, $\Delta F'p$, from ΔF.

3. Discussion

An examination of the curvature of J versus Q_{fn} in fig. 2 indicates that there is more than one characteristic Q_{fn}, and therefore, σ_D, for the whole detrapping process. This can also be seen from the ΔF versus Qfn curve, as shown in fig. 3. To describe the features of this curve by equation (3), it was found that two detrapping mechanisms (i.e., M=2) should be used. The values of the extracted detrapping efficiencies and their corresponding electrical field strength shifts are listed in Table 1 for an average field strength 8.44 MV/cm.

From AEI results it was found that there are two cross sections for electron capture. Fig. 3 revealed that there are also two mechanisms for the detrapping process. The question naturally arises as to whether there is any connection between a capture cross section and a detrapping efficiency for a given oxide. Fig. 5 plots the gate voltage shift ΔVg corresponding to each detrapping efficiency. Also shown are the midgap voltage shifts corresponding to the capture cross sections. Clearly the curves for the detrapping and capture processes are by no means coincident, and it can be seen that ΔVmg2 (= 10^{-17} cm2) reaches its saturation at a smaller value of Qaei than that obtained for ΔVg1 or ΔVg2. Furthermore, the magnitude of ΔVmg2 is less than either ΔVg1 or ΔVg2. Thus, the amount of trapped electrons in traps with capture cross sections of 10^{-17} cm2 is insufficient to account for ΔVg1 or ΔVg2. Therefore the electrons trapped in sites with capture cross sections of 10^{-18} cm2 contribute to both ΔVg1 and ΔVg2. Thus, within the accuracy and range of the experiments reported here, no correspondence between a detrapping mechanism with a specific efficiency and a trapping mechanism with a particular capture cross section was found. This was also confirmed by tests on MOSC not subjected to a PMA.

TABLE 1

σ_{D1} (cm^2)	ΔF_{sat1} (MV/cm)	σ_{D2} (cm^2)	ΔF_{sat2} (MV/cm)
4.54x10^{-16}	0.228	1.52x10^{-17}	0.147

4. Conclusions

The detrapping of trapped electrons under FN stress is quantitatively investigated in this study. The following conclusions are drawn:
(i) There is more than one mechanism in operation for the detrapping process each with a different reaction rate.
(ii) For the same trapping site, there can exist two detrapping mechanisms. One of them has an efficiency in the order of 5 x 10^{-16} cm2, which agrees with that reported by Nissan-Cohen et al (1985). The other is in the order of 10^{-17} cm2, which has not been reported before.
(iii) For the experimental conditions considered here, the detrapping efficiencies are not sensitive to the capture cross sections of the traps.

References
M. Lenzlinger and E.H. Snow, J.Appl.Phys., 1969, 40 (1), 278.
C.C.H. Hsu and C.T. Sah, Solid-State Electronics, 1988, 31 (6), 1003.
D.R. Young, E.A. Irene, D.J. DiMaria, R.F. DeKeersmaeker and
   H.Z. Massoud, J.Appl.Phys., 1979, 50 (10), 6366.
T.H. Ning, J.Appl.Phys., 1978, 49 (7), 4077
Y. Nissan-Cohen, J.Shappir and D. Frohman-Bentchkowsky, J.Appl.Phys.,
   1985, 58 (6), 2252.

Paper presented at INFOS '91, Liverpool, April 1991
Contributed Papers, Section 7

Characterization of hot-carrier effects in short channel NMOS devices using low frequency noise measurements

M.J. Deen and C. Quon
Engineering Science, Simon Fraser University
Burnaby, British Columbia, Canada V5A 1S6

ABSTRACT: This paper presents results on the use of low frequency noise measurements $S_{VG}(f)$ to characterize hot-carrier stressing effects in short channel NMOS devices. It also presents results of $S_{VG}(f)$ for both the linear and saturation modes of NMOSFET operation after d.c. stress at peak substrate current generation. From experiments, we found that $S_{VG}(f)$ increased with stressing time and that it was larger for linear mode of operation than for the saturation mode. In the saturation mode, $S_{VG}(f)$ was higher for reverse mode than for forward mode of operation. We also found that $S_{VG}(f)$ was predominantly due to charge-carrier number fluctuations, and the 'noise profile' along the device channel indicated that the noise was highest near the source and drain junction edges for virgin devices, but it increased significantly near the drain junction for stressed devices. The noise results are also compared to d.c. results for the stressed devices.

1 INTRODUCTION:

Hot-carrier degradation in short channel MOS devices poses a serious limitation for their continued miniaturization. In NMOS devices, electrons flowing from the source to the drain gain energy from the longitudinal channel electric field and if their energies are large enough, the electrons can be injected into the SiO_2 gate insulator. It is now agreed among many researchers that hot-electron stress in NMOS devices results in both trapped charges and interface states in the gate oxide near the drain. The creation of interface states near the drain has been proven through charge pumping measurements, subthreshold and current-voltage characteristics, and by capacitance-voltage measurements in hot-carrier stressed MOS transistors. The hot-carrier degradations in MOSFETs are typically manifested as changes in the device transconductance, gain, threshold voltage, and increased low frequency noise, and are therefore important for the long term reliability and proper operation of the devices.

Low frequency noise in MOS transistors has been studied extensively in the past, for example, by Backensto and Viswanathan (1980), Fang et al (1986), Kung et al (1990), Park and Van Der Ziel (1982) and Stegherr (1984). Several low frequency noise theories and models have been proposed to explain the nature and behavior of the noise. While many of the theories differ in details and assumptions, most of them attempt to relate the low frequency noise to oxide traps in the $Si-SiO_2$ interface. The most prominent low frequency noise models are the charge-carrier number fluctuation model; the carrier mobility fluctuation model; and the combined carrier number and mobility fluctuation model. For the devices studied in this paper, we found that the charge-carrier fluctuation model is most relevant with the gate referred noise voltage spectral density given by the usual expression

$$S_{VG}(f) = \frac{q^2 kTN_{ss}}{fWLC_{ox}^2 \ln(\tau_2/\tau_1)} \quad (V^2/Hz) \quad \text{where} \quad \ln(\tau_2/\tau_1) \approx 40. \tag{1}$$

© 1991 IOP Publishing Ltd

In this paper, we use low frequency noise measurements to study the relative interface state density characteristics in unstressed and stressed short-channel NMOS devices at room temperature.

2 EXPERIMENTAL DETAILS:

For the experiments, an automated noise profiling system was developed to determine the low frequency voltage noise spectral density of short channel MOS devices as a function of gate and drain voltages. By varying the drain bias at a fixed gate voltage when the device is saturated, different amounts of interface oxide traps are 'sampled'. Consequently, these noise results provide information regarding the spatial distribution of interface states along the transistor channel for both virgin and stressed devices. Experiments were also performed by varying the gate voltage at a fixed drain bias.

The devices used in the experiments have a common width of 10μm, gate oxide thickness of 25nm, and effective channel lengths from 0.7μm to 2.9μm. Low frequency noise measurements were performed in both the forward mode and the reverse for several devices before and after d.c. stress. When the device is in saturation, the pinch-off region near the drain junction edge is depleted and it screens the traps above it. This pinch-off region ΔL_D near the drain junction was varied by changing the drain bias voltage, thus varying the effective device area 'sampled' by the channel inversion layer current for saturated mode of operation. In this way, a 'noise profile' near the junction edges could be made since the area under the gate 'sampled' for each drain bias is given by $W \cdot (L_{eff} - \Delta L_D)$ where the drain pinch-off distance is given by

$$\Delta L_D = \sqrt{\frac{2\varepsilon_{si}}{qN_A}(V_D - V_{D,SAT})} \quad (\mu m) \qquad (3)$$

Here, $V_{D,SAT} \approx (V_{GS} - V_T)/(1+\delta)$, $\delta = \gamma/(2\sqrt{2 \cdot \phi_F})$ and $\gamma = \sqrt{2\varepsilon qN_A}/C_{ox}$. For reverse mode of operation, the 'noise profile' near the source junction edge could similarly be determined with a similar expression for ΔL_S to that given in equation (3) above. After the initial noise and threshold voltage measurement on virgin devices, they were then electrically stressed for up to twenty hours. During the stress, the drain of the devices were fixed at either 7V or 8V, and the gate bias is set to a voltage that yields the maximum substrate current ($I_{SUB,P}$) of ~3.5V and ~4V respectively.

3 RESULTS AND DISCUSSION:

The effect of gate biasing of interface state creation and charge trapping has recently been investigated by Doyle at el (1990) who showed that interface state generation N_{ss} was the dominant degradation mechanism in NMOS devices when stressed at bias voltages corresponding to the peak substrate current generation. Results from our experiments, shown in Figure 1, confirm the conclusion of Doyle et al and we have found a power law relation between ΔV_T and time (t) in seconds given by

$$\Delta V_T \approx A \cdot t^n \qquad (3)$$

where the exponent n varied between 0.55 and 0.7 for our devices for stressing times up to 72000s, and V_D=7V and V_{GS}~3.4V at $I_{SUB,P}$. Here, the threshold voltage was determined by extrapolating the linear I_D-V_{GS} measurement data at the maximum transconductance point.

The best straight line is drawn through the maximum transconductance point in the I_D-V_{GS} curve and the threshold voltage is the intercept on the V_{GS} axis.

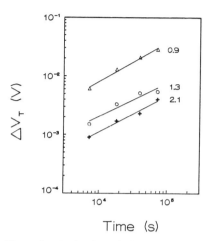

Figure 1: ΔV_T for three devices at 296K. Note the power law relationship between ΔV_T and t and n varied between 0.55 and 0.7.

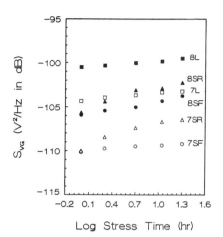

Figure 2: Variation of S_{VG} with *Log Stress Time (hours)* for a 2.1µm device at $V_D = 7V$ and $8V$. S_{VG} is shown for both linear and saturation modes.

Figure 2 shows the variation of S_{VG} at *10Hz* with *Log Stress Time (hours)* for a 2.1µm device in which the gate voltage was adjusted to give the maximum substrate current (peak hot-electron aging biasing condition). This figure shows that while there is a small increase in S_{VG} for the linear mode of operation ($V_D = 0.2V$ and $V_G = 2V$) of ~1dBV, S_{VG} for the saturation reverse mode ($V_D = 3V$ and $V_G = 2V$) increased by ~3.5dBV for both stressing drain voltages, and by ~1dBV for the saturation forward mode of operation. This result clearly indicates that the damage in the stressed device is localized near the drain so that in the forward mode, N_{SS} near the drain is screened by the channel depletion region, resulting in a lower S_{VG} for the forward saturation mode of operation, than for the reverse saturation mode.

An indication of the 'lateral distribution' of the noise voltage can be obtained from the difference between S_{VG} for two

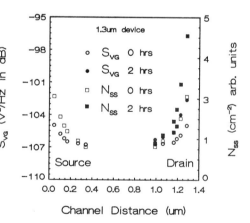

Figure 3: Variation of S_{VG} and N_{SS} with channel distance for a 1.3µm device in which $V_{GS} - V_T = 0.5V$ and V_D was varied from 0.4 to 3V. The measurements were made at *296K* and *10Hz*. This device was stressed for 2 hours. N_{SS} is given in arbitrary units here.

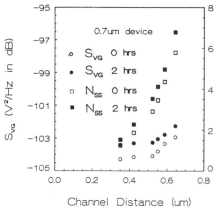

Figure 4: S_{VG} variation with channel position (corresponding to V_D from 0.4 to 3V and $V_{GS} - V_T = 0.5V$ near the drain junction edge for the 0.7μm device at 296K, t=2hrs and 10Hz. Note that N_{SS} is given in arbitrary units here to indicate its variation with channel position.

successive drain biases, that is $S_{VG}(V_{D1}) - S_{VG}(V_{D2})$ is the noise contribution from the region $X_{D2} - X_{D1}$ since the increased drain bias results in the channel 'pinch-off' point moving further away from the drain junction edge. Results using this technique to calculate both S_{VG}, and N_{SS} calculated from equation (1), and are shown in figure 3 for an unstressed and stressed device at 10Hz. Note that both S_{VG} and N_{SS} are largest near the junction edges, indicating that N_{SS} is non-uniformly distributed in the channel. This non-uniform N_{SS} distribution has been confirmed by charge pumping experiments of Li and Deen (1990) for similar devices. Results from stressed devices indicated that both S_{VG} and N_{SS} increased primarily near the drain junction edge, as shown in figure 3 previously.

Figure 4 shows the variation of S_{VG} and N_{SS} with channel position near the drain depletion region before and after 2hours of d.c. stress. The S_{VG} and N_{SS} profiles before stress had similar shapes near the source junction edge, as shown in detail for a longer device in figure 3. Here, the effect of hot-electron stress is to increase N_{SS} near the drain junction edge, but almost no change in N_{SS} was measured near the source junction edge.

4 CONCLUSIONS:

In this paper, we presented results on the use of low frequency noise to characterize hot-electron stressing in NMOS devices. We found that for virgin MOSFETs, the noise has peaks near the source and drain junctions, and that after hot-electron stress, the noise increased mostly near the drain junction. Results from our devices also indicated the noise is predominantly due to charge-carrier fluctuations. Changes in the threshold voltages were also measured and these changes followed a power law dependence on stressing times with the exponent in the 0.55 to 0.7 range.

5 ACKNOWLEDGEMENTS:

We are pleased to acknowledge the assistance of J. Ilowski, R. Hadaway and X.M. Li. This research was supported in part by grants from the Natural Sciences and Engineering Research Council of Canada, Northern Telecom Electronics Ltd, Ottawa, and the Center for Systems Science, Simon Fraser University.

6 REFERENCES:

Backensto W, and Viswanathan C 1980 *IEE Proceedings*, **27**, **Part I**, 237.
Doyle B, Bourcerie M, Marchetaux J-C and Boudou A 1990 *IEEE Trans Elect. Dev.*, **ED-37** 744.
Fang ZH, Cristoloveanu S, and Chovet A 1986 *IEEE Electronics Device Letters*, **EDL-7** 371.
Kung K, Ko P K, Hu C, and Cheng Y C 1990 *IEEE Trans Electron Devices*, **ED-37** 654.
Li X M and Deen M J 1990 *IEDM Technical Digest*, **IEDM-90** 85.
Park H and Van Der Ziel A 1982 *Solid-State Electronics*, **25** 213.
Stegherr M 1984 *Solid-State Electronics*, **27** 1055.

Paper presented at INFOS '91, Liverpool, April 1991
Contributed Papers, Section 7

A search for protons in irradiated MOS oxides

J.T. Krick, J.W. Gabrys, D.I. Semon, and P.M. Lenahan

The Pennsylvania State University, University Park, PA 16802 USA

ABSTRACT: The technique of electron spin resonance (ESR) has been used in an attempt to detect protons in irradiated Si/SiO$_2$ structures. We searched for ESR signals of hydrogen in oxides which had been heavily irradiated at low temperature (T≈210K) and then subjected to electron photoinjection at 77K.

1. INTRODUCTION

In the past twenty years, a great deal of progress has been made in characterizing radiation-induced damage in the metal-oxide-semiconductor (MOS) transistor. More recently, several studies regarding the time, temperature and field dependence of interface state generation following electron beam irradiation[1-4] have been used to explore the kinetics of the radiation damage process. The results of these LINAC studies have been rather convincingly interpreted in terms of a two-step process proposed by McLean[5].

In the McLean model, holes created during irradiation are thought to release positive ions -- almost certainly protons -- which then drift to the interface under the influence of an applied electric field. Upon reaching the interface, the protons are thought to react with hydrogen groups (ie. by breaking Si-H or Si-O-H bonds to form H$_2$ or H$_2$O) and leave behind a silicon dangling bond. These interfacial dangling bonds, known as P$_b$ centers, have been shown using the technique of electron spin resonance (ESR) to be the dominant radiation-induced interface state in MOS devices[6]. In this study, we have used ESR in an attempt to verify an important aspect of the model proposed by McLean, the creation of protons in the oxide.

The idea behind our experiment is relatively straightforward. The radiation-induced interface state buildup is known to last many hours at a relatively low temperature[7]. Thus, if proton drift to the interface is the rate limiting step in this process, it should be possible to 'freeze' the protons in place by maintaining the MOS sample at such a temperature both during and after irradiation. Unfortunately, protons, without an unpaired electron, are undetectable by ESR techniques. However, they could be rendered paramagnetic by capturing an electron to form atomic hydrogen; a species which has a distinctive and easily identifiable ESR lineshape consisting of two very narrow ($\Delta H_{pp} \approx 2G$) lines separated by 503 gauss[8].

2. EXPERIMENTAL DETAILS

The samples used in this study were 0.2mm thick 400 Ωcm <111> oriented p-type silicon wafers polished on both sides. The 1200Å wet oxides examined were grown on both sides of the wafers at 1100°C and subjected to a 1150°C Argon anneal for 90 minutes. A 20 minute anneal in forming gas at 450°C completed the processing sequence. ESR samples of dimensions 0.35x2cm^2 were then cut from the processed wafers. These radiation "soft" oxidizing and annealing conditions were specifically chosen in order to maximize the anticipated hydrogen ESR signal. Furthermore, the use of thin, high resistivity silicon

© 1991 IOP Publishing Ltd

wafers allowed us to 'stack' three slices in the spectrometer in an effort to maximize the sample oxide volume.

The Si/SiO$_2$ samples were irradiated to doses of 15 to 50 Mrad using sequences of 125 Krad LINAC pulses. The temperature of the samples during irradiation (T≈210K) was such that any atomic hydrogen created during irradiation would quickly disappear[8]; at this temperature, however, the interface state buildup is known to last many hours. After irradiation, the samples were then cooled to 100K and exposed to ultraviolet illumination. The UV illumination was used to photoinject electrons into the oxide from the silicon substrate; The electrons would then presumably be captured by protons to form atomic hydrogen. Finally, electron spin measurements were performed on the samples and the resulting ESR traces were averaged for a period of several hours.

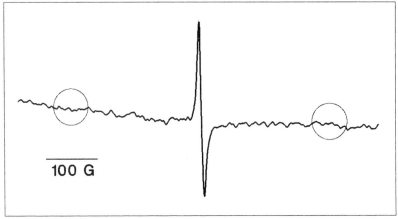

Figure 1: ESR spectrum of a sample which has been LINAC irradiated and exposed to UV illumination at low temperature. The scan width is 700G and is centered on g=2.000. The trace shown is the average of 25 3.2 minute sweeps performed at T=100K. Signals corresponding to atomic hydrogen would appear within the circles shown. However, no atomic hydrogen signal is detected.

The ESR trace shown in figure 1 is typical of the ten measurements performed on Si/SiO$_2$ samples which have undergone pulsed LINAC irradiation and subsequent UV illumination. The trace was performed on three samples irradiated to a total dose of 15 Mrad at a temperature of 213K. The samples were then exposed to intense UV illumination while immersed in liquid nitrogen for a total of 20 minutes (10 minutes on each side). The UV source was a focused 100W mercury-xenon lamp which emitted photons of energy ≤5eV. After transferring the samples to the "cold finger" of an ESR cavity in a test tube filled with liquid nitrogen, electron spin resonance measurements were performed at 100K. The magnetic field sweep width used was 700G and the signal shown is the average of 25 separate 3.2 minute sweeps. Signals corresponding to the presence of atomic hydrogen would appear in the circles shown. However, no atomic hydrogen signal is observed. Comparing this irradiated sample trace to that which we obtained with a calibrated spin standard, we estimate that the total concentration of atomic hydrogen is less than 1×10^{11}/cm^2; we believe that this limit is, in turn, accurate to a factor of two. As a result, we conclude with a high degree of certainty that the density of atomic hydrogen is less than 2×10^{11}/cm^2 in these irradiated samples.

3. DISCUSSION

On the basis of the model proposed by McLean, we had -- perhaps naively -- anticipated a *huge* hydrogen ESR signal. Our oxides were intentionally processed to be radiation "soft."

Since they were subjected to such high radiation doses, we would expect average interface state densities of $\geq 10^{12}/cm^2 eV$. Assuming that one proton is required to create an interface state defect, we would, therefore, expect *at least* the same proton concentration -- though one several times greater seemed quite likely -- in the oxide following irradiation (if protons drift to the interface and interact with an existing imperfection (ie. H-Si≡Si$_3$) to create an interface state, one might guess that the efficiency would be far lower than one defect per proton). Thus, with a spectrometer sensitivity of $\approx 1 \times 10^{11}/cm^2$, we had expected to observe an atomic hydrogen signal.

The success of our experiment, in terms of evaluating the validity of the proton model proposed by McLean, relies on two factors: (1) protons must be present in the oxide following irradiation and (2) photoinjected electrons must be captured by protons which, presumably, have a large coulombic capture cross section. It is also conceivable that errors in temperature control could affect our ability to see atomic hydrogen. However, in two runs we exposed the irradiated oxides to UV illumination only after placing the samples into the chilled cold finger inside the ESR cavity. The outcome was identical in each case -- *no detectable hydrogen signal.*

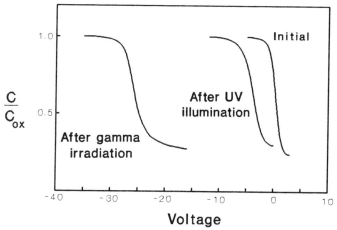

Figure 2: Capacitance-voltage measurements of a gamma irradiated 1000Å oxide which has been exposed to UV illumination while immersed in liquid nitrogen. The illumination scheme is identical to that of the LINAC irradiated samples. Photoinjection of electrons annihilates the majority of the positive oxide charge.

To test the effectiveness of our photoinjection technique, capacitance-voltage (CV) measurements were performed on a 1000Å oxide which had been first gamma irradiated and then exposed to UV illumination while immersed in liquid nitrogen (figure 2). The illumination process used here was identical to that of the LINAC samples shown in figure 1. Exposure to UV illumination clearly annihilates most of the positive charge; electrons are photoinjected into the oxide and are captured by the large capture cross section hole centers. Presumably, protons present in the oxides of samples subjected to LINAC irradiation would have a similarly large capture cross section and would thus be equally effective in trapping injected electrons. Therefore, we provisionally assume that the number of protons present in the oxides of our samples following irradiation is quite small ($\leq 10^{11} cm^2$).

If the proton model is correct, the flaw in our experiment would have to be the lack of an applied electric field during irradiation. It is well known that the amount of radiation-induced interface state buildup is dependent on the magnitude and direction of the applied electric field[2]. If an electric field in the oxide were required for substantial proton release, we would be unable to see atomic hydrogen in our experiment. Secondly, the presence of an electric field would affect the probability of electron capture by positively charged species (holes or protons). At the temperatures used in our irradiation experiments, atomic hydrogen formed by electron trapping on a proton would quickly dissipate from the oxide and thus would not be detected in our ESR measurements.

4. CONCLUSION

We have used the technique of electron spin resonance in an attempt to examine the role of protons in the degradation of irradiated MOS structures. Based on our interpretation of the proton model proposed by McLean, we expected to generate a large atomic hydrogen signal by irradiating MOS oxides and subsequently photoinjecting electrons at low temperatures. We did not. From our results, we conclude that after irradiation in the absence of an applied electric field, the concentration of protons in the oxide must be less than $\approx 1 \times 10^{11} cm^2$. Our findings do not invalidate the proton drift models. However, our results do show that protons are not generated in large numbers in unbiased soft oxides subjected to high levels of irradiation. Nevertheless, further ESR experiments in which an electric field is applied during irradiation would provide a more conclusive test to determine the role of protons in radiation-induced degradation in MOS structures.

REFERENCES

1. P.S. Winokur, H.E. Boesch, J.M. McGarrity and F.B. McLean, *J. Appl. Phys.* **50**, 3492 (1979).
2. Earlier work in this area is summarized by P.S. Winokur in "Radiation Induced Interface Traps" of Ionizing Radiation in MOS Devices, T.P. Ma and P.V. Dressendorfer editors, Wiley Interscience, New York (1989).
3. N.S. Saks and D.B. Brown, *IEEE Trans. Nuc. Sci.* **NS-36**, 1848 (1989).
4. N.S. Saks and D.B. Brown, *IEEE Trans. Nuc. Sci.* **NS-37**, 1624 (1990).
5. F.B. McLean, *IEEE Trans. Nuc. Sci.* **NS-27**, 1651 (1980).
6. P.M. Lenahan and P.V. Dressendorfer, *J. Appl. Phys.* **54**, 1457 (1983) and references contained therein.
7. P.S. Winokur, H.E. Boesch, J.M. McGarrity and F.B. McLean, *IEEE Trans. Nuc. Sci.* **NS-24**, 2113 (1977).
8. T.E. Tsai, D.L. Griscom and E.J. Friebele, *Phys. Rev. B* vol. **40** no. 9, 6374 (1989)

Paper presented at INFOS '91, Liverpool, April 1991
Contributed Papers, Section 7

Prediction of hot electron degradation in MOSFETs: a comparative study of theoretical energy distributions

C.C.C. Leung & P.A. Childs

School of Electronic & Electrical Engineering, University of Birmingham, P.O. Box 363, Birmingham B15 2TT.

ABSTRACT: We compare the hot electron energy distribution functions derived by Keldysh and Ridley with monte carlo simulations and find the Keldysh result to be extremely accurate. We show that the Keldysh theory can be extended to take full account of dead space in the channel of MOSFET.

1. INTRODUCTION

Hot carrier degradation presents long term reliability problems for small geometry MOSFETs. In order to predict the susceptibility of a device to this degradation an accurate description of the hot electron distribution within the channel is required. The field dependence of hot electron distribution functions has been established for many years by the rigorous solution of the Boltzmann Transport Equation (BTE) due to Keldysh (1965). However, it is not obvious that this theory can be applied to MOSFETs where the electric field in the channel is strongly divergent. The relatively new lucky drift theory developed by Ridley (1983) has found wide acceptance precisely because of its adaptability to a range of situations. Both theories are superior to existing lucky electron models and can be adopted for use within a device simulator.

In this paper we compare the theories of Keldysh and Ridley for accuracy and ease of use. We find that although both theories are accurate in predicting the impact ionisation rate their dominant exponent terms differ by a factor 3/2 in the high field limit. We conclude that the normalisation procedure adopted is fundamental to the successful application of the theories within a device simulator. Monte carlo simulations show the exponent term derived by Keldysh to be accurate. Furthermore, the Keldysh result is shown to provide all the information required on dead space.

2. COMPARISON OF DISTRIBUTION FUNCTIONS

In comparing the distribution functions we examine initially the dominant exponential terms and treat other terms as normalising factors. The distribution function derived by Keldysh has the form

$$f_0(E) = \text{const} \cdot \left[\frac{E}{q\varepsilon\lambda}\right]^\nu \exp\left(-\frac{sE}{q\varepsilon\lambda}\right) \qquad (1)$$

© 1991 IOP Publishing Ltd

where ε is the electric field, λ is the mean free path between optical phonon collisions and v is a slowly varying function of field. Neglecting acoustic phonon scattering s is given by

$$\frac{\cosh\left(\frac{\hbar\omega}{2KT}\right)}{\cosh\left(\frac{\hbar\omega}{2KT} - s\frac{\hbar\omega}{q\varepsilon\lambda}\right)} + \frac{1}{2s}\ln\left(\frac{1-s}{1+s}\right) = 0 \qquad (2)$$

In the lucky drift model the distribution can be expressed as

$$f_0(E) = \text{const} \cdot \exp\left(-\frac{E}{q\varepsilon\lambda_E}\right) \qquad (3)$$

where

$$\lambda_E = \frac{q\varepsilon\lambda^2(2n+1)}{2\hbar\omega} \qquad (4)$$

$\hbar\omega$ is the optical phonon energy and n is the Bose-Einstein number. In equation (3) we are assuming that the field is strong enough that transport is dominated by lucky drift and ballistic transport is negligible.

A true comparison of distribution functions can only be made when the assumptions about phonon scattering are the same. We have therefore modified (2) and (4) allowing only emission of optical phonons and neglecting acoustic scattering. Figure 1 shows a comparison of the distribution functions with monte carlo simulations. The agreement between monte carlo simulation and the Keldysh result is excellent. The important difference between lucky drift and the Keldysh theory lies in the exponent terms, $1/\lambda_E$ and s/λ respectively. Figure 2 shows λ_E and λ/s plotted as a function of electric field. These terms diverge strongly in the high field region where the ratio $\lambda_E/(\lambda/s) \to 3/2$. Comparison of these terms with monte carlo simulation shows good agreement with the Keldysh theory throughout the range

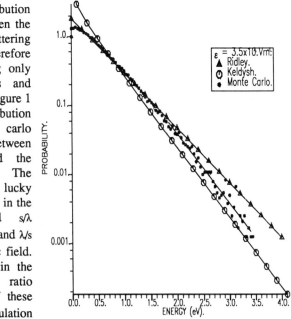

figure 1. Comparison of energy distribution functions

of fields examined.

Despite the large difference in the exponent terms both theories successfully predict the impact ionisation rate in silicon. This results from a difference in the pre-exponential terms which in both theories are slowly varying functions of energy. Hence, although agreement can be achieved at the threshold energy for impact ionisation, lucky drift would overestimate the probability of electrons acquiring higher energies and consequently would overestimate the gate current in a MOSFET.

3. DEAD SPACE

The distribution functions described in the previous section are applicable when the electric field is uniform. Childs (1987) showed that lucky drift theory can be extended to include divergent fields providing the rate of variation is not too rapid. The physical arguments used in his analysis are equally applicable to the Keldysh theory. Hence the hot electron energy distribution in the channel of a MOSFET can reasonably be described by a function of the form

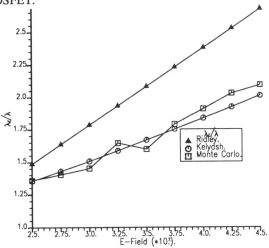

figure 2 Variation of λ_E and λ/s with electric field

$$f_0(E) = \text{const} \cdot \exp\left[\int_0^E f(\varepsilon)dE'\right] \qquad (5)$$

where $f(\varepsilon)$ depends upon the theory used and ε is function of position within the channel.

However, it is not obvious over which path in the channel the integral in (5) should be taken. Figure 3 illustrates this point. Path (a) is applicable to lucky electron and lucky drift models where collisions with phonons are assumed to be absent or elastic. Path (b) is a path that an electron might actually travel in arriving at energy E. Clearly, $f_0(E)$ will give different values according to which path is used.

We have shown (Childs and Leung, 1991) that the optimum path (b) can be derived directly from the Keldysh theory. In our analysis the mean energy gained between collisions is given by

$$\langle E-E'\rangle = \left[\frac{\frac{2}{(1-s^2)} - \frac{1}{s}\ln\left(\frac{1+s}{1-s}\right)}{\ln\left(\frac{1+s}{1-s}\right)}\right] q\varepsilon\lambda \qquad (6)$$

Indeed path (b) in figure 3 has been derived using (6). It is expected that this result will provide a more accurate prediction of substrate and gate currents than can be achieved at present, particularly at low voltages.

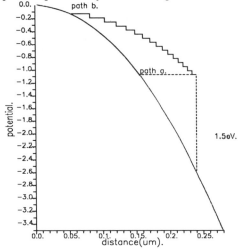

figure 3 Effect of dead space on the evaluation of $f_0(E)$

4. CONCLUSIONS

We have compared the hot electron distribution functions derived by Keldysh and Ridley with a view to application within a device simulator. Monte carlo simulations have shown the Keldysh theory to be accurate over a wide range of fields. Lucky drift theory is very flexible and physically instructive but we expect it would overestimate the number of carriers in the high energy tail and, therefore, the gate current in a MOSFET.

Both theories can be extended to include the divergent electric fields found in the channel of a MOSFET. However, as the Keldysh theory includes energy loss due to phonon emission we have shown that it can take full account of the dead space within the channel. We would expect models that do not include the full dead space to overestimate the number of electrons in the high energy tail, particularly at low voltages.

5. ACKNOWLEDGMENTS

The authors would like to thank the Science and Engineering Research Council for funding this work.

6. REFERENCES

Keldysh L V   1965 Soviet Physics-JETP **21** 1135
Ridley B K 1983 J Phys. C:Solid State Phys. **16** 3373
Childs P A 1987 J Physics C: Solid State Phys. **20** L243
Childs P A, Leung C C C, to be published

Anneal characteristics of E_1' centres in buried oxide layers and oxide precipitates in silicon

[1]R C Barklie, [1]T J Ennis, [2]K J Reeson and [2]P L F Hemment

[1]Physics Department, Trinity College, Dublin 2, Ireland.
[2]Electronic and Electrical Engineering Dept, University of Surrey, UK.

ABSTRACT: Electron Paramagnetic Resonance (EPR) measurements have been made on samples of (100) silicon wafers implanted at a temperature between 500°C and 620°C with 200 keV $^{16}O^+$ ions in the dose range 0.5 to 2.2 x $10^{18}O^+$ cm^{-2}. Amongst the defects present revealed by EPR are E_1' centres. It is found that samples in the as-implanted state only contain E_1' centres in oxide precipitates but that their precursors remain in the buried oxide layer and can be converted to E_1' centres by small doses of ionising radiation. The anneal behaviour of these centres is both precipitates and buried a-SiO_2 layer is reported.

1. INTRODUCTION

One of the main types of defect in either crystalline or amorphous SiO_2 is the E' centre which may be characterised as an unpaired electron in a dangling orbital of a silicon bonded to just three oxygens (Griscom 1984). Of the several variants the one most widely observed, especially as a result of ion or γ-irradiation, is the E_1' centre (or E_γ' centre as it sometimes called in a-SiO_2); Feigl et al (1974) have shown the E_1' centre in quartz to be a bridging oxygen vacancy with a hole trapped on one of the silicon atoms nearest the vacancy and an unpaired electron on the other, $\equiv Si^+ \ldots \cdot Si \equiv$, which is identical to the model for E_γ' (Griscom 1984). These centres are the main source of trapped positive charge in the thermal oxides of irradiated MOS devices (Lenahan and Dressendorfer 1984).

Although the model for the E_1' centre is well established the same cannot be said of the anneal mechanism(s). Several have been proposed including:

$$Si^+ \ldots \cdot Si + O_2 \rightarrow Si^+ \cdot O\text{-}O\text{-}Si \quad (1)$$
$$Si^+ \ldots \cdot Si + H_2O \rightarrow Si^+ OH\text{-}Si + H° \quad (2)$$
$$Si^+ \ldots \cdot Si + e^- \rightarrow Si \cdot \cdot Si \quad (3)$$

Reaction (2) has been suggested to account for the annealing of E_1' in high-OH silicas (Griscom 1985) whereas the growth of the peroxy radical, $\equiv Si\text{-}O\text{-}O\cdot$, along with the decay at $T \gtrsim 100°C$ of E_1' centres in low-OH silicas (Stapelbroek et al 1979, Devine 1987, Pfeffer 1988) strongly supports reaction (1), which was suggested by Edwards and Fowler (1982). Many authors (for example Oldham et al 1986, Saks et al 1989) have attributed the annealing of trapped positive charge in the oxide of irradiated MOS structures to a process involving the tunnelling of electrons from the silicon, in which case reaction (3) may be the appropriate reaction.

We have investigated the nature, location and anneal behaviour of E' centres in silicon wafers implanted with high doses (~ 10^{18} O$^+$ cm^{-2}) of 200keV oxygen ions. Such samples are unusual in that not only do they contain a buried a-SiO$_2$ layer (if the dose is > 1.4 x 10^{18} O$^+$ cm^{-2}) but they also contain many oxide precipitates. We find that E_1' centres can exist in both the precipitates and the buried oxide layer. However their anneal behaviour in these two locations is different and this provides a clue to the nature of their anneal mechanisms which we discuss.

2. SAMPLES AND TECHNIQUES

Samples were prepared at the University of Surrey by implanting n type (100) silicon wafers with 200 keV ^{16}O$^+$ ions with doses of order 10^{18} O$^+$cm^{-2} at implantation temperatures of about 500-600°C. The samples were heated by the ion beam. It has been found by Hemment et al (1983) that for ions of this energy the critical dose above which a continuous buried layer of a-SiO$_2$ is formed in the as-implanted state is 1.4 x 10^{18} O$^+$cm^{-2}; for a dose of 1.8 x 10^{18} O$^+$cm^{-2} this layer has a thickness of about 0.28 μm. All samples contain oxide precipitates. We do not know their size but in similar samples they have been found (Krause et al 1986) to be mostly less than about 10 nm in diameter; their size and number is greatest where the volume concentration of implanted oxygen is a maximum.

EPR measurements were made at Trinity College at room temperature using 100 kHz field modulation and a microwave frequency of about 9.6 GHz.

3. LOCATION AND ANNEAL BEHAVIOUR

Figure 1 shows the EPR spectrum of a sample implanted with 1.8 x 10^{18} O$^+$ cm^{-2} at an implantation temperature of 560°C; the figure also shows the effect of a further irradiation, at room temperature, with a dose of 7 Mrad of 30 keV electrons. Two features A and C are apparent. The defects responsible for feature A are D centres (i.e. amorphous silicon centres) in the silicon above and below the buried oxide layer and P_b centres (i.e. $Si_3 \equiv Si\cdot$) mostly at the Si/SiO$_2$ interfaces of the oxide precipitates; these are discussed elsewhere (Barklie et al 1986) and are not the concern of this paper. We are

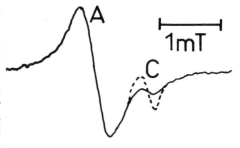

Fig 1: EPR Spectrum of sample implanted with 1.8 x 10^{18} O$^+$ cm^{-2} before (—) and after (- - -) electron irradiation. Magnetic Field //[100]

instead concerned with feature C. It is an isotropic, slightly asymmetric, line with a g value of 2.0003(4) (evaluated at the point where the signal crosses the base line) which saturates easily with increasing microwave power; these characteristics together with its anneal behaviour, discussed later, strongly suggest that it is due to E_1' centres. The only slight asymmetry of the lineshape is characteristic of E_1' centres produced by ion implantation - at least at fairly high doses (Devine and Fiori 1985). The question that arises is where are these defects located. We consider first the E_1' centres present in the as-implanted samples before any further irradiation. It might be supposed that they are in the buried oxide layer but we have shown (Barklie et al, 1988, 1989) that they are still present even when the dose, φ, is reduced below the critical value, φ_c, of 1.4 x 10^{18} O$^+$ cm^{-2} above which a continuous buried oxide layer is formed and this shows that they must exist in the oxide precipitates. Furthermore the E_1' areal concentration, $N(E_1')$ (i.e. population per unit area of implanted surface) depends strongly on dose : for $\varphi < \varphi_c$ $N(E_1')$ increases as φ increases but for $\varphi > \varphi_c$ it decreases as φ increases (Barklie et al 1989). One reason for this might be that the E_1' centres in the as-implanted samples are all or nearly all in the oxide precipitates so that as φ increases above φ_c the reduction in the number of precipitates, as a result of their incorporation into the growing buried oxide layer, causes a reduction in $N(E_1')$. If this is so, then it raises the question of what has happened to the E_1' centres in the buried layer. Devine (1984) and Golanski et al (1984) have shown that E_1' centres produced in thermally grown a-SiO_2 by ion implantation anneal irreversibly for $T > 500°C$ but that the annealing in the range $100 < T < 300°C$ is reversible in the sense that the E_1' centres can be "reactivated" by ionising radiation. It occurred to us therefore that the buried oxide layer might still contain E_1' centre precursors and so we irradiated the sample with an approximate dose of 7 Mrad of 30 keV electrons; figure 1 shows how this changes the EPR spectrum. The intensity of feature C is increased and since this is the only way it is changed we conclude that more E_1' centres have been produced. Furthermore, from measurements with different electron doses, we find that the increase saturates at a dose of about 7 Mrad and this indicates that the new E_1' centres arise from the transformation of precursors. To see whether the precursors are in the buried layer or precipitates or both we measured the increase for samples implanted with doses in the range 0.75 to 2.2×10^{18} O^+ cm^{-2}. An increase only occurs for samples with a buried oxide layer which strongly suggests that the precursors are only in this layer.

The results provide strong evidence that the E_1' centres in the samples in their as-implanted state are only in the oxide precipitates but that precursors of these centres remain in the buried oxide layer. This implies that the anneal characteristics of the centres must be different in these two locations. The E_1' centres in the as-implanted state of samples implanted with 1.4×10^{18} O^+ cm^{-2} using 200 keV O^+ ions and at various implantation temperatures, T_i, in the range 250-600°C all begin to anneal at about 450°C (Barklie et al 1988) whereas those in one implanted with 1.8×10^{18} O cm^{-2} using 400 keV $^{32}O_2^+$ ions at $T_i = 520°C$ begin to anneal at about 350°C (Barklie et al 1986). For samples implanted with 1.8×10^{18} O^+ cm^{-2} using 200 keV $^{16}O^+$ ions at $T_i = 560°C$ we now report that the E_1' centres, in the as-implanted state of the sample, also begin to anneal at about 350°C and that this anneal process is irreversible. For this latter sample, we have performed a series of electron irradiation-anneal-electron irradiation ... steps in which the same sample is annealed for 10 min in air at successively higher temperatures from room temperature up to 700°C and its EPR spectrum recorded at room temperature after each anneal and each irradiation. This series of measurements shows that the "reactivated" E_1' centres in the buried oxide layer anneal reversibly (in the sense previously defined) in the temperature range $100 \lesssim T \lesssim 350°C$ and irreversibly for $T \gtrsim 350°C$.

4. DISCUSSION

We consider now the anneal mechanisms. Since the annealing at $T \gtrsim 350°C$ is irreversible it presumably involves the restoration of the Si-O-Si network but what of the low temperature reversible annealing? Several papers (Devine 1984, Golanski et al 1984, Devine and Fiori 1985) consider models for this in the case of E_1' centres in ion implanted thermally grown a-SiO_2. Models (1)-(3) are all considered and model (1) is suggested to be the most likely but no conclusive proof is obtained. We believe model

(1) could account for the behaviour we observe; the annealing of the E_1' centres in the buried layer would involve diffusion of O_2 through a-SiO_2 which occurs with an activation energy of 0.85 - 1.26 eV (Edwards and Fowler 1982) whereas the oxide precipitates may well be deficient in oxygen and so the annealing of E_1' centres within them would then require the O_2 to first diffuse through the surrounding silicon - a process requiring a higher activation energy of 2.4 - 3.5 eV (Wilkes 1988) and hence a higher temperature. If this higher temperature is close to or greater than that at which the irreversible annealing begins then no reversible stage would be observed, as is the case. As regards model (3) it is hard to see why it should occur in the buried layer and not in the precipitates. There is the additional question - where does the electron come from? Many authors (for example Oldham et al 1986, Saks et al 1989) have given evidence that positive charge trapped in the SiO_2 within a few nm of the Si/SiO_2 interface in MOS devices is removed by electrons tunnelling from the silicon. This mechanism should play a role in removing E_1' centres from at least the outer layer of the oxide precipitates but the buried layer is too thick for it to remove more than a small fraction of the centres. Perhaps the electrons are thermally released from traps in the oxide and their high mobility allows them to escape from the small precipitates but not from the thicker buried layer. This does not seem very plausible and we feel that model (1) is more likely to apply.

In summary we find that the as-implanted state of the samples contain E_1' centres in oxide precipitates and that precursors of these centres remain in the buried oxide layer. E_1' centres in both locations anneal irreversibly at $T \gtrsim 350°C$ (at least for an oxygen dose of 1.8×10^{18} cm^{-2}) and those in the latter location also anneal reversibly for $100 \lesssim T \lesssim 350°C$.

REFERENCES

Barklie R C, Hobbs A, Hemment P L F and Reeson K 1986 J Phys C: Solid State Phys 19, 6417.
Barklie R C, Ennis T J, Hemment P L F and Reeson K 1988 Nucl Inst and Methods B32, 433.
Barklie R C, Ennis T J, Reeson K and Hemment P L F 1989 Appl Surf Science 36, 400.
Devine R A B 1984 J Appl Phys 56, 563.
Devine R A B and Fiori C 1985 J Appl Phys 57, 5162.
Devine R A B 1987 Phys Rev B35, 9783.
Edwards A H and Fowler W B 1982 Phys Rev B26, 6649.
Feigl F J, Fowler W B and Yip K L 1974 Solid State Commun 14, 225.
Golanski A, Devine R A B and Oberlin J C 1984 J Appl Phys 56, 1572.
Griscom D L 1984 Nucl Instr and Methods B1, 481.
Griscom D L 1985 J Non-Crystalline Solids 73, 51.
Hemment P L F, Maydell-Ondrusz E, Stephens K G, Butcher J, Ioannou D and Alderman J 1983 Nucl Instr and Methods 209/210, 157.
Krause S J, Jung C O, Wilson S R, Lorigan R P and Burnham M E, 1986 Mat Res Soc Symp Proc 53, 257.
Lenahan P M and Dressendorfer P V 1984 J Appl Phys 55, 3495.
Oldham T R, Lelis A J and McLean F B 1986 IEEE Trans Nucl Sci NS-33, 1203.
Pfeffer R L 1988 The Physics and Technology of Amorphous SiO_2 (New York: Plenum) pp 181-186.
Saks N S, Ancona M G and Modolo J A 1989 IEEE Trans Nucl Sci NS-31, 1249.
Stapelbroek M, Griscom D L, Friebele E J and Sigel Jr G H 1979 J Non-Crystalline Solids 32, 313.
Wilkes J 1988 The Properties of Silicon, EMIS Data Reviews Series No 4 (London: INSPEC) p 281.

Paper presented at INFOS '91, Liverpool, April 1991
Contributed Papers, Section 8

An application of a new chemical etching process for oxidation induced stacking faults in SIMOX structures; comparison with silicon

C.TSAMIS[1], D.TSOUKALAS[2], N.GUILLEMOT[1], J.STOEMENOS[3], J.MARGAIL[4]

[1] Lab.de Physique des Composants à Semiconducteurs , ENSERG, BP 257, 38016 Grenoble Cedex, FRANCE.
[2] NCSR "Democritos", Instit. of Microelectronics, 15310 Aghia Paraskevi, GREECE.
[3] University of Thessaloniki, Physics Dept., 54006 Thessaloniki, GREECE.
[4] LETI, Centre d' Etudes Nucleaires de Grenoble , 38041 Grenoble Cedex, FRANCE.

ABSTRACT: The observation of Oxidation Induced Stacking Faults (OISF) in SIMOX structures with an optical Microscope is achieved with the aid of a new method consisting of chemical etching and HF decoration. With the aid of this method we compare the length of S.F obtained in SIMOX structures with those obtained in bulk silicon under the same oxidation conditions.

1. INTRODUCTION

Thermal oxidation of Silicon is a frequently encountered high–temperature processing step in the fabrication of any device on silicon and on Silicon–on–Insulator structures. Such oxidations have been known to induce stacking faults in the silicon (called OISF,i.e. oxidation induced stacking faults). Since these defects can degrade device performance, knowledge of their growth kinetics is of great importance.Several techniques have been developed for the observation of these defects (chemical etching, X–ray techniques and Transmission Electron Microscopy). The use of chemical etching allowed us to establish easily laws concerning the growth kinetics of the Stacking Faults. However this versatile technique cannot be applied to SIMOX (Separation by IMplanted OXygen technique) structures: The high etch rates of the existing etchants (Wright Etch,Sirtl Etch,etc) and the fact that the Si–overlayer is very thin (0.2μm), makes their use impossible.

In the present work we propose a new method for the observation of Stacking Faults in SIMOX. The method consists of chemical etching using a solution with a low etch rate and subsequent decoration of the defect with an HF solution. The defect is observed with an optical microscope.

2.EXPERIMENTAL.

In order to obtain a solution with a low etch rate, the Wright Etch (Wright 1977) was diluted down by adding both acetic acid (CH_3COOH) and water (H_2O). The addition of water lowers the etch rate by decreasing the concentration of oxidizing species (Ghandi 1983) while the addition of acetic acid results in good surface quality preventing the formation of bubbles during etching (Wright 1977). Several solutions were prepared. The etch rate was estimated by measuring a step created at the surface (using a surface profiling device) of unoxidized silicon samples (<100>, n–type,4–6 Ω cm) after exposing

© 1991 IOP Publishing Ltd

part of the surface. The concentration of the solution selected (named Diluted Wright Etch [DWE]) exhibiting an etch rate of 0.15 μm/min is listed below:
60 ml conc. HF (50%), 30 ml conc.HNO_3 (70%), 30 ml of 5M CrO_3, 2g $Cu(NO_3)_2 3H_2O$, 250 ml conc. acetic acid (glacial) and 250 ml H_2O (deionized)
In order to examine the ability of the DWE to reveal the crystal defects we compared it with WE using oxidized silicon samples. The results obtained after etching similar amounts of silicon were identical. Also experiments performed with SIMOX samples, using the DWE solution, gave the same etch rate (0.16 μm/min).

For the study of the growth kinetics and the comparison between SIMOX and Silicon both types of samples were used. The SIMOX samples were fabricated with a standard procedure (by LETI/CEA).The silicon overlayer thickness was 210 nm. The silicon samples used were <100>, p–type,4–6 Ω cm. Prior to oxidation, Boron was implanted (8 10^{13} cm^{-2},50 KeV) in order to increase the density of the Stacking Faults by creating nucleation cites. The samples were oxidized in dry oxygen for various temperatures (1050°C,1100°C,1150°C) and times. After the removal of the oxide the SIMOX samples were dipped in the DWE solution. The thickness of the oxide and of the etched silicon was measured using the surface profiling device. The time of etching is limited since silicon overlayer is very thin. Subsequently the samples were exposed to HF solution (50%) for different times in order to decorate the faults. Observation of the surface with an optical microscope was performed after every step.For the revealing of OISF in silicon the Wright Etch was used.

3. Results and Conclusions

The method described previously consists of two phases. We explain what happens during each phase : The DWE etches the region with the fault faster than surrouding region. Due to this windows are created that extend down to the SiO_2 buried layer. However the contrast of the produced image is not adequate. This means that the decoration with the HF solution is always necessary During this,the HF solution passing through the windows opened during etching reaches the buried layer and etches it in every direction. An exposure to HF for 20 to 30 sec is sufficient to provide very good image contrast. This results to a good definition of the defects. However the HF solution will have no effect if the windows opened during etching do not extend down to the oxide buried layer. We have estimated that this will not happen when the thickness of the Si–overlayer is of the order of 10 nm or less This critical value results from the difference of the etch rates between the region with the fault and the defect free. TEM measurements have confirmed our results (Tsamis et al).

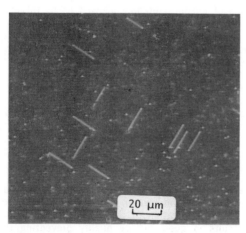

Fig. 1. Stacking Faults on a SIMOX wafer. The average length is 14 μm.

Fig.1 shows Stacking Faults revealed on a SIMOX sample. The defect delineation and the surface quality are excellent. Comparison of the length and the density obtained with TEM measurements shows that chemical etching gives higher mean size (5_10%)and the lower density (10%) for the faults. This is attributed to the presence of very small SF that are not detected by the optical microscope due to the uncertainty HF introduces. The width of the slot which HF opens along the SF in the SiO_2 buried layer is over 1 μm. This is even larger at the ends of the fault, where the partial dislocation that bound the fault are situated. The hole formed is comparable to the diameter of the holes which delineate the presence of threading dislocations. Thus OISF smaller than 2 μm cannot be distinguished from the threading dislocations. However chemical etching is most appropriate for the determination of low Stacking Faults densities (less than 10^6 cm^{-2}) due to the large field of view that can be scanned.

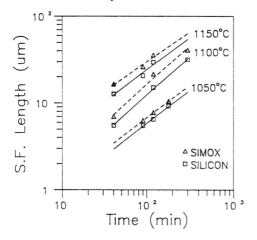

Fig. 2. Time dependence of the Stacking Faults length in Silicon and SIMOX oxidized at different temperature in dry oxygen.

With the aid of this method we compare the length of OISF in silicon and in SIMOX structures oxidized under the same conditions To the best of our knowledge this is the first time such a comparison is reported. Fig.2 shows the OISF length dependence on the oxidation time .It can be seen that although OISF in SIMOX are larger than those induced in silicon by 20_30% the growth kinetics is determined by the known relation $L \sim t^n$ (where L is the length of the Faults and n a number exponent). The values of n are listed in Table 1.We notice that the values obtained for n are the same for the same temperature. A possible explanation of the fact that OISF are larger in SIMOX could be the following: The growth of the OISF is due to the injection of Si—interstitials from the Si/SiO_2 interface. In bulk silicon the growth of the faults is not limited in depth, resulting in semi_elliptical shapes. In SIMOX structures however the Si—overlayer thickness is about 0.2μm, the OISF reaches the back Si/SiO_2 interface very quickly and then it can growth only in a direction parallel to the surface.

T (°C)	Silicon	Simox
1050	0.75	0.74
1100	0.86	0.88
1150	0.73	0.70

Table 1 lists the values obtained for the n coefficient.

The dependence of Stacking Faults length in silicon with temperature is given by the relation $L \sim \exp(-Ea/kT)$, where Ea is the activation energy and T the oxidation temperature. From fig.3 it it seen that this relation holds also for Simox. The activation energies for Silicon and Simox are found to be 2.45 eV and 2.48 eV respectively.

In conclusion we have presented a new chemical etching process capable of revealing the Oxidation

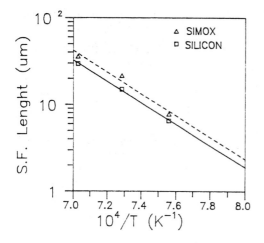

Stacking Faults in SIMOX material. Stacking Faults are found to be larger in Simox by 20-30%, although they have the same dependence on oxidation time and temperature with those induced in silicon.

Fig.3. Temperature dependence of the Stacking Faults length in Silicon and SIMOX after oxidation in dry oxygen.

REFERENCES

Ghandi S.K., VLSI Fabrication Principles, J.Wiley & Sons edition 1983

Tsamis C., Tsoukalas D., Guillemot N., Stoemenos J., Margail J.,"A chemical etching for the delineation of oxidation stacking faults in silicon implanted with oxygen structures" J. Electrochem. Soc.(to be published).

Wright Jenkins M. 1977, J. Electrochem. Soc. 124, 757

Paper presented at INFOS '91, Liverpool, April 1991
Contributed Papers, Section 8

Infrared microscopic spectroscopy analysis of silicon on insulator bevelled samples

J. Samitier[1], A. Pérez-Rodríguez[1], B. Garrido[1], J.R. Morante[1] and P.L.F. Hemment[2]

(1) L.C.M.M., Departament de Física Aplicada i Electrònica, Universitat de Barcelona, Diagonal 645-647, 08028-Barcelona, SPAIN
(2) Department of Electronic and Electrical Engineering, University of Surrey, Guildford Surrey GU2 5XH, UK

ABSTRACT: The analysis of SOI/SIMOX structures obtained by different processes has been performed by FTIR reflection spectroscopy measurements on low angle bevelled samples. This has allowed to directly observe the characteristics of the different buried oxide layers. The simulation of the experimental data with the theoretical parameters reported for a thermal oxide points out the high quality of the buried layers, although in some cases a higher contribution in the TO_4-LO_4 region is observed, which has been related to the existence of disorder induced effects determined by the technological process.

1. INTRODUCTION

FTIR (Fourier Transform Infra-Red) spectroscopy constitutes a powerful non destructive technique for the analysis of multilayered structures for VLSI applications, such as Silicon on Insulator (SOI) substrates. It is specially well suited to the characterization of the SiO_2 buried layers present in these structures, as the measurements are made in the mid-infrared spectral region, where the spectra are characterized by the presence of absorption peaks related to the different transverse and longitudinal optical (LO and TO) vibrational modes of the Si-O bond unit. The position and shape of these peaks is very sensitive to the structural characteristics of the oxide, being very much dependent on any microstructural feature affecting the local random network order (Lange 1989).

Different authors have investigated the quality of the buried oxide layers on SOI structures by IR spectroscopy, specially in the case of SIMOX structures (Harbecke et al 1987, Yu et al 1989). SIMOX (Separation by Implanted Oxygen) technique constitutes nowadays one of the leading methods of obtention of high quality SOI substrates for VLSI applications (Sturm et al 1988). In this technique, the substrates are obtained by high dose oxygen ion implantation, followed by a high temperature anneal (HTA) treatment. In these previous works, IR spectra were obtained from transmission measurements performed with normal light incidence, and the direct evaluation of the peaks has been used to characterize the buried oxide layers. However, for these kind of structures the existence of a Si layer over the SiO_2 one

© 1991 IOP Publishing Ltd

can modify the shape and position of the different peaks in the spectra, as they also depend on the morphological characteristics of the structures. This dependence is enhanced when using reflection or transmission measurements under variable angle of incidence. Such measurements have a special interest, as they provide with additional significant information than the usual normal transmission ones (due to the excitation of both kinds of vibrational modes, LO and TO).

In this work we present the results obtained by FTIR microscopic reflection spectroscopy measurements on bevelled SIMOX samples obtained by different processes. These measurements have been compared with those obtained from ZMR (Zone Melting Recrystallization) structures. Measurements performed on the bevel surface have allowed us to directly observe the buried oxide layers, as well as the evolution of the spectra as the different layers in the structure are included in the measured region. The theoretical simulation of these spectra -taking into account the different geometric conditions- has enabled their correct interpretation to be made, yielding significant information regarding the oxide structure for the different buried layers.

2. EXPERIMENTAL RESULTS AND DISCUSSION

SIMOX wafers were produced by Ibis Technology Corp. Samples studied were obtained by a standard process and by sequential implantation and annealing (SIA) and, according to the obtention process, have been labelled SS1, SIA1 and SS2, respectively. SS1 and SS2 wafers were prepared by a single step oxygen implantation at an energy of 200 keV with doses of 1.7×10^{18} cm^{-2} and 1.8×10^{18} cm^{-2} (implantation temperatures 640°C and 600°C, respectively). Subsequently, wafers were annealed at high temperature (1300°C for the SS1 and 1320°C for the SS2) during 6 hours. The SIA1 wafer is basically equivalent to the SS1: one third of the total dose was implanted at 640°C and the wafer was annealed at 1300°C for 2 hours. This cycle was carried out three times to give a total dose of 1.5×10^{18} cm^{-2} and a total annealing time of 6 hours at 1300°C. The thicknesses of the top Si and buried SiO$_2$ layers from these structures have been determined by cross section TEM and are the following: (SS1) 295nm and 310nm, (SIA1) 435nm and 210nm and (SS2) 300nm and 480nm.

Samples have been bevelled by mechanical polishing. This has been made by mounting the samples with wax on bevelling fixtures having the desired bevel angle. Then, samples are polished with fine diamond abrasive in an oil-based slurry on a frosted glass plate. Bevel angle has been optically determined, being for all the samples tg(α) = 0.0060. FTIR measurements have been performed with a Bomen MB-120 spectrometer. The use of a Spectra Tech IR-Plan microscope has allowed to obtain a light spot with a diameter of about 30μm. The measurements have been done at room temperature, with an angle of incidence of 28° and a spectral resolution of 1 cm^{-1}.

In figure 1 are plotted the absorbance spectra obtained from the different SIMOX samples with the spot directly located on the buried oxide, together with the theoretical one. All these spectra are characterized by the presence of a positive peak and a negative one which have been identified as the due to the LO$_3$ and TO$_3$ vibrational modes (Lange 1989). The position of these peaks is very similar for all the samples: for the LO$_3$ one the positions are 1258 (SS1, SIA1) and 1257 cm^{-1}

Silicon on Insulator

(SS2) and, for the TO_3 one, 1086 (SS1), 1084 (SIA1) and 1082 cm^{-1} (SS2).

The main difference between these spectra is given by the full width at half height of the TO_3 peak, which has values of 118 cm^{-1} (SS1), 93.5 cm^{-1} (SIA1) and 88 cm^{-1} (SS2). These differences are due to the presence in the spectra from SS1 and SIA1 samples of a shoulder in the TO_4-LO_4 region, which indicates a higher contribution of these vibrational modes in these samples (specially in the SS1 one). According to previous works (Lange 1989, Kirk 1988) the increase of these modes is due to disorder induced vibrational coupling effects, and this points out the presence of these disorder effects in the samples related to the technological process.

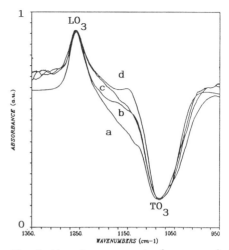

Fig. 1. Absorbance spectra from samples (a) SS1, (b) SIA1, (c) SS2. (d) is the theoretical simulation

So, the lower intensity of these effects from SIA1 sample when compared to SS1 would corroborate the ability of the SIA technique for the obtention of higher quality structures (Sturm et al 1988). On the other side, the higher presence of disorder effects in the SIA1 sample than in the SS2 is consistent with the fact that SIA1 does not correspond to an optimized process. Previous observations have shown the presence in this sample of a region with a high concentration of defects (oxide precipitates and dislocations) in the top Si layer close to the buried oxide and, according to this, this sample is not typical of SIA material.

The theoretical simulation of these spectra is also plotted in figure 1 and has been performed assuming a model of different layers corresponding to the layers present in the measured region of the bevelled sample (in this case, a SiO_2 layer on a Si substrate). For each material, the dielectric function has been estimated according to the oscillator model for a crystal (Baker 1964), assuming for the amorphous material the existence of a gaussian distribution of bonds. According to this model, each oscillator mode is characterized by 4 parameters. The values used for these parameters are essentially the reported by different authors for thermally oxidized SiO_2 (Naiman et al 1984, Grosse et al 1986). Only the parameter related to the strength of the TO_4 mode has been increased. As it is shown, there is a good agreement between the simulated spectrum and the experimental ones, specially in the case of SS2 sample, which indicates the higher quality of the buried oxide layer from this structure, with characteristics very similar to those of a thermal oxide.

In figure 2 are plotted the spectra corresponding to the SS2 sample and a ZMR structure, together with the theoretical one. In the case of ZMR material, the buried oxide is obtained by a high temperature thermal oxidation process. The spectra shown in this figure are very similar which indicates that the characteristics of these layers are very similar to those of a thermal oxide. According to these results, there seems not to be an effect of the bevelling procedure on the layer characteristics.

Finally, measurements performed on different positions of the bevel surface when the top Si layer is included in the measured region show a strong dependence of spectra on the geometric conditions. In general, the theoretical simulation of these data -which has to take into account the morphological characteristics of the structure under study- with the parameters correspondent to a thermal oxide seems not to agree so much as the previous simulation. This suggests the existence of higher differences in the oxide structure from a thermal oxide in the upper region of the buried layers, although this needs further research to be done.

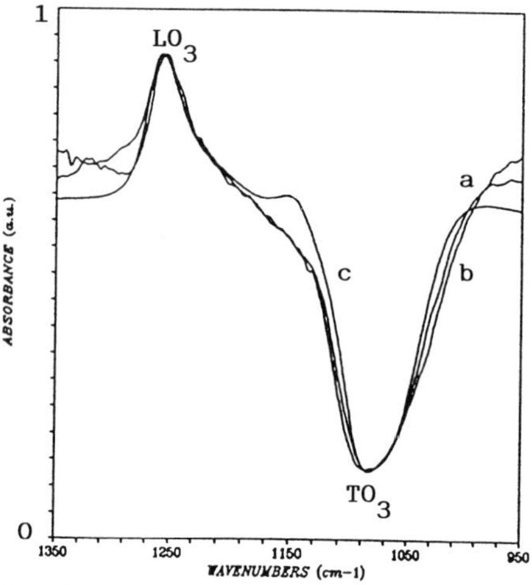

Fig. 2. Absorbance spectra from samples (a) SS2 and (b) ZMR. (c) is the theoretical simulation.

CONCLUSIONS

The analysis performed has allowed us to directly observe the characteristics of the buried oxide from SIMOX structures obtained by different processes. The comparison of these results and those from ZMR structures, together with their theoretical simulation with the parameters corresponding to a thermal oxide has allowed us to confirm the high quality of the synthesized buried SiO_2 layers, which have structural characteristics similar to a thermal oxide. However, in some cases a higher contribution in the TO_4-LO_4 region is observed, which has been related to the existence of disorder induced effects determined by the technological process. These results confirm the utility of the FTIR technique for the analysis of SOI structures and contribute to a better understanding of the physical nature of the SiO_2 layers obtained in the SIMOX technology.

REFERENCES

Baker A S 1964 Phys. Rev. **136** 1290
Grosse P, Harbecke B, Heinz B, Meyer R and Offenberg M 1986 Appl. Phys. A **39** 257
Harbeke G, Steigmeier E F, Hemment P, Reeson K J and Jastrzebski L 1987 Semicond. Sci. Technol. **2** 687
Kirk C T 1988 Phys. Rev. B **38** 1255
Lange P 1989 J. Appl. Phys. **66** 201
Naiman M L, Kirk C T, Aucoin R J, Terry F L, Wyatt P W and Senturia S D 1984 J. Electrochem. Soc.: Solid-St. Scien. Techn. **131**
Sturm J C, Chen C K, Pfeiffer L and Hemment P L F (eds) 1988 Silicon on Insulator and Buried Metals in Semiconductors (Mat. Res. Soc. Symp. Proc. Vol. 107) (Pittsburgh: Mat. Res. Soc.)
Yu Y, Fang Z, Lin C, Zou S and Hemment P L F 1989 Materials Letters **8** 95

Paper presented at INFOS '91, Liverpool, April 1991
Contributed Papers, Section 8

Silicon-on-insulator waveguides

N Mohd Kassim, T M Benson, D E Davies[*], A McManus[+]

Dept of Electrical and Electronic Engineering, University of Nottingham, UK.
[*] European Office of Aerospace Research and Development, US Air Force.
[+] Dept of Electronics and Computer Science, University of Southampton, UK.

ABSTRACT: Optical waveguiding at a wavelength of 1.3 µm has been observed in silicon on insulator material formed using the SIMOX (Separation by IMplanted OXygen) process. Measured optical losses are related to both the material structure and its quality.

1. Introduction.

The development of a reliable silicon optical waveguide technology compatible with electronic device processing is essential if targets such as silicon based optoelectronic integration or wide bandwidth on-chip optical networks are to be met. In the III-V semiconductors high quality lattice matched multi-layers such as GaAs/GaAlAs or InGaAsP/InP can conveniently be used to form low-loss optical waveguides (McIlroy et al 1987, Kapon and Bhat 1987, Angenent et al 1989). For silicon, the layered material commonly encountered with it is SiO_2 so a $Si-SiO_2$ configuration is worth exploring for optical waveguide fabrication. A silicon equivalent to a planar optical waveguide formed using III-V semiconductor heterostructures requires a buried-oxide confining layer. The SIMOX (Separation by IMplanted OXygen) process, in which a buried oxide layer is created by multiple ion-implantation stages each followed by thermal annealing (Van Ommen 1989), is one technology being developed for silicon-on-insulator MOS electronic components. SIMOX material has provided our most successful SOI optical waveguides to date (Mohd Kassim et al 1990). Here we describe the use of optical waveguide measurement to assess the quality and structure of SOI material.

2. Theoretical considerations for SOI waveguides.

The material refractive indices and dimensions of SIMOX material as illustrated in Figure 1, fortuitously provide single mode slab waveguiding at 1.3 µm. The thickness of the implanted oxide layer must be sufficient to ensure the confined modes do not leak into the silicon substrate. This is an important point, since there is a limit to the thickness of the oxide layer that can be found in practical structures fabricated by the SIMOX technique. A surface oxide can be produced relatively easily by chemical vapour deposition or alternatively by epitaxial growth followed by thermal oxidation.

© 1991 IOP Publishing Ltd

A guided TE$_0$ mode supported by the structure shown in Figure 1 satisfies the eigenvalue equation

$$\tan(q_3 h) = \frac{z p_4 q_3 - q_2 q_3}{z q_3^2 + q_2 p_4} \quad (1)$$

where

$$z = -\frac{p_1 \tanh(q_2 t) + q_2}{p_1 + q_2 \tanh(q_2 t)}$$

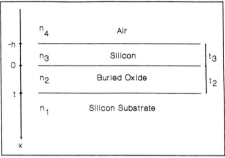

Figure 1: Schematic diagram of SIMOX structure.

and

$$p_i^2 = \beta^2 - n_i^2 k_0^2 \;,\; q_2^2 = \beta^2 - n_2^2 k_0^2 \;,\; q_3^2 = n_3^2 k_0^2 - \beta^2 \quad \text{with } i=1,4$$

To determine the attenuation that results from substrate leakage of this structure, we solve the usual four medium waveguide dispersion equation above. The solution to the eigenvalue equation is in general complex and the imaginary part of associated propagation constant describes loss by leakage into the substrate. The TE$_0$ mode leakage loss at 1.3 µm shown in Figure 2 as a function of the buried oxide layer thickness for the case where the thickness of guiding silicon is 0.2 µm. Note that at a buried oxide layer thickness of 0.4 µm, the leakage substrate loss is ≈ 3.6 dBcm^{-1}. This loss can be made negligible

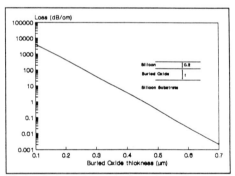

Figure 2: Loss due to leakage into substrate as a function of buried oxide thickness.

for buried oxide layer thickness above 0.5 µm. Losses for the TM$_0$ mode and higher order modes of both polarisations are substantially higher.

In SIMOX wafers the interface between the buried oxide layer and the superficial silicon guiding layer can be rough because of the presence of precipitates. It is known that surface scattering loss can be significant even for relatively smooth surfaces because the propagating waves interact strongly with the surfaces of the waveguide. A new scattering loss formula applicable to scattering from a single rough interface in a slab waveguide has been developed with the assistance of Kendall (to be submitted for publication). The expression derived for the attenuation

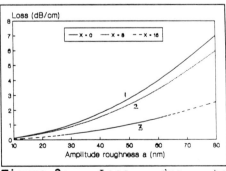

Figure 3: Loss due to scattering as a function of a.

constant of a corrugated three layer slab waveguide is

$$\alpha_S = \frac{K_3 [2K_2^2 - (K_2^2 - K_3^2) \sin^2 K_2 h] a^2 E_0^2 (k_2^2 - k_3^2)^2}{8\omega\mu [K_2^2 (K_1 - K_3)^2 + (K_2^2 - K_1^2)(K_2^2 - K_3^2) \sin^2 K_2 h]} \quad (2)$$

where

$$K_i^2 = k_i^2 - (p - \beta)^2 \;, \quad k_i^2 = n_i^2 k_0^2$$

p is the spatial frequency of the roughness and i = 1, 2 or 3.

Figure 3 shows the attenuation coefficient α_S in dBcm^{-1} for a slab waveguide at 1.3 µm plotted as a function of a. The curves numbered 1 to 3 are for values of $X = 20(p-\beta)/k_3$ of 0, 8 and 16.

3.0 Experimental techniques.

The slab waveguide structures mentioned above only provide confinement of fields in one direction. However many applications require optical confinement in two dimensions. The additional lateral confinement can be obtained by etching the surface oxide to provide strip-loaded waveguides 3 to 5 µm wide.

TE-like optical waveguiding at 1.3 µm was observed by end-fire coupling into the cleaved end of a waveguide via a x45 microscope objective. The overall thickness of the silicon wafer was reduced to about 100 µm to assist the cleaving process. The light transmitted was focused on to an IR camera and a Ge photodiode via another x45 objective. Because of the small guide thickness encountered coupling problems were anticipated and whilst excitation of both slab and pencil-beam guided modes was clear it was also inefficient. All the guides tested were single moded in both vertical and horizontal directions as expected.

The cleaved input and output facets form a Fabry-Perot cavity and the transmitted power I_T varies as (Regener and Sohler 1985)

$$I_T = \frac{\eta I_0 T^2 \exp^{-\alpha L}}{(1 - R\exp^{-\alpha L})^2 + 4R\exp^{-\alpha L}\sin^2 \beta L} \quad (3)$$

where I_0 is the incident intensity, R the mode reflectivity, T the transmittance, ß the propagation constant of the guided mode, L the sample length and η the coupling efficiency to the mode. The optical phase difference (ßL) can be tuned by gently heating the sample to produce a periodic variation in transmission. I_T is a maximum (I_{TMAX}) when $\sin^2\beta L = 0$ and a minimum (I_{TMIN}) when $\sin^2\beta L = 1$. It follows (Walker 1985) that

$$f(u) = \ln\left[\frac{1+\sqrt{u}}{1-\sqrt{u}}\right] = \alpha L - \ln R \quad (4)$$

where u is the ratio of I_{TMIN} to I_{TMAX}.

4. Result and discussion.

If R and L in equation (4) are known, measurement of u gives a non-destructive means of calculating α. In our initial experiments f(u) was measured for various length L of each sample and α and R found from the slope and intercept of a plot of f(u) against L. The smallest TE_0 attenuation value obtained to date is 4.67 dBcm^{-1} with a facet reflection coefficient of 0.26. This particular sample was implanted at a temperature of 640°C and beam energy of 200 keV to a total dose of 1.8×10^{18} cm^{-2} in three sequential stages. Intermediate and final annealing was under taken for 6 hours each at 1300°C.

This experimentally measured loss includes contributions from both substrate leakage and interface scattering. For the measured buried oxide layer thickness of 0.4 μm, Figure 2 shows that the substrate leakage loss is approximately 3.6 dBcm^{-1}. Since the quality of the Si-SiO$_2$ interface is good, as shown in the XTEM photograph of Figure 4, the loss due to surface scattering is small and is estimated at 0.43 dBcm^{-1} from Figure 3. The leakage and surface scattering loss mechanisms thus seem to account for most of the measured loss.

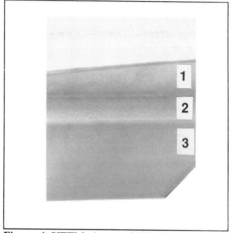

Figure 4: XTEM picture of SIMOX structure. 1 Surface oxide (0.32μm), 2 Silicon (0.21μm) and 3 Buried oxide (0.40μm).

We have previously reported that optical loss varies significantly between multiple implant wafers fabricated using slightly different process parameters (Mohd Kassim et la 1990). Since the buried oxide layer thickness were measured to be 400 ± 10 nm for all samples, we have been able to relate this variation to roughness scattering. Excellent agreement between experimental losses and theoretical calculations made using XTEM measurement of precipitate size has been achieved (Kendall to be submitted for publication).

Acknowledgement.

The authors are grateful to Drs Prewitt and Curran for their assistance with device processing and to the various laboratories who supplied the SIMOX wafers used in our studies.

References

Angenent J H et al, Elec Lett 25, p 629, 1989
Van Ommen D H, Nucl Inst and Methods in Phys Res B39, p 194, 1989
Mohd Kassim N et al, Proc ESSDERC 90, p 5, 1990
Kendall P C et al, To be submitted for publication.
Regener R and Sohler W, Appl Phys B39, 143, 1985
Walker R G, Elec Lett 21, 581, 1985

Evaluation of SOI/SIMOX structures by Raman scattering measurements obtained at different excitation powers

A. Pérez-Rodríguez[1], F. Coromina[1], J.R. Morante[1], J. Jiménez[2], P.L.F. Hemment[3] and K.P. Homewood[3]

(1) L.C.M.M., Departament de Física Aplicada i Electrònica, Universitat de Barcelona, Diagonal 645-647, 08028-Barcelona, SPAIN
(2) Departamento de Física de la Materia Condensada, Universidad de Valladolid, 47011 Valladolid, SPAIN
(3) Department of Electronic and Electrical Engineering, University of Surrey, Guildford, Surrey, GU2 5XH, UK

> ABSTRACT: In this work we report the evaluation of SIMOX structures obtained by different processes performed by Raman scattering measurements made at different excitation powers and wavelengths. The comparison between the spectra obtained from different samples has revealed the existence of significant differences related to the technological process. The electrical evaluation of samples with spectra very similar to those from bulk Si has confirmed the value of the SIMOX technique for the preparation of high quality structures with thin Si films free of strain and with good electrical properties.

1. INTRODUCTION

Silicon on Insulator (SOI) structures made by high dose oxygen implantation and annealing (technique known as SIMOX: Separation by Implanted Oxygen) are very promising materials for VLSI applications (Sturm et al 1988). However, during the preparation of these substrates different processes can affect the electrical performance of the devices. Therefore, the optimization of these technologies requires a wide range of material characterization to be carried out, in order to evaluate the influence of the fabrication steps on the SOI wafer characteristics as well as on the final device performance.

Of the different characterization techniques, special interest has been given to the optical methods such as Raman spectroscopy, due to their non destructive character as well as their applicability for the analysis of multilayered materials. So, different authors have reported the analysis by Raman spectroscopy of SIMOX structures obtained under different conditions (Harbecke et al 1987, Olego et al 1988, Takahashi et al 1988). For as implanted samples, they have observed the existence of a tensile strain in the top silicon layer which can be relieved by high temperature anneal (HTA) treatments, and has been attributed to the presence of oxide precipitates. In these works measurements were performed in conditions were sample temperature effects could be neglected.

In the present work SIMOX structures made by different processes have been analized by Raman spectroscopy measurements obtained under different excitation powers and wavelengths. This has allowed us to investigate the dependence of the spectra on the temperature gradient present in the scattering regions. The measurements have allowed to deduce the existence of significative differences related to the particular technological processes. Moreover, the electrical evaluation of the samples has confirmed that thin Si films free of strain and with good electrical properties can be formed in SIMOX substrates.

2. RESULTS AND DISCUSSION

SIMOX wafers were produced by Ibis Technology Corp. Samples were obtained from three different wafers which were prepared by standard processes (single step implantation and annealing) and by sequential implantation and annealing (SIA) (Van Ommen 1989). According to their preparation, samples have been labelled SS1, SS2 and SIA1. SS1 and SS2 wafers were prepared by a single step oxygen implantation at an energy of 200 keV with doses of 1.7×10^{18} cm^{-2} and 1.8×10^{18} cm^{-2} and at temperatures of 640°C and 600°C, respectively. Subsequently, wafers were annealed at high temperature (1300°C for the SS1 and 1320°C for the SS2) during 6 hours. The SIA1 wafer is basically equivalent to the SS1: one third of the total dose was implanted at 640°C and the wafer was annealed at 1300°C for 2 hours. This cycle was carried out three times to give a total dose of 1.5×10^{18} cm^{-2} and a total annealing time of 6 hours at 1300°C. The thicknesses of the top Si and buried SiO_2 layers from these structures have been determined by cross section TEM and are the following: (SS1) $d(Si) = 295$nm $d(SiO_2) = 310$nm, (SS2) $d(Si) = 300$nm $d(SiO_2) = 480$nm, (SIA1) $d(Si) = 435$nm $d(SiO_2) = 210$nm.

Moreover, in the case of SIA1 sample the TEM images show the presence in the top Si layer of a region with SiO_2 precipitates and dislocations pinned to them, close to the buried oxide layer and with an average thickness of 200nm. This indicates that this sample does not correspond to an optimized process. The SIA method has been reported as one of the leading techniques of production of high quality SOI material (Van Ommen 1989) and, so, this wafer is not typical of SIA material. However, it has been included in order to observe the sensitivity of the performed analysis on the defects present in the structure.

Raman scattering measurements have been performed with a DILOR XY spectrometer. The samples were excited with the 457.9, 488 and 514nm lines from an Ar$^+$ laser, using microscope optics in backscattering configuration. For these lines, the penetration depths for the scattered light have been estimated to be approximately 300, 600 and 800nm, respectively (Takahashi et al 1988). Detection has been made in multichannel mode with photon counting electronics.

As previously reported, the existence of temperature gradients in the scattering volume due to heating by the excitation laser can determine both a shift and a broadening of the Raman line, essentially on the side of small frequencies shifts (Lo et al 1980, Raptis et al 1984, Liarokapis et al 1988), due to the non uniform heating of the lattice by the laser. According to this, Raman measurements have been performed using different excitation powers for all the wavelengths and have been compared with those obtained at the same conditions in bulk silicon. The comparison between the spectra obtained in the different samples indicates that such an effect is largely determined by the characteristics of the region of the sample under study as well as the presence and quality of interfaces in this region. From the Stokes to Antistokes intensity ratios, the temperature in the laser spot has been estimated to be lower than 150°C for the highest excitation power, which gives a temperature gradient higher than 100°C/μm. We have to remark that at low excitation powers, similar to those used in previous works, the spectra are very similar to those from bulk Si and these effects are scarcely observed.

In figures 1 and 2 are plotted the spectra obtained from the different samples using the highest excitation power and with wavelengths of 457.9 and 488nm. Each spectrum is normalized to its maximum value, and those spectra coming from different samples have been arbitrarily shifted. The spectra obtained with the wavelength of 514nm show a behaviour very similar to that plotted in figure 2. In almost all the spectra, there is a shift towards the low frequency side and a broadening of the peaks in relation to that obtained in bulk silicon. This broadening tends to occur in the low frequency side, and increases with the excitation power. Moreover, significant differences are observed between the different samples.

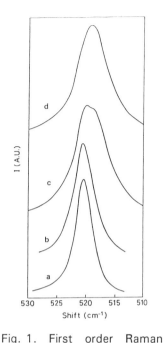

Fig. 1. First order Raman peaks from samples of bulk Si (a), SS2 (b), SS1 (c) and SIA1 (d). $\lambda = 457.9$nm, $P = 0.75$Mw/cm^2

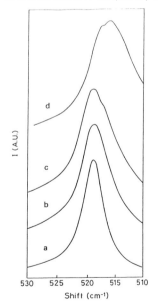

Fig. 2. First order Raman peaks from samples of bulk Si (a), SS2 (b), SS1 (c) and SIA1 (d). $\lambda = 488$nm, $P = 0.25$Mw/cm^2.

So, for the SS2 sample, and for all the wavelengths, the spectra are very similar to those obtained from bulk silicon. Only at the higher excitation powers is a certain broadening observed. In the case of SS1 and SIA1 samples, higher changes in the peaks are observed. However, the spectra obtained with the lower wavelength behave in a different way than those obtained with the higher wavelengths. At the lower wavelength (penetration depth 300nm) the spectra from the SS1 sample give higher shifts and broadening of the Raman lines than those from the SIA1 sample. At higher wavelengths (penetration depths 600 and 800nm) the higher changes occur in the SIA 1 sample. For the spectra obtained at the higher powers, a certain splitting of the peaks occur.

The changes observed in the Raman peaks indicate

the existence in the samples of a distribution of tensile strains, dependent on the temperature gradient in the scattering volume. The higher changes observed in the SS1 sample in relation to the SIA1 for the lower penetration depth indicate a higher strain in the surface region of this structure. At greater penetration depths, the contribution to the spectra of the region with precipitates in the SIA1 sample increases. Then, the increase in the changes of the Raman peaks from this sample in relation to the other ones suggests the existence of a significant strain component related to this region, which is in agreement with previous observations by Olego et al (1988) and Takahashi et al (1988). The splitting of some of the peaks would be determined by the different contributions of the surface region and the region below the buried oxide (with lower strains) and the precipitates or interfaces region. These data indicate the higher quality of the surface region from the SIA1 structure when compared to SS1 and would corroborate the potential of the SIA process for obtaining higher quality structures.

Finally, an electrical evaluation of samples from wafer SS2 has also been performed by Photoconductive Frequency Resolved Spectroscopy (PCFRS, Homewood et al 1988). PCFRS measurements made on these samples indicate the existence of an exponential excess carrier recombination mechanism (corresponding to a first order recombination kinetics) with a value of the excess carrier lifetime of 11 μs. This data confirms the good electrical quality of the Si films under study, free of strain.

3. SUMMARY

The analysis performed has corroborated the value of the SIMOX technique for obtaining high quality SOI structures, with Si films almost free of strain and with good electrical properties. Although the dependence of the changes of the Raman peaks on the temperature gradient for this kind of structures needs further research - to obtain quantitative information about strain distributions on the different layers - the strong dependence of these effects on the nature and characteristics of the layers suggest this technique as a powerful tool for their non destructive analysis. Moreover, and in spite of the fact that the studied SIA samples do not correspond to optimized structures, the data obtained from the comparison with an equivalent SS structure seem to corroborate the potential of the SIA process for obtaining higher quality SOI material.

REFERENCES

Harbecke G, Steigmeier E F, Hemment P and Reeson K J 1987 Semicond. Sci. Technol. **2** 687
Homewood K P, Wade P G, and Dunstan D J 1988 J.Phys E **21** 84
Liarokapis E and Anastassakis E 1988 Physica Scripta **38** 84
Lo H W and Compaan A 1980 J. Appl. Phys. **51** 1565
Olego D J, Baumgart H and Celler G K 1988 Appl. Phys. Lett. **52** 483
Raptis J, Liarokapis E and Anastassakis E 1984 Appl. Phys. Lett. **44** 125
Sturm J C, Chen C K, Pfeiffer L and Hemment P L F (eds) 1988 Silicon on Insulator and Buried Metals in Semiconductors (Mater. Res. Soc. Symp. Proc. Vol. 107) (Pittsburgh: Mat. Res. Soc.)
Takahashi J. and Makino T. 1988 J. Appl. Phys. **63** 87
Van Ommen A H 1989 Nucl. Inst. and Meth. in Phys. Res. B **39** 194

Paper presented at INFOS '91, Liverpool, April 1991
Contributed Papers, Section 8

Investigation of hysteresis and floating-body effects in SOI-MOSFETs

T Ouisse[1], G Ghibaudo[2], J Brini[2], S Cristoloveanu[2] and G Borel[1]

[1] Thomson–TMS, BP 123, 38521 Saint-Egrève Cedex, France.
[2] Laboratoire de Physique des Composants à Semiconducteurs (UA-CNRS)
Institut National Polytechnique, ENSERG, BP 257, 38016 Grenoble Cedex, France.

Abstract. An analytical model of floating body effects is proposed and experimentally confirmed in SIMOX MOSFET's. It is demonstrated that the conductance and transconductance become simultaneously negative. Floating body effects are shown to affect the device reliability under irradiation exposure.

1. Introduction

Floating body effects represent a specific and essential feature of partially-depleted SOI devices. There are undesirable consequences, such as the kink effect in the $I_D(V_D)$ characteristics (Tihanyi and Schlötterer 1975) or hysteresis in the $I_D(V_G)$ curves (Chen et al 1988). The related reliability problems cannot be solved without an adequate modeling. In this paper, an original model of floating body effects is proposed which is validated by experimental data obtained on SIMOX–MOSFET's.

2. Model

For the hysteresis effects to be accurately modeled, we must consider the combination of several physical mechanisms : (i) drain current flow in the saturation mode, (ii) gate voltage influence, (iii) continuity from weak to strong inversion, (iv) relation between the body potential V_B, body current I_B and reverse junction current (given by the usual diode law), (v) drain voltage dependence of I_B and (vi) threshold voltage variation. Using available mathematical equations for these mechanisms, an *analytical relationship* can easily be derived. This allowed us to fit the hysteresis which occur in the $I_D(V_G)$ characteristics (Fig.1(a)).

Although this fully analytical approach gives a quantitative estimation of the drain current, it requires the use of an approximate model for weak inversion. The interesting point is that more general formulas, which stand whatever the current model is, can be derived with a small-signal analysis. Considering *generic* expressions for the drain and substrate currents and using the gate charge conservation

© 1991 IOP Publishing Ltd

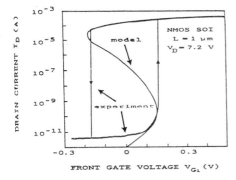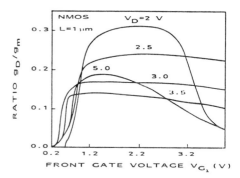

Figure 1: (a) Experimental and simulated hysteresis in $I_D(V_G)$ characteristics and (b) ratio between conductance and transconductance versus gate voltage.

equation, it is easy to demonstrate the following relationship which correlates the conductance g_d and transconductance g_m:

$$g_d = \frac{C_D}{C_{ox}} \frac{\partial V_B}{\partial I_B} \frac{\partial I_B}{\partial V_D} g_m \qquad (1)$$

where C_D and C_{ox} are the capacitances of the depletion layer and gate oxide. This expression is valid from weak to strong inversion, and can be explicited according to the precise models chosen for the drain current or impact ionization current.

3. Experiment

It is important to keep in mind that, although the device operates in the saturation mode, the conductance is not constant. Note also that g_d and g_m are not directly proportional, since the correlating coefficient is a function of the surface and body potentials, which in turn depend on V_G and V_D. Plotted in Fig.1(b) is the experimental variation of this coefficient (i.e. the ratio g_d/g_m) which well illustrates the evolution of the floating body effects with V_G. It can be seen that, in strong inversion, the body potential still increases for high gate voltages, since the ratio g_d/g_m does not vanish. The shape of the curves of Fig.1(b) can easily be reproduced with the above analytical model. Equation (1) predicts that hysteresis phenomena must occur *simultaneously* in the $I_D(V_G)$ characteristics (i.e. $g_m < 0$) and $I_D(V_D)$ curves (i.e. $g_d < 0$). Indeed, a *negative conductance* region can be found experimentally in the "kink region" when the device is current-controlled (Fig.2(a)). It does not involve any avalanche breakdown mechanism and, although unstable, it corresponds to a steady state regime. We have verified that the addition of an external capacitor on the drain terminal induces relaxation oscillations. The domains where g_m and g_d are negative have been experimentally determined and look very similar.

For a given surface potential Ψ_s, the critical body potential V_{Bc} corresponding to the onset of negative transconductance/conductance values can be roughly approximated by:

$$V_{Bc} = \Psi_s - \frac{q\epsilon_s N_A}{2}\left(\frac{n-1}{C_{ox}}\right)^2 \tag{2}$$

where N_A is the film doping and n is the emission coefficient of the body-to-source diode. It is now possible to define the main parameters of the floating body effects:

- The *doping level* N_A is an essential factor, since it appears at each stage of the modeling. A previously unnoticed effect is the dependence of the reverse current of the body-to-source diode on N_A. Thus, when reducing the doping, floating body effects are reduced because (i) the impact ionization decreases, (ii) the reverse junction current is increased and (iii) an electrical coupling appears between the opposite interfaces.

- *Body-to-source diode* properties: The quality of the junction is important because a slight increase of the emission factor n may substantially amplify the floating body potential (see Eq.(2)). In the specific case of SOI devices, the experimental current-voltage characteristics of the body-to-source diode may correspond to empirical values of n greater than 3 !

- *Oxide thickness*: The thicker the oxide, the larger the floating body effects.

- *Silicon film thickness*: In partially-depleted SOI films, an increase of the Si film thickness attenuates the static floating body effects, because the junction area is increased. Hence, for rad-hard or high speed applications, a trade-off must be found.

- *Longitudinal electric field*: The higher the peak of the electric field, the worse the floating body effects become. This parameter can be adjusted by drain engineering. Experimental results show that minor modifications of this parameter, induced by variations of the series resistances or LDD doping, may dramatically increase floating body effects.

4. Radiation effects

The threshold voltage is defined here as the gate voltage corresponding to a given current level. Since the self-biasing of the body induces an important threshold voltage lowering at high V_D, the $I_D(V_G)$ characteristics during irradiation should preferentially be monitored in the saturation mode. Comparison between the ohmic and saturation regimes demonstrates that the aggravation of the floating body effects leads to an additional lowering of the threshold voltage measured at high V_D. A slight increase of the transconductance peak occurs in the earliest stage of exposure and indicates that the series resistances are reduced. This variation of the

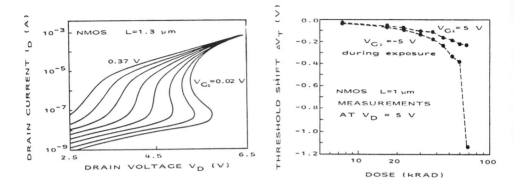

Figure 2: (a) Experimental negative conductance curves and (b) threshold voltage shift under irradiation for various biases.

series resistances is due to a positive charge trapping into the spacer which increases the electron density in the low doped region. The worst case of biasing under exposure surprisingly corresponds to a negative gate voltage (Fig.2(b)). Indeed, for a positive gate bias, the ohmic threshold voltage decreases more significantly. This causes an increase of the saturation drain voltage, which partially compensates the increase of the electric field due to the lowering of the series resistances.

5. Conclusion

An original model has been proposed to account for the floating body effects and relevant experimental data obtained on SIMOX MOSFET's have been presented. It has been demonstrated that the conductance and transconductance are strongly correlated even in the saturation mode. A negative conductance takes place in the "kink region" of the $I_D(V_D)$ characteristics, *simultaneously* with hysteresis phenomena in the $I_D(V_G)$ characteristics. The main parameters of the floating body effects have been reviewed. It has been shown that the worst case for irradiating LDD N–MOSFET's, fabricated in partially-depleted SOI films, does not necessarily correspond to a positive applied gate bias during exposure, but possibly to a negative gate voltage, due to increasing impact ionization.

References

Tihanyi J and Schlötterer H 1975 *IEEE Trans. Electron Dev.* **ED-22** 1017
Chen C E, Matloubian M, Sundaresan R, Mao B Y, Wei C C and Pollack G P 1988 *IEEE Electron Device Lett.* **9** 636

A comparison of the relative merits of n+ or p+ polysilicon gates for ultra thin SOI MOSFETs

G.A.Armstrong and W.D.French
Department of Electrical and Electronic Engineering
Queen's University Belfast, N.Ireland

Abstract: The factors involved in the choice of gate material for thin film sub–micron SOI transistors are considered. Control of threshold voltage, subthreshold slope and bipolar holding voltage becomes progressively more significant as the gate length is reduced. Two–dimensional simulation indicates that with a carefully optimised design, a maximum operating voltage of 5V is feasible, for a 0.5 micron gate length.

Introduction

Thin film, or fully depleted, silicon–on–insulator (SOI) CMOS transistors offer the benefits of low parasitic capacitance, reduced short channel effects, higher drive currents, together with the inherent advantages of high packing density and simplified processing for sub–micron VLSI. To extend the design of n–channel SOI transistors with conventional n+ polysilicon gates to the submicron regime requires careful scaling of both the gate oxide and SOI film thickness[1]. For example, for a gate length of $0.5\mu m$, the film thickness should be scaled to between 50–70nm and the gate oxide to around 14nm. Transistors with these dimensions however, require a film doping greater than $2.5 \times 10^{17} cm^{-3}$, in order to give a threshold voltage in excess of 0.6V. However, such a large film doping is undesirable, as it makes the threshold voltage more difficult to control because of the high implant dose. An alternative approach has been suggested, where utilisation of the lower work function of p+ polysilicon as the gate material, necessitates a very low film doping to achieve the same threshold voltage. This has been shown to give improved threshold voltage control, higher gain and higher breakdown voltage, for a $1\mu m$ process[2]. However, the lower film doping can result in degraded punchthrough resistance, together with significant short channel effects, for submicron devices.

Two–dimensional simulation

In this paper we consider an investigation into the performance of thin film SOI transistors with n+ or p+ polysilicon as the gate material, using a two–dimensional finite difference device simulator[3]. This simulation package has been based on MINIMOS4[4], but contains some significant enhancements for SOI transistors, which are fully described in [3]. In particular, a capability exists to compare sub–micron gate transistors with different types of source/drain structure, including lightly doped drain (LDD) and gate overlap drain(GOLD). Silicidation of source and drain can be modelled as an ideal ohmic contact which extends into the source and drain region[5]. Full details of parameter models and solution algorithms are given in [4]. The significance of the choice of coefficients to model impact ionisation and bandgap narrowing in heavily doped regions is described in more detail in another paper in the proceedings of this conference[6].

Control of threshold voltage and subthreshold slope

Control of punchthrough in p+ polysilicon gate transistors requires optimisation of a threshold adjustment boron implant. The effect of variation in dose of a 60 keV boron implant on threshold voltage and subthreshold slope is shown in Fig.1, for three different gate lengths. For gate lengths longer than $1\mu m$, near ideal subthreshold slopes (<75mV/decade) and low threshold voltages (<0.8V) may be obtained.

However, as the gate length is reduced to 0.5μm, short channel effects tend to degrade the subthreshold slope of devices with a low dose implant. A heavier boron implant dose is therefore required to suppress this degradation in slope, but this tends to counteract the benefits of utilising a fully depleted SOI film. Alternatively, if the SOI film thickness is reduced to 50 nm and the oxide thickness to 14 nm, the lowest achievable threshold voltage is 0.8V, but the subthreshold slope can be reduced to 80 mV per decade, as shown in Fig.2. Fig.3 shows that it is much easier to obtain both a low threshold voltage and subthreshold slope, using an n⁺ polysilicon gate. Clearly, for a given threshold voltage, an n⁺ polysilicon gate provides a lower subthreshold slope over the entire range of film thickness, as the utilisation of higher film doping eliminates short channel effects. The disadvantage of the n⁺ polysilicon gate for ultra thin films is the very large implant dose that is required to raise the threshold voltage above 0.6V. Care must be taken to ensure that the doping becomes large enough to reintroduce a 'kink' in the output characteristics.

Bipolar snapback

Transistors with p⁺ polysilicon gate lengths of 1 micron have been found to give enhanced resistance to bipolar snapback, as a result of a buried channel mode of operation[2]. The onset of bipolar snapback is determined by the magnitude of the rate of generation of hole electron pairs due to impact ionisation in the drain region. The peak impact ionisation rate is governed by the magnitude of the lateral current density in the vicinity of the peak electric field. In a transistor with a relatively thick film, the peak electric field occurs close to the surface, while the current flows in a buried channel. If, however, the film is thinned from 150nm down to 50nm, not only is the peak electric field increased from 3.7×10^5 V.cm⁻¹ to 5.5×10^5 V.cm⁻¹, but the drain current is constrained to flow through the region of peak field, with a significant increase in peak ionisation rate. Fig.4 shows the dependence of peak impact ionisation rates of 0.5 μm n⁺ and p⁺ polysilicon gate transistors on film thickness for the bias condition, $V_g = 1V$ and $V_d = 4V$. In each case the film doping was chosen to give a threshold voltage of 0.8V, and no LDD region was included. At high drain bias, the peak electric field in an n⁺ polysilicon device tends to exceed that of p⁺ polysilicon device because of the higher doping. However, when the film thickness is reduced, the peak field occurs in a region where there is significant current flow. For this reason the peak ionisation rate, and hence the holding voltage, of the p⁺ polysilicon gate transistors may well exceed that of the n⁺ polysilicon gate transistor. For the specific bias condition shown this occurs when when the film thickness is less than 75 nm.

The bipolar holding voltage is often used as a measure of the maximum operating voltage of an SOI transistor. Although this voltage may be determined precisely by simulation[1], it proves to be more efficient to simply determine the magnitude of the drain voltage which gives a drain current of 1 μA per μm width, when $V_g = 0$. This has proved to be satisfactory in the simulation of LDD transistors and is used in the definition of a nominal breakdown voltage.

This voltage can be increased by the use of a lightly doped drain, with silicidation of the n⁺ region[7]. The choice of the n⁻ dose is largely governed by considerations of device degradation due to charge injection into the gate oxide. It has been suggested[8] that a dose of 1×10^{13} cm⁻² is acceptable. Fig.5 and Fig.6 permit a direct comparison of the simulated output characteristics for 0.5 micron n⁺ and p⁺ polysilicon gate transistors. The use of 1×10^{13} cm⁻² 50keV n⁻ doses with a 0.25 μm spacer maximises the breakdown voltage in both cases. The p⁺ polysilicon gate transistor suffers from punchthrough which can be cured by the use of a low dose boron implant as suggested by Fig.3. The negative resistance region at high gate voltage is due to a temperature rise resulting from joule heating and has been observed experimentally.

Conclusions

We conclude that there is no significant advantage in using p^+ polysilicon as the gate material, when the gate length is reduced to 0.5 micron. In order to counter the degradation in subthreshold slope due to short channel effects, films must be thinned to around 50nm. The choice then lies between the 'kink' degraded output characteristic of the n^+ polysilicon gate, and the increased punchthrough of the fully depleted p^+ polysilicon gate transistor. In both cases the minimum achievable subthreshold slope is around 80mV per decade. Simulations indicate that it is possible to achieve an operational voltage of 5V for both types of gate material if a n^- implant of 1×10^{13} cm^{-2} is used.

Acknowledgement

The authors wish to acknowledge financial support from the Procurement Executive, Ministry of Defence (RSRE) under contract RP009/335.

References

1. H.Lifka and P.H.Woerlee, IEDM Tech Dig., 1989, pp. 821–824.
2. J.R.Davis et al, IEEE Trans. ED–38, 1991, pp.32–38.
3. G.A.Armstrong, J.R.Davis and A.Doyle, IEEE Trans. ED–38, 1991, pp. 328–336.
4. W.Hansch and S.Selberherr, IEEE Trans. ED–34,1987, pp.1074–1080.
5. G.A.Armstrong, W.D.French and J.R.Davis, Proc ESSDERC 90,pp. 425–428.
6. G.A.Armstrong and W.D.French, proc INFOS 91
7. P.H.Woerlee et al, IEDM Tech Dig., 1990.
8. H.Katto et al, IEDM Tech Dig., 1984, pp.774–777.

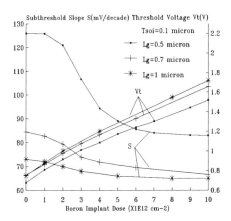

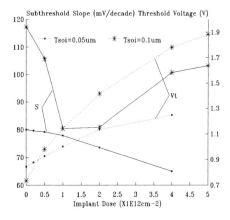

Fig.1   Dependence of subthreshold slope and threshold voltage on boron implant dose of p+ poly gate non LDD SOI transistors of film thickness 0.1 microns, gate oxide 20 nm and three gate lengths.

Fig.2   Dependence of subthreshold slope and threshold voltage on boron implant dose of 0.5 micron gate length non LDD p+ poly gate transistors with film thicknesses of 0.1 and 0.05 microns and a 14 nm gate oxide.

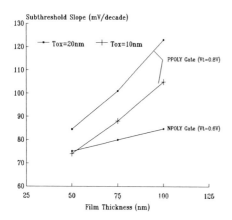

Fig.3 Dependence of subthreshold slope on film thickness and gate type of 0.5 micron non LDD SOI transistors.

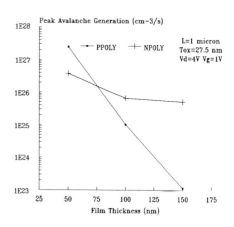

Fig.4 Dependence of peak impact ionisation rate of n^+ and p^+ poly gate transistors on film thickness.

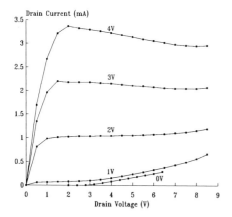

Fig.5 Simulated output characteristics for a 0.5 micron p^+ poly gate LDD transistor with a film thickness of 0.05 microns, a gate oxide of 14 nm and a threshold voltage of 0.74 V.

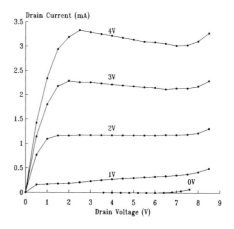

Fig.6 Simulated output characteristics for a 0.5 micron n^+ poly gate LDD transistor with a film thickness of 0.05 microns, a gate oxide of 14 nm and a threshold voltage of 0.6 V.

A study of heavy doping effects on the performance of thin film SOI MOSFETs

G.A.Armstrong and W.D.French
Department of Electrical and Electronic Engineering
Queen's University Belfast, N.Ireland

Abstract: To model bipolar snapback in thin film SOI transistors correctly, it is necessary to model the bandgap narrowing in the heavily doped source region. Two dimensional device simulation is used to quantify the dependence of the holding voltage on bandgap narrowing in the source.

Introduction

In the design of silicon–on–insulator transistors, the maximum operating voltage is limited by bipolar snapback, which reduces the holding voltage to a value less than that which can be achieved by conventional bulk technology. This problem is caused by holes, generated by impact ionisation at the drain junction, raising the potential of the floating region sufficiently to forward bias the source junction. When scaling a design based on thin film SOI to sub–micron gate lengths, this constraint becomes a severe limitation if the holding voltage is required to exceed 5V.

Parameter models

It has been shown that two–dimensional device simulation[1] can be helpful in predicting the onset of bipolar snapback, and hence determining the holding voltage. Previous work, however, ignored heavy doping effects in source and drain, such as doping dependent carrier lifetime and bandgap narrowing. Inclusion of these effects is essential for accurate simulation of bipolar transistors, but proves to be insignificant for simulation of MOS devices, as long as the current flow is unipolar. When snapback occurs however, the source junction becomes forward biased and current flow is dominated by the minority carrier injection characteristic of a bipolar transistor.

A model for the effective lifetime in heavily doped silicon may be approximated by an effective lifetime τ_{eff} of the form[2]

$$\tau_{eff} = \tau_{n,p} / (1 + |N_d - N_a| / N_{ref}) \qquad (1)$$

where $\tau_{n,p}$ represents the lifetime of electrons or holes in lightly doped material and N_{ref} is a doping reference level. Typically it it has been found that the base lifetime for both holes and electrons in thin film SOI should be of the order of 10^{-6}s and a value of $5 \times 10^{16} cm^{-3}$ is appropriate for N_{ref}.

As pointed out in [3], band gap narrowing describes the increase in the equilibrium pn product in silicon at high doping levels, such that

$$pn = n_{ie}^2 = n_i^2 \exp(\Delta E_g / kT) \qquad (2)$$

where ΔE_g in meV is obtained by fitting to measured data[3]

$$\Delta E_g = 18.7 \ln(|N_d - N_a| / 7 \times 10^{17}) \qquad (3)$$

for $N_d - N_a > 7 \times 10^{17}$. Band gap narrowing is not significant in the SOI film, since the net doping is normally of the order of 10^{17} or less. The variation in bandgap modifies the driving force E in the drift term in the carrier continuity equations, such that for electrons and holes

$$E_n = -\nabla(\psi + \ln(n_{ie})) \quad (4a) \qquad\qquad E_p = -\nabla(\psi - \ln(n_{ie})) \quad (4b)$$

Impact ionisation is modelled as a Chywoneth law where the generation rate G is given by

$$G = (\alpha_n |J_n| + \alpha_p |J_p|)\, q \quad (5)$$

where the respective coefficients are defined as a function of electric field in the direction of current flow E_j by

$$\alpha_{n,p} = A_{n,p} \exp(-B_{n,p}/E_j) \quad (6)$$

The choice of appropriate coefficients A and B is vital in order to obtain accurate simulations. In practice, the significant parameters for electron current flow in an n–channel transistor are A_n and B_n. Because of the exponential term in (6), it is particularly important to specify B_n accurately. One way to determine appropriate values for A and B is to model substrate current in a conventional MOSFET. Fig.1 shows a comparison of simulated and measured substrate current in a conventional LDD MOSFET, as a function of gate length. The ionisation coefficients used were obtained from experimental measurements of surface breakdown[4]. Excellent agreement has been achieved for 3 different n⁻ LDD doses, providing additional confirmation of the validity of these coefficients.

Bandgap narrowing

The significance of the inclusion of a model for bandgap narrowing is shown in Fig.2. This shows a simulation of the $V_g = 0$ breakdown characteristic of a 0.5 μm (non LDD) SOI transistor. An approximate one–dimensional analytical model [5] predicts that bipolar induced breakdown is initiated when the product of bipolar current gain and impact ionisation rate, βM, tends to unity. If bandgap narrowing is omitted from the simulation, the predicted bipolar current gain is over–emphasized and the lateral bipolar transistor turns on at a much lower value of M, (i.e. lower drain voltage), as shown in curve (a). With bandgap narrowing included, however, the bipolar gain is reduced, and a much less realistic breakdown characteristic results, if the same impact ionisation coefficients are used, as shown in curve (b). However, experimental measurements [1] indicate that the breakdown voltage of a 0.5 μm (non LDD) transistor is typically less than 3V. Hence we conclude that it is possible to obtain a correct estimate for the breakdown voltage, by over–emphasising the effect of the bipolar gain β, and underestimating the effect of the impact ionisation rate M. If the coefficients in the equation for impact ionisation are altered, according to the values reported in [4], a more realistic breakdown characteristic is obtained, as shown in (c). More significantly, a consistent set of ionisation parameters have been demonstrated, equally valid for both bulk and SOI technologies.

Recently, the importance of both source and drain engineering for successful sub–micron SOI design has been stressed[6]. Drain engineering increases the breakdown voltage by the inclusion of a lightly doped n⁻ region to reduce the peak electric field, as described in an accompanying paper[7]. With source engineering there are two main factors to be taken into account. Firstly it is desirable to utilise as low a doping as is feasible, in order to reduce the emitter efficiency of the parasitic bipolar transistor. Fig.3 shows the simulated increase in holding voltage which can be achieved in a non–LDD transistor by lowering the source implant dose, with the drain implant dose kept constant. As the source implant dose is increased up to $5 \times 10^{16} \text{cm}^{-2}$, the additional bandgap narrowing does not appear to give rise to a significant reduction in current gain of the parasitic bipolar transistor. This suggests a lightly doped source is preferable. Indeed, it appears that the holding voltage is more sensitive to a low source doping than to a low drain doping, suggesting that the emitter efficiency is the dominant factor in determining the maximum operating voltage.

Furthermore, it is desirable to minimise the effective distance, L_s between the

source contact and the edge of the gate contact. In practice this can be achieved by silicidation, as this tends to minimise the diffusion length for holes in the source by bringing the contact as close as possible to the edge of the junction. Fig.4 shows the increase in holding voltage achieved by reducing L_s, for a 0.5 micron non−LDD SOI transistor with a film thickness of 0.1 micron. A similar 0.5V increase has been observed experimentally[6]. Fig.5 confirms that the increased breakdown voltage can be attributed to the reduced emitter efficiency of the parasitic bipolar transistor.
The emitter efficiency has been calculated by calculation of the ratio of diffusion current to total emitter current from the two dimensional simulation.

Conclusions

In order to correctly estimate the bipolar holding voltage of thin film SOI transistors with sub−micron gate lengths, it is necessary to model correctly both the bandgap narrowing in the heavily doped source region and impact ionisation at the drain. Ionisation parameters which have been validated for bulk technology have been shown to be equally applicable to SOI technology. The bipolar holding voltage is more sensitive to a lightly doped source than a lightly doped drain. Silicidation of source and drain minimises the minority carrier diffusion length and hence the emitter efficiency.

Acknowledgements

The authors acknowledge financial support from the Procurement Executive, Ministry of Defence(RSRE) and the UK Science and Engineering Research Council.

References

1. G.A.Armstrong, W.D.French and J.C.Alderman, Proc. IEEE SOS/SOI Technology Conf., Key West Florida, 1990, pp.17−18.
2. A.S.Grove,"Physics and technology of semiconductor devices", Wiley, 1967.
3. J.A. del Alamo, S.Swirhun and R.M.Swanson, IEDM Tech. Dig.,1985, pp.290−294.
4. J.W.Slotboom et al, IEDM Tech. Dig., 1987, pp.494−497.
5. K.K.Young and J.A.Burns, IEEE Trans ED−35, pp. 426−430, 1988.
6. H.Lifka and P.H.Woerlee, Proc. ESSDERC90, Nottingham, 1990,pp.453−456.
7. G.A.Armstrong and W.D.French, Proc INFOS 91.

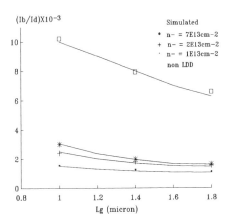

Fig.1 Ratio of bulk to drain current vs. gate length at Vd=5V, Vg= 2V for measured (solid lines) and simulated (points) LDD and non LDD transistors.

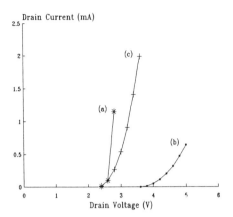

Fig.2 Simulated output characteristics at Vg=0V for 0.5 micron non LDD transistors of 0.1 micron film thickness and 20 nm gate oxide.

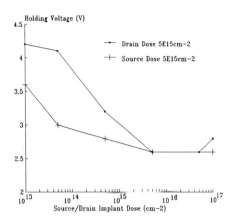

Fig.3 Dependence of the holding voltage on source and drain implant doses for an asymmetric non LDD transistor.

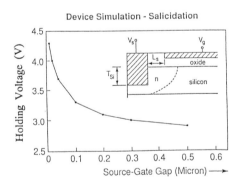

Fig.4 Variation of holding voltage with source-gate separation for a 0.5 micron non LDD transistor with a film thickness 0f 0.1 microns and gate oxide of 20 nm.

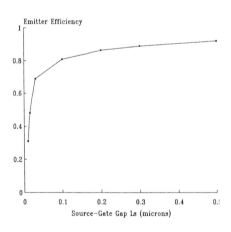

Fig.5 Variation of emitter efficiency with source - gate separation for a 0.5 micron non LDD transistor with a 0.1 micron film thickness and 20 nm gate oxide.

Paper presented at INFOS '91, Liverpool, April 1991
Contributed Papers, Section 8

Determination of generation lifetime in thin film silicon-on-insulator (SOI) material using capacitance time, charge time and gated diode measurements

L.J. McDaid, S. Hall and W. Eccleston, Department of Electrical Engineering and Electronics, University of Liverpool, P.O. Box 147, Liverpool, L69 3BX, U.K.
J.C. Alderman, National Microelectronic Research Centre (NMRC), University College, Lee Maltings, Prospect Row, Cork, Ireland.

We demonstrate that a two terminal charge-time measurement followed by a three terminal capacitance-time measurement, performed on an SOI capacitor, can yield the bulk body generation lifetime (τ_{gb}) and the surface generation velocity at the body/buried oxide interface. We highlight that for the former measurement, the depletion approximation is violated over an appreciable thickness of the body region and this is taken into account in the analysis. For the latter measurement, charge storage occurs at the body/buried oxide interface and a simple analysis is presented. The value obtained for τ_{gb} compares well with that obtained from a gated diode measurement.

1. INTRODUCTION

We have developed new techniques for the measurement of generation lifetime in the silicon overlayer or body, and the surface generation velocity at the back interface, of thin-film SOI wafers. The lifetime in the bulk of the body can be obtained from the charge transient (Q-t) measurement and this allows the effect of interface generation to be separated from the capacitance-time (C-t) measurement. A gated diode measurement provides an independent check of the technique. The capacitors are 200μm square, produced in SIMOX wafers annealed at 1250°C [1]. Thicknesses of oxide and silicon regions were obtained from C-V measurements [1]. Problems relating to charge storage together with the use of the depletion approximation in the interpretation of generation rate, are discussed.

2. THE CHARGE TRANSIENT MEASUREMENT

Figure (1) shows a SOI capacitor with an n-type body and substrate and the interfaces I1, I2 and I3 are identified. Because the body region is open circuit, the SOI capacitor, can be considered as a conventional MOS capacitor with an equivalent "oxide" of thickness, WT (= Wof + Wb ($\varepsilon ox/\varepsilon s$) + Wob). This "oxide" contains "oxide-charge" associated with the inversion charge (Qinb), accumulation charge (Qacb) and the ionised donors charge (Qb) arising from the application of a negative substrate voltage. Each of these three components will contribute a charge centroid effect and influence the distribution of charge between the top gate and the substrate at I3. If a step voltage is now applied to the substrate, the potential at I3 will reach equilibrium almost immediately and any subsequent charge fluctuations at I3 will not alter ϕI3. The electric field across the system has increased and electron/hole pairs must be generated within the body, to maintain the continuity of the dielectric displacement. Generated electrons (holes) drift to interfaces I1 (I2) respectively and the associated time varying centroids result in a current in the external circuit. It is easy to show that this current is related to the generation lifetime in the body region τ_{gb}, by:

$$dQ_g/dt = (C_{oT}/C_b)\ dQ_{inb}/dt = (C_{oT}/C_b)\ (qn_i W_b'/2\tau_{gb}) \tag{1}$$

© 1991 IOP Publishing Ltd

CoT, Cb are the capacitances associated with the thickness tT and tb respectively. Maximum generation only occurs over the region Wb' where both electron (n) and hole (p) concentrations are less than ni and this distance is estimated as follows. Noting that the field in the two dielectrics must be equal and that for sufficiently high inversion layer density (Qinb >> Qb), then Qinb~Qacb. It also follows that the distance over which Qinb and Qacb fall to a level equal to the body doping can be assumed equal (~3nm) and this is taken as the distance over which the potential drops by (Eg/2-ϕb) as we assume that the quasi Fermi levels are flat and almost coincident with the appropriate band edges at I1 and I2. (Eg is the bandgap, ϕb is the fermi potential). To calculate the distance Wn, over which the electron concentration falls from a level equal to the background doping Nd, to ni, we solve Poisson's equation in 1-D. The distance xp associated with the hole concentration at interface I2 is found similarly. The point at which n,p=ni is defined where the intrinsic energy level crosses the respective quasi-fermi levels. The values of Wn, Wp obtained for Nd=10 cm and Vstep=10V are 30nm and 40nm respectively yielding Wb'=0.65Wb. These widths were obtained for the condition just after the application of the step voltage therefore the initial gradient of transient must be used in equation (1), as Wb' (and hence the generation rate) varies with time. Figure (2) shows a typical charge transient for bias and step voltages both equal to -10V and equation (1) gives a lifetime, τgb=270ns. The temperature dependence of the charge transient was also measured and a straight Arrhenius plot gave an activation energy of 0.6 eV which provides strong evidence for thermal generation. Also, the reverse leakage current in a gated diode close to the capacitor has been measured. Table (1) summarises values for lifetime.

3. THE CAPACITANCE TRANSIENT MEASUREMENT

The capacitor is biased as shown in figure (3) with I1 accumulated, the body region and I2 fully depleted. A depletion region is formed at I3 also. Application of a voltage step causes deep depletion in the substrate followed by generation in both body and substrate. Holes generated in the body drift to I1, electrons drift to I2 and subsequently diffuse laterally to the n+ body contact, which is grounded. The diffusion process gives rise to a concentration gradient whereby the majority carrier density at I2 n_{I2}, may exceed ni. If this condition

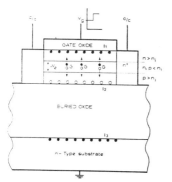

Figure 1: Cross section of the SIMOX SOI capacitor for the charge transient measurement: positive gate bias showing interfaces I1, I3 accumulated, I2 inverted and surface potential ϕI3. The region of maximum generation, We, is indicated.

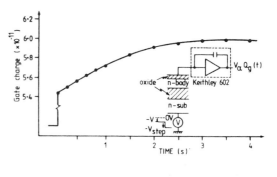

Figure 2: A charge transient obtained for an SOI capacitor of area 4x10⁻⁸ m , with Vsub=Vstep= -10V, giving Tgb=270ns, allowing the reduced generating volume. The insert shows the measuring circuit with bias and step voltages applied to the substrate. Parameters required for the analysis obtained from [1], were tof=20nm, tb=220nm, tob=320nm.

exists, then surface generation at I2 is reduced. It can be shown [2] that the capacitance-time relationship in the presence of this charge storage is

$$C = C_o[((1+(1+C_{of}/C_b+C_{of}/C_{ob})^2) - t/T_b)^{1/2} - (1+C_{of}/C_b+C_{of}/C_{ob})]^{-1} \quad (2)$$

where

$$T_b = 2\tau_{gbe} N_s/n_i = 2[1/\tau_{gb} + s_{I2}/W_b]^{-1} N_s/n_i \quad (3)$$

Cob is the buried oxide capacitance Ns is substrate doping and sI2 is the surface generation velocity. τgbe is an effective generation lifetime which includes the contribution from surface generation, assumed constant, along the interface I2 and this is discussed further below. Note that we have also assumed that dnI2/dt = 0. This is because the thermal generation rate attains it's maximum value within a time negligible compared to the total relaxation time. The magnitude of the voltage step is crucial as it sets the potential at I3 and determines the level of minority carrier charge arising from electron/hole generation in the substrate. A large voltage step results in a potential at I3 of sufficient magnitude to maintain an inversion layer for part of the recovery whereas for a small voltage step, an inversion layer cannot be maintained and minority carriers recombine with no effect upon the electrostatics and therefore the observed C-t response. This is illustrated in the results of figure (4) which have the voltage step as a parameter. For the case of the -2V step, the generation in the body dominates and therefore generation in the substrate is ignored leading to the C-t expression of equation (2). Note that because of the approximation of constant generation rate, equation (2) is only valid over the initial part of the transient. We have used equation (2) to obtain the lifetime τgbe, from the -2V transient and hence sI2 from equation (3), because τgb is known from the Q-t measurement. To obtain the maximum value, soI2, the density of free charge at I2 (nI2) must be calculated. By approximating the square capacitor to an equivalent circular device of radius ro and assuming for simplicity, that the majority carriers at I2 arise from bulk generation only, we can relate the charge in the centre of the device, to the lifetime in the body yielding:

$$n_{I2} = n_i W_b E_o r_o^2/(8V_t D_e \tau_{gb}) \quad (4)$$

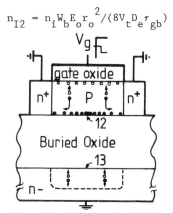

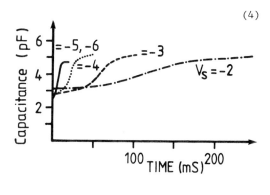

Figure 3: Cross section of the SIMOX SOI capacitor for the capacitance transient measurement; negative gate bias showing interface I1 inverted and I2, I3 depleted. The charge storage effect at I2 is indicated by the increase of electron density towards the centre of the capacitor.

Figure 4: Capacitance-time response with step voltage Vs as a parameter. The decrease in recovery time with increasing voltage step amplitude is of note. The effective generation lifetime Tgbe is 110ns, calculated from the Vs=-2V curve.

De is the diffusion constant at I2, found from the measured low field mobility of a nearby transistor as 10cm s. Vt is thermal voltage and Eo is the electric field due to the voltage step. From SRH statistics, sI2 is related to soI2 by

$$s_{I2} = s_{oI2}\, 2n_i/(n_{I2} + 2n_i) \tag{5}$$

Note that the maximum value for nI2 is calculated from (4) and hence the value for soI2 is obtained from equation (5). Further details of this calculation will be published soon [2]. It is of note that τ_{gbe} is about half τ_{gb} indicating that surface generation is small in agreement with the charge storage effect.

Method	τ_{gbe} (ns)	τ_{gb} (ns)	s_{oI2} (cm/s)	W_b' (nm)
Q-t	-	270	-	144
C-t	110	-	19	-
G-D	-	260	-	160

Table 1: Generation lifetime (T_{gb}) and velocity (SOI_2) values obtained by the charge transient (Q-t), three terminal capacitance transient (C-t) and gated diode (G-D) techniques. The calculated effective generating thickness t_b' is also shown for the structure with $t_b=200$ and $N_{body}=10^{16} cm^{-3}$.

4. DISCUSSION

Central to the analysis of the Q-t technique is the validity of the depletion approximation in defining generation rates. If maximum generation rate n_i/τ_{gb}, it to be used, both electron and hole concentrations must be less than ni. We have shown that this is only true over a certain fraction of the body region in these thin films. For the C-t measurement, the concentration gradient of majority carriers at I2 means that the generation velocity is varying laterally and only reaches it's maximum value close to the body contact. Therefore the effective generation lifetime measured by the C-t technique arises from a constant bulk generation and a distance dependent surface generation component. It is difficult to analyse completely this effect, but if we assume that majority carriers at I2 arise from bulk generation only, then it can be shown that for generation lifetimes less than 1ms, the majority carrier density at I2 will exceed ni over most of the interface. Lifetimes in these SOI films are typically less than 1µs and therefore surface generation will be reduced significantly over most of the interface. This charge storage effect is present also in the gated diode and interpretation of that data is also not straightforward. We maintain therefore, that values for lifetimes obtained previously for both SOI and bulk are likely to be in error if charge storage effects are neglected.

5. ACKNOWLEDGEMENTS

The work was funded by the Procurement Executive, Ministry of Defence (RSRE) under contract RP009/341. One of the authors (JCA) would like to acknowledge Plessey Research (Caswell) where some of the work was carried out.

6. REFERENCES

[1] L.J. McDaid, S. Hall, W. Eccleston and J.C. Alderman, Solid-St. Electron., Vol. 32, No. 1, pp. 65-68, Jan. 1989.
[2] L.J. McDaid, S. Hall, W. Eccleston and J.C. Alderman, submitted to IEEE Trans. Electron Devices, April 1991.

Author Index

Abraham, P, *207*
Afshar-Hanaii, N, *183*
Aizenberg, I A, *159*
Alderman, J C, *339*
Ancona, M G, *139*
Andersson, M O, *147*
Andrianov, A V, *159*
Armstrong, G A, *331, 335*
Asenov, A, *247*
Aymerich, X, *243*

Balk, P, *167*
Balland, B, *211*
Barklie, R C, *307*
Barlow, K, *171*
Bauer, A, *175*
Bekkaoui, A, *207*
Bellens, R, *271*
Benson, T M, *319*
Berger, J, *247*
Bik, W M A, *191*
Bollu, M, *131*
Borel, G, *327*
Botton, R, *211*
Bouchakour, R, *279*
Boyd, I W, *163*
Brini, J, *323*
Brotherton, S D, *117, 231*
Brox, M, *267*
Brozek, T, *227*
Buchwald, W R, *215*
Bureau, J C, *211*
Burte, E P, *175, 199*
Busch, M C, *223*

Carter, J C, *183*
Cartier, E, *43*
Cerva, H, *187*
Chen, Wenliang, *139*
Childs, P A, *303*
Conley, J F, *259*
Coromina, F, *323*
Coxon, P, *235*
Cristoloveanu, S, *53, 251, 327*

Davies, D E, *319*
De Keersmaecker, R F, *179*
Deen, M J, *295*
DiMaria, D J, *65*
Dietl, J, *93*

Dooryhee, E, *223*

Eccleston, W, *183, 235, 275, 291, 339*
El-Hdiy, A, *279*
Engström, O, *147*
Ennis, T J, *307*
Esser, A, *167*
Evans, A G R, *183*

Farmer, K R, *1, 147*
Farrés, E, *243*
Flietner, H, *151*
Fogarassy, E, *195*
French, W D, *331, 335*
Fuchs, C, *195*

Gabrys, J W, *299*
Garrido, B, *315*
Gendry, M, *207*
Gerardi, G J, *215*
Ghibaudo, G, *327*
Gill, A, *231*
Groeseneken, G, *271*
Gueorguiev, V K, *239*
Guillemot, N, *311*

Habraken, F H P M, *191*
Hall, S, *339*
Haond, M, *107*
Hartnagel, H L, *207*
Hassein-Bey, A, *251*
Hemment, P L F, *307, 315, 323*
Heremans, P, *271*
Heyers, K, *167*
Heyns, M M, *73, 179, 263, 283*
Homewood, K P, *323*
Hönlein, W, *187*
Hurley, P K, *235*

Iwai, H, *83*

Jiménez, J, *323*
Jourdain, M, *279*

Karmann, A, *143*
Kassabov, J, *33*
Keeble, D J, *215*
Khvostov, V A, *159*
Kiblik, V Y, *227*
Koch, F, *131, 247, 255*

Author Index

Köster jr, H, *135*
Krafft, F, *207*
Krawczyk, S K, *207*
Krick, J T, *299*
Kurz, H, *167*
Kux, A, *255*

Le Néel, O, *107*
Lemiti, M, *211*
Lenahan, P M, *259, 299*
Leung, C C C, *303*
Logush, O I, *227*

Maes, H E, *271*
Marée, C H M, *191*
Margail, J, *311*
McDaid, L J, *339*
McFeely, F R, *43*
McManus, A, *319*
Mohd Kassim, N, *319*
Monteil, Y, *207*
Morante, J R, *315, 323*
Mühlhoff, H M, *93*
Murray, D C, *183*

Nafría, M, *243*
Nayar, V, *163, 171*
Nosenko, S V, *159*
Novikov, S V, *203*

Ouisse, T, *327*

Paneva, R, *239*
Pérez-Rodríguez, A, *315, 323*
Pham, V V, *155*
Poindexter, E H, *215*
Popova, L I, *239*
Poppe, T, *131*
Prudon, J, *211*

Quon, C, *295*

Rausch, N, *199*
Razafindratsita, R, *155*
Reeson, K J, *307*
Reisinger, H, *187*
Richter, R, *207*
Riemenschneider, R, *207*
Roitman, P, *259*

Romanova, G F, *227*
Rong, F C, *215*

Saks, N S, *139*
Salace, G, *279*
Samitier, J, *315*
Sarrabayrouse, G, *155*
Sassi, Z, *211*
Schels, M, *255*
Schmidt, M, *135*
Schulz, M, *127, 143*
Schütz, R, *207*
Schwerin, A V, *73, 263, 283*
Semon, D I, *299*
Siffert, P, *195, 223*
Simonne, J J, *155*
Slaoui, A, *195, 223*
Sokolov, N S, *203*
Speckbacher, P, *247*
Spitzer, A, *187*
Steimle, M, *93*
Stesmans, A, *219*
Stoemenos, J, *311*
Stoneham, A M, *19*
Suñé, J, *243*

Tardy, J, *207*
Taylor, S, *183, 275, 291*
Toulemonde, M, *223*
Tsamis, C, *311*
Tsoukalas, D, *311*

Uren, M J, *287*

Van Gorp, G, *219*
Vassilev, P, *239*
Verhaverbeke, S, *179*
Viktorovitch, P, *207*
Vuillaume, D, *279*

Weber, W, *247, 255, 267*
Willer, J, *187*

Yakovlev, N L, *203*
Young, N D, *231*

Zhang, J F, *235, 275, 291*
Zhang, J, *183*
Zorn, G, *187*